Tassilo Pellegrini, Sören Auer, Klaus Tochtermann, and Sebastian Schaffert (Eds.)

Networked Knowledge – Networked Media

Studies in Computational Intelligence, Volume 221

Editor-in-Chief

Prof. Janusz Kacprzyk
Systems Research Institute
Polish Academy of Sciences
ul. Newelska 6
01-447 Warsaw
Poland
E-mail: kacprzyk@ibspan.waw.pl

Further volumes of this series can be found on our
homepage: springer.com

Vol. 200. Dimitri Plemenos and Georgios Miaoulis *Visual
Complexity and Intelligent Computer Graphics Techniques
Enhancements*, 2009
ISBN 978-3-642-01258-7

Vol. 201. Aboul-Ella Hassanien, Ajith Abraham,
Athanasios V. Vasilakos, and Witold Pedrycz (Eds.)
Foundations of Computational Intelligence Volume 1, 2009
ISBN 978-3-642-01081-1

Vol. 202. Aboul-Ella Hassanien, Ajith Abraham,
and Francisco Herrera (Eds.)
Foundations of Computational Intelligence Volume 2, 2009
ISBN 978-3-642-01532-8

Vol. 203. Ajith Abraham, Aboul-Ella Hassanien,
Patrick Siarry, and Andries Engelbrecht (Eds.)
Foundations of Computational Intelligence Volume 3, 2009
ISBN 978-3-642-01084-2

Vol. 204. Ajith Abraham, Aboul-Ella Hassanien, and
André Ponce de Leon F. de Carvalho (Eds.)
Foundations of Computational Intelligence Volume 4, 2009
ISBN 978-3-642-01087-3

Vol. 205. Ajith Abraham, Aboul-Ella Hassanien, and
Václav Snášel (Eds.)
Foundations of Computational Intelligence Volume 5, 2009
ISBN 978-3-642-01535-9

Vol. 206. Ajith Abraham, Aboul-Ella Hassanien,
André Ponce de Leon F. de Carvalho, and Václav Snášel (Eds.)
Foundations of Computational Intelligence Volume 6, 2009
ISBN 978-3-642-01090-3

Vol. 207. Santo Fortunato, Giuseppe Mangioni,
Ronaldo Menezes, and Vincenzo Nicosia (Eds.)
Complex Networks, 2009
ISBN 978-3-642-01205-1

Vol. 208. Roger Lee, Gongzu Hu, and Huaikou Miao (Eds.)
Computer and Information Science 2009, 2009
ISBN 978-3-642-01208-2

Vol. 209. Roger Lee and Naohiro Ishii (Eds.)
*Software Engineering, Artificial Intelligence, Networking and
Parallel/Distributed Computing*, 2009
ISBN 978-3-642-01202-0

Vol. 210. Andrew Lewis, Sanaz Mostaghim, and
Marcus Randall (Eds.)
Biologically-Inspired Optimisation Methods, 2009
ISBN 978-3-642-01261-7

Vol. 211. Godfrey C. Onwubolu (Ed.)
Hybrid Self-Organizing Modeling Systems, 2009
ISBN 978-3-642-01529-8

Vol. 212. Viktor M. Kureychik, Sergey P. Malyukov,
Vladimir V. Kureychik, and Alexander S. Malyoukov
Genetic Algorithms for Applied CAD Problems, 2009
ISBN 978-3-540-85280-3

Vol. 213. Stefano Cagnoni (Ed.)
Evolutionary Image Analysis and Signal Processing, 2009
ISBN 978-3-642-01635-6

Vol. 214. Been-Chian Chien and Tzung-Pei Hong (Eds.)
*Opportunities and Challenges for Next-Generation Applied
Intelligence*, 2009
ISBN 978-3-540-92813-3

Vol. 215. Habib M. Ammari
*Opportunities and Challenges of Connected k-Covered
Wireless Sensor Networks*, 2009
ISBN 978-3-642-01876-3

Vol. 216. Matthew Taylor
Transfer in Reinforcement Learning Domains, 2009
ISBN 978-3-642-01881-7

Vol. 217. Horia-Nicolai Teodorescu, Junzo Watada, and
Lakhmi C. Jain (Eds.)
Intelligent Systems and Technologies, 2009
ISBN 978-3-642-01884-8

Vol. 218. Maria do Carmo Nicoletti and
Lakhmi C. Jain (Eds.)
*Computational Intelligence Techniques for Bioprocess
Modelling, Supervision and Control*, 2009
ISBN 978-3-642-01887-9

Vol. 219. Maja Hadzic, Elizabeth Chang,
Pornpit Wongthongtham, and Tharam Dillon
Ontology-Based Multi-Agent Systems, 2009
ISBN 978-3-642-01903-6

Vol. 220. Bettina Berendt, Dunja Mladenic,
Marco de de Gemmis, Giovanni Semeraro,
Myra Spiliopoulou, Gerd Stumme, Vojtech Svatek, and
Filip Zelezny (Eds.)
*Knowledge Discovery Enhanced with Semantic and Social
Information*, 2009
ISBN 978-3-642-01890-9

Vol. 221. Tassilo Pellegrini, Sören Auer, Klaus Tochtermann,
and Sebastian Schaffert (Eds.)
Networked Knowledge - Networked Media, 2009
ISBN 978-3-642-02183-1

Tassilo Pellegrini, Sören Auer, Klaus Tochtermann, and Sebastian Schaffert (Eds.)

Networked Knowledge – Networked Media

Integrating Knowledge Management, New Media Technologies and Semantic Systems

Springer

Tassilo Pellegrini
SemanticWeb School
Zentrum fürWissenstransfer
Lerchenfelder Gürtel 43
1160Wien Top 5/2
Austria
E-mail: t.pellegrini@sematic-web.at

Prof. Klaus Tochtermann
Know-Center GmbH
Inffeldgasse 21
8010 Graz
Austria
E-mail: ktochter@know-center.at

Sören Auer
University of Leipzig
Institute for Applied Informatics
Johannisgasse 26
04109 Leipzig
Germany
E-mail: auer@uni-leipzig.de

Sebastian Schaffert
Salzburg Research
ForschungsgesellschaftmbH
Jakob Haringer Straße
5/III 5020 Salzburg
Austria
E-mail: sebastian.schaffert@salzburgresearch.at

ISBN 978-3-642-02183-1   e-ISBN 978-3-642-02184-8

DOI 10.1007/978-3-642-02184-8

Studies in Computational Intelligence   ISSN 1860949X

Library of Congress Control Number: Applied for

Typeset & Cover Design: Scientific Publishing Services Pvt. Ltd., Chennai, India.

Printed in acid-free paper

9 8 7 6 5 4 3 2 1

springer.com

Preface

The book at hand reflects on the increasing convergence of Social Media and Semantic Web technologies. It was the editors' intention to collect up-to-date and high quality contributions that illustrate various approaches to this young and emerging technology area. To guarantee this we invited best-of authors from the international conferences I-SEMANTICS and I-KNOW from the years 2007 & 2008 that took place in Graz / Austria to prepare extended contributions that illustrate this agile and fast-evolving development and hint at trends that might accompany us for the years to come. More than 60 authors from 12 countries with either academic or industrial background provided us with 20 articles covering a broad range of topics related to frameworks, applications and concrete use cases.

After more than a year of preparation we would like to thank all authors for sharing their experiences and insights with the reader. We also would like to thank them for their motivation, dedication and creativity uncomplainingly including our feedback into their contributions thereby ensuring the high quality of the book. Special thanks also go to Gisela Granitzer from the Know Center in Graz who accompanied the editorial process with valuable comments and contributions and helped us to finally bringing this volume to you.

Have an inspiring reading!

Vienna, March 23, 2009

Tassilo Pellegrini
Sören Auer
Sebastian Schaffert
Klaus Tochtermann

Contents

Introduction

Networked Knowledge - Networked Media: - Bringing the
Pieces Together 1
Tassilo Pellegrini, Sören Auer, Sebastian Schaffert,
Klaus Tochtermann

Frameworks and Infrastructures

RDF Support in the Virtuoso DBMS 7
Orri Erling, Ivan Mikhailov

Semantic Task Management Framework: Bridging
Information and Work 25
Ernie Ong, Uwe V. Riss, Olaf Grebner, Ying Du

AUTOMS-F: A Framework for the Synthesis of Ontology
Mapping Methods 45
Alexandros G. Valarakos, Vassilis Spiliopoulos, George A. Vouros

Developing Semantic Web Applications with the OntoWiki
Framework 61
Norman Heino, Sebastian Dietzold, Michael Martin, Sören Auer

Conceptual Foundations for a Service-oriented Knowledge
and Learning Architecture: Supporting Content, Process
and Ontology Maturing 79
Andreas Schmidt, Knut Hinkelmann, Tobias Ley,
Stefanie Lindstaedt, Ronald Maier, Uwe Riss

ARS/SD: An Associative Retrieval Service for the
Semantic Desktop 95
Peter Scheir, Chiara Ghidini, Roman Kern, Michael Granitzer,
Stefanie N. Lindstaedt

GRISINO - A Semantic Web Services, Grid Computing
and Intelligent Objects Integrated Infrastructure 113
*Tobias Bürger, Ioan Toma, Omair Shafiq, Daniel Dögl,
Andreas Gruber*

Application Areas

Collaborative Web-Publishing with a Semantic Wiki 129
Rico Landefeld, Harald Sack

Collaborative Wiki Tagging 141
Milorad Tosic, Valentina Nejkovic

O'CoP, an Ontology Dedicated to Communities of
Practice 155
Amira Tifous, Adil El Ghali, Alain Giboin, Rose Dieng-Kuntz

Incremental Approach to Error Explanations in
Ontologies 171
Petr Křemen, Zdeněk Kouba

Using Ontologies Providing Domain Knowledge for Data
Quality Management 187
Stefan Brüggemann, Fabian Grüning

Semantic Search and Visualization of Time-Series Data 205
Tatiana von Landesberger, Viktor Voss, Jörn Kohlhammer

An Evaluation Framework and Adaptive Architecture for
Automated Sentiment Detection 217
Stefan Gindl, Johannes Liegl, Arno Scharl, Albert Weichselbraun

Managing Ontology Lifecycles in Corporate Settings 235
Markus Luczak-Rösch, Ralf Heese

A Semantic Policy Management Environment for
End-Users and Its Empirical Study 249
*Anna V. Zhdanova, Joachim Zeiß, Antitza Dantcheva, Rene Gabner,
Sandford Bessler*

Use Cases: From Digital TV to Health Care Systems

User-Driven Semantic Wiki-Based Business Service
Description 269
Heiko Paoli, Andreas Schmidt, Peter C. Lockemann

Facilitating Knowledge Management in Pervasive Health Care Systems 285
Bo Hu, Srinandan Dasmahapatra, Paul Lewis, David Dupplaw, Nigel Shadbolt

Integrating Semantic Technologies with Interactive Digital TV 305
Antonis Papadimitriou, Christos Anagnostopoulos, Vassileios Tsetsos, Sarantis Paskalis, Stathes Hadjiefthymiades

Marrying Game Development with Knowledge Management: Challenges and Potentials 321
Jörg Niesenhaus, Steffen Lohmann

Author Index 337

Subject Index 339

Networked Knowledge - Networked Media: - Bringing the Pieces Together

Tassilo Pellegrini, Sören Auer, Sebastian Schaffert, and Klaus Tochtermann

The book title Networked Knowledge - Networked Media reflects on the convergence of Social Media and the Semantic Web. When these developments became popular a few years ago it was a simple co-existence between the two, but in the meantime they have increasingly melted making it impossible to think of knowledge technologies without thinking of the Semantic Web.

Semantic Web principles have not only proven applicable at the technological level solving problems such as data integration and data quality management, but have also proved their applicability at the service level networking knowledge and people across diverse media, supporting search and retrieval, recommendation, collaboration and even trust and accountability issues. Although basic frameworks and infrastructures are essential prerequisites for this, it is not the technology per se but the added value derived from it that convinces about the practical relevance of the Semantic Web. But how good can networked knowledge be without people creating, using and sharing it?

Since the emergence of the Web 2.0 actively generating and sharing knowledge has become a common practice. The technological enabler to this are Social Media including Wikis, Weblogs or Social Networks. Initially almost exclusively used by experts and geeks, today Social Media are widely accepted and common in educational, research and business contexts. However, as we can witness at the moment only the combination of Social Media and the Semantic Web brings both developments to their full potential.

Conceptual Foundations

The term "networked knowledge" exemplifies several important facets of knowledge: first, in a "knowledge society" knowledge needs to be connected in order to generate new knowledge or innovation, which can be realised by Semantic Web technologies. Second, knowledge also needs to be shared among people in order to be used effectively, and much of this sharing is based on collaboration, social software, and social networks. And third, knowledge is never isolated but always embedded in a context, connected with other information.

Where "networked knowledge" mostly describes networks on a conceptual level, the corresponding term "networked media" addresses the technological feasibility of integrated information and communication environments that connect and explore knowledge distributed over several systems and locations. Additionally the term also indicates that nowadays, we are not only speaking about connecting textual content but about connecting different media like video and audio content, images, documents, and facts to create novel software applications that support diverse knowledge processes.

Both notions are deeply rooted in the understanding of the computer as a medium and its multiple connotations. Networked media refers to several epistemological concepts that have accompanied the evolution of the computer as an information and communication device. First, pioneers like Vanevar Bush (1945), Douglas Engelbart (1962) or J.C.R. Licklider & Robert W. Taylor (1968) created the idea of the "scientific workplace", where the computer aids complex workflows and enhances the cognitive capabilities of its users. Second, thinkers like Sherry Turkle (1984), Donna Haraway (1991) and Howard Rheingold (1993) drove the attention to the human-computer-interface and the role a networked computer plays in constructing identities and virtual communities. Third, theoreticians like Manuel Castells (1996) created the vision of a networked society, in which connected computers provide the viral infrastructure - the Information Superhighway (Al Gore 1994) or the European Information Society (Martin Bangemann 1994) - for all sorts of social transactions. And finally we can witness a new branch of discourse that reflects on the ongoing integration and transformation of private and public life in terms of normative changes of privacy, property, labour and social relations by authors like Yochai Benkler (2006) or Rishab Ayer Gosh (2005).

Given this vast richness of theoretical reflexivity it is interesting to observe that one of the still dominating notions in contemporary computer science is rooted in the "scientific workplace" tradition. It seems that with the ongoing technological progress old ideas finally become technologically feasible and ready for the masses, opening new challenges of technological, political and economic nature. This trend is best exemplified by the discourse about the Semantic Web, where proponents like Tim Berners-Lee (2009) regularly refer to aspects of knowledge media and cognitive enhancement through human-computer-interaction to explain the practical value of it.

Contributions of This Book

The volume aims at supplying practitioners as well as academic and industrial researchers with fundamental knowledge about technologies, methodologies and tools for building networked knowledge applications. Consequently, contributions of the book are grouped into the chapters Frameworks and Infrastructure, Application Areas and Use Cases.

In the first part of the book on *Frameworks and Infrastructure* the individual chapters present tools and technologies, which lay the foundation for the implementation of networked knowledge and media applications. RDF and RDF-based knowledge representation techniques evolved into industry standards for developing networked knowledge and media applications. Hence, Orri Erling and Ivan Mikhailov present current developments of one of the most advanced RDF knowledge stores - Virtuoso. A framework, which uses Virtuoso as a persistence layer is the Semantic Task Management Framework (STMF) presented in a paper by Ernie Ong, Uwe V. Riss, Olaf Grebner and Ying Du from SAP research. It represents a platform for establishing a task-oriented ecosystem for desktop applications built on top of it. A framework which addresses the heterogeneity and diversity of networked knowledge representations is the AUTOMS-F framework for the synthesis of ontology mapping methods as presented by Alexandros G. Valarakos, Vassilis Spiliopoulos, and George A. Vouros. AUTOMS-F facilitates rapid prototyping and adapts some well established programming design patterns for the development of synthesized mapping methods. The OntoWiki framework for the development of Semantic Web applications is presented in a chapter by Norman Heino, Sebastian Dietzold, Michael Martin, and Sören Auer. Besides being a Semantic Wiki it supports the rapid development of customized Web applications based on Semantic Web standards such as RDF, RDFa and SPARQL. ARS/SD: An Associative Retrieval Service for the Semantic Desktop aims at improving retrieval performance in a setting where resources are sparsely annotated with semantic information and is introduced in a chapter by Peter Scheir, Chiara Ghidini, Roman Kern, Michael Granitzer and Stefanie N. Lindstaedt. The first part of the book is concluded by a chapter on GRISINO - an infrastructure integrating Semantic Web Services, Grid Computing and Intelligent Objects. GRISINO aims at facilitating next generation distributed, networked applications and is presented by Tobias Bürger, Ioan Toma, Omair Shafiq, Daniel Dögl, and Andreas Gruber.

In the second part on *Application Areas* selected applications of networked knowledge and networked media are introduced. In the first chapter of this part Rico Landefeld and Harald Sack discuss how collaborative Web-Publishing can be realized using a Semantic Wiki. They showcase their Maariwa architecture and implementation, which particularly aims at dramatically simplifying the semantic annotation and textual Wiki content. In the next chapter Milorad Tosic and Valentina Nejkovic present an approach how Wiki technology can be employed for collaborative semantic tagging. The use of semantic representations for facilitating the collaboration of communities of practice is discussed in a chapter by Amira Tifous, Adil El Ghali, Alain Giboin and Rose Dieng-Kuntz from INRIA. The problem of explaining modeling errors in description logic ontologies is addressed by a chapter by Petr Kremen and Zdenek Kouba. The quality of instance data is addressed by a contribution from Stefan Brüggemann and Fabian Grüning, which uses ontologies providing domain background knowledge. Tatiana von Landesberger, Viktor Voss and Jörn Kohlhammer investigate how networked knowledge based semantic search and visualization can support the exploration of time-series data in the conweaver system. The sentiment dimension of networked knowledge is explored by Stefan Gindl, Johannes Liegl, Arno Scharl and Albert Weichselbraun in a chapter, which presents an evaluation framework as well as an adaptive architecture for automated sentiment detection. Better ontology management support in corporate lifecycles is tackled by the contribution of Markus Luczak-Rösch and Ralf Heese, which incorporates adaptive knowledge engineering techniques such as Wikis, Weblogs and the like. Managing the end-user access to networked knowledge by means of a semantic policy management environment is the concern of a chapter by Anna V. Zhdanova, Joachim Zeiß, Antitza Dantcheva, Rene Gabner, Sandford Bessler.

The *Use Cases part* of the book showcases some concrete application scenarios of networked knowledge and media technologies. The first chapter by Heiko Paoli, Andreas Schmidt, and Peter C. Lockemann explores how business service descriptions can be obtained in user-driven ways based on semantic Wiki technology. A contribution by Bo Hu, Srinandan Dasmahapatra, Paul Lewis, David Dupplaw and Nigel Shadbolt applies networked knowledge management in the context of pervasive health care systems. The chapter by Antonis Papadimitriou, Christos Anagnostopoulos, Vassileios Tsetsos, Sarantis Paskalis and Stathes Hadjiefthymiades presents the POLYSEMA approach for integrating semantic technologies with Interactive Digital TV. Last but not least, a chapter by Jörg Niesenhaus and Steffen Lohmann is devoted to the application of networked knowledge in the domain of computer game development.

Outlook and Future Challenges

After more than 10 years of research and development aiming at transforming the Web of documents into a Web of interconnected knowledge we observe that this transition will rather be an long-running evolutionary process than a rapid technological revolution. While the contributions in this volume already address many of the upcoming issues, there are several key areas that we think will need to receive significant attention in the future:

Web-scale data integration. Current Semantic Web applications run mostly in isolated environments and are concerned with building consistent knowledge structures in limited domains. The big promise of the "Web of Data", however, still remains to a large extent unaddressed. Initiatives like Linking Open Data offer first ideas how such web-scale data integration can be realised, but work in this area is currently still in its infancy and many issues are not yet tackled. Challenges in this area include efficient distributed querying and reasoning with several sources, dealing with inconsistencies, uncertainties and contradictions, finding relevant linked data sources, or performance in working with huge amounts of data.

Reasoning and querying. Current reasoning on the Semantic Web is mostly focused on a very particular scenario: checking consistency of the formalised knowledge. However, not only will data on the web scale inevitably be always inconsistent, users will also demand for other kinds of reasoning that is actually much more useful for them: deriving new knowledge, representing rule-based knowledge, presenting relevant knowledge, etc. are all reasoning tasks that have not been addressed much in research and even less in applications.

Semantic search. Searching for content is still either classical full-text search or a structural search over the formalized knowledge. Arguably, none of these kinds of searches are very semantic (at least from a user perspective), and they also cannot really be combined. Semantic search is a big challenge indeed, because it is one of the few points where semantics inevitably need to be exposed to the ordinary user, but it still needs to be easy to use. There are a number of different approaches from different research fields towards this issue, e.g. statistical approaches like Latent Semantic Indexing or NLP approaches like POS tagging, but up until now none of them really reaches the goal of an easy-to-use semantic search.

References

1. Bangemann, M.: Europe and the Global Information Society. Bangemann Report (1994), http://www.cyber-rights.org/documents/bangemann.htm (last viewed February 20, 2009)
2. Benkler, Y.: The Wealth of Networks: How Social Production Transforms Markets and Freedom. Yale University Press, Yale (2006)
3. Berners-Lee, T.: The next Web of open, linked data. TED Talk (Feburary 2009), http://www.ted.com/index.php/talks/tim_berners_lee_on_the_next_web.html (last viewed March 10, 2009)
4. Bush, V.: As We May Think (1945), http://www.theatlantic.com/doc/194507/bush (last viewed February 20, 2009)
5. Castells, M.: The Rise of the Network Society. Blackwell Publishers, London (1996)
6. Engelbart, D.: Augmenting Human Intellect: A Conceptual Framework (1962), http://www.bootstrap.org/augdocs/friedewald030402/augmentinghumanintellect/ahi62index.html (last viewed February 20, 2009)
7. Gore, A.: Remarks on the Information Superhighway Prepared for Delivery by Vice President Al Gore, Royce Hall, UCLA Los Angeles, California (January 11, 1994), http://www.ibiblio.org/icky/speech2.html (last viewed February 20, 2009)

8. Haraway, D.: A Cyborg Manifesto: Science, Technology, and Socialist-Feminism in the Late Twentieth Century. In: Simians, Cyborgs and Women: The Reinvention of Nature, pp. 149–181. Routledge, New York (1991)

9. Licklider, J.C.R., Taylor, R.W.: The Computer as Communication Device. Science & Technology (April 1968), `http://memex.org/licklider.pdf` (last viewed February 20, 2009) (reprinted)

10. Rheingold, H.: The Virtual Community: Homesteading on the Electronic Frontier. Addison-Wesley, Reading (1993)

11. Turkle, S.: The Second Self: Computers and the Human Spirit. Simon & Schuster, New York (1984)

12. Gosh, R.A.: Collaboration, Ownership and the Digital Economy. MIT Press, Cambridge (2005)

RDF Support in the Virtuoso DBMS

Orri Erling and Ivan Mikhailov

Abstract. This paper discusses RDF related work in the context of OpenLink Virtuoso, a general purpose relational / federated database and applications platform. The use cases are dual 1. large RDF repositories 2. making arbitrary relational data queriable with SPARQL and RDF by mapping on demand. We discuss adapting a relational engine for native RDF support with dedicated data types, bitmap indexing and SQL optimizer techniques. We discuss adaptations of the query engine for running on shared nothing clusters, providing virtually unbounded scalability for RDF or relational warehouses. We further discuss mapping existing relational data into RDF for SPARQL access without converting the data into physical triples. We present conclusions and metrics as well as a number of use cases, from DBpedia to bio informatics and collaborative web applications.

1 Introduction and Motivation

Virtuoso is a multi-protocol server providing ODBC/JDBC access to relational data stored either within Virtuoso itself or any combination of external relational databases. Besides catering for SQL clients, Virtuoso has a built-in HTTP server providing a DAV repository, SOAP and WS* protocol end points and dynamic web pages in a variety of scripting languages. Given this background and the present emergence of the semantic web, incorporating RDF functionality into the product is a logical next step. RDF data has been stored in relational databases since the inception of the model [2][14]. Performance considerations have however led to the

Orri Erling
OpenLink Software, 10 Burlington Mall Road Suite 265 Burlington, MA 01803 U.S.A
e-mail: oerling@openlinksw.com
http://www.openlinksw.com

Ivan Mikahilov
OpenLink Software,
e-mail: imikhailov@openlinksw.com

S. Schaffert et al. (Eds.): Networked Knowledge - Networked Media, SCI 221, pp. 7–24.
springerlink.com　　　　　　　　　　　　　　© Springer-Verlag Berlin Heidelberg 2009

development of custom RDF engines, e.g. RDF Gateway [13], Kowari [15] and others. Other vendors such as Oracle and OpenLink have opted for building a degree of native RDF support into an existing relational platform.

For a production strength DBMS, we need a balanced set of capabilities. Failure with any of the below may cost orders of magnitude in performance.

- Doing the right things — A bad query plan can destroy any possibility of performance, no matter how good the rest is.
- Doing things in the right place — Processing must take place close to the data. For example, if an unmodified RDBMS is used as back end for SPARQL, impedance mismatch between type systems may cause the SPARQL front end to do things that belong to the back end. If SPARQL is mapped to SQL, we must get a single SQL statement with all joins inside, so that the back end database can optimize the query. Failure to do either will kill performance by requiring client-server round trips.
- Doing things in memory — A single disk read takes the time of thousands of table lookups in memory. Inefficient use of space leads to needless disk access. This is specially bad with RDF, where the data model is not geared to application specific disk layout.
- Scale — If RDF is the means of turning the web into a database, then scale is important. This means that a scale out approach becomes inevitable at some point. When we move from a single server to multiple servers, performance dynamics change qualitatively.

We shall discuss our response to all these challenges in the course of this paper.

2   Triple Storage

Virtuoso's initial storage solution is fairly conventional: a single table of four columns holds one quad, i.e. triple plus graph per row. The columns are G for graph, P for predicate, S for subject and O for object. P, G and S are IRI ID's, for which we have a custom data type, distinguishable at run time from integer even though internally this is a 32 or 64 bit integer. The O column is of SQL type ANY, meaning any serializable SQL object, from scalar to array or user defined type instance. Indexing supports a lexicographic ordering of type ANY, meaning that with any two elements of compatible type, the order is that of the data type(s) in question with default collation.

Since O is a primary key part, we do not wish to have long O values repeated in the index. Hence O's of string type that are longer than 12 characters are assigned a unique ID and this ID is stored as the O of the quad table. For example Oracle [16] has chosen to give a unique ID to all distinct O's, regardless of type. We however store short O values inline and assign ID's only to long ones.

Generally, triples should be locatable given the S or a value of O. To this effect, the table is represented as two covering indices, G, S, P, O and O, G, P, S. Since both indices contain all columns, the table is wholly represented by these two indices and no other persistent data structure needs to be associated with it. Also there is never a need for a lookup of the main row from an index leaf.

Using the Wikipedia data set [19] as sample data, we find that the O is on the average 9 bytes long, making for an average index entry length of 6 (overhead) + 3 * 4 (G, S, P) + 9 (O) = 27 bytes per index entry, multiplied by 2 because of having two indices.

We note however that since S is the last key part of P, G, O, S and it is an integer-like scalar, we can represent it as a bitmap, one bitmap per distinct P, G, O. With the Wikipedia data set, this causes the space consumption of the second index to drop to about a third of the first index. We find that this index structure works well as long as the G is known. If the G is left unspecified, other representations have to be considered, as discussed below.

For example, answering queries like

```
graph <my-friends> {
    ?s sioc:knows people:John , people:Mary }
```

the index structure allows the AND of the conditions to be calculated as a merge intersection of two sparse bitmaps.

The mapping between an IRI ID and the IRI is represented in two tables, one for the namespace prefixes and one for the local part of the name. The mapping between ID's of long O values and their full text is kept in a separate table, with the full text or its MD5 checksum as one key and the ID as primary key. This is similar to other implementations.

The type cast rules for comparison of data are different in SQL and SPARQL. SPARQL will silently fail where SQL signals an error. Virtuoso addresses this by providing a special QUIETCAST query hint. This simplifies queries and frees the developer from writing complex cast expressions in SQL, also enhancing freedom for query optimization.

Other special SPARQL oriented accommodations include allowing blobs as sorting or distinct keys and supporting the IN predicate as a union of exact matches. The latter is useful for example with FROM NAMED, where a G is specified as one of many.

Compression

We have implemented compression at two levels. First, within each database page, we store distinct values only once and eliminate common prefixes of strings. Without key compression, we get 75 bytes per triple with a billion-triple LUBM data set (LUBM scale 8000). With compression, we get 35 bytes per triple. Thus, key compression doubles the working set while sacrificing no random access performance. A single triple out of a billion can be located in less than 5 microseconds with or without key compression. We observe a doubling of the working set when using 32 bit IRI ID's. Going from 32 bit IRI ID's to 64 has hardly any effect with key compression since most ID's are not stored at full length.

When applying gzip to database pages, we see a typical compression to 40% of original size, even after key compression. This is understandable since indices are by nature repetitive, even if the repeating parts are shortened by key compression. Over 99% of 8K pages filled to 90% compress to less than 3K with gzip at default compression settings. This does not improve working set but saves disk. Detailed performance impact measurement is yet to be made.

Alternative Index Layouts

Most practical queries can be efficiently evaluated with the GSPO and OGPS indices. Some queries, such as ones that specify no graph are however next to impossible to evaluate with any large data set. Thus we have experimented with a table holding G, S, P, O as a dependent part of a row id and made 4 single column bitmap indices for G, S, P and O. In this way, no combination of criteria is penalized. However, performing the bitmap AND of 4 given parts to check for existence of a quad takes 2.5 times longer than the same check from a single 4 part index. The SQL optimizer can deal equally well with this index selection as any other, thus this layout may prove preferable in some use cases due to having no disastrous worst case.

In practice, we find it preferable to use many covering indices. If queries must be made against the union of all graphs, the preferred index layout is SPOG, GPOS, POGS, OPGS. The three last are bitmap indices.

Using this index layout plus full text index on all literals, The Billion Triples Challenge data set, 1150M triples, including DBpedia, Freebase, US Census and numerous web crawls took 120GB of allocated database pages.

3   SPARQL and SQL

Virtuoso offers SPARQL inside SQL, somewhat similarly to Oracleś RDF_MATCH table function. A SPARQL subquery or derived table is accepted either as a top level SQL statement of wherever a subquery or derived table is accepted. Thus SPARQL inherits all the aggregation and grouping functions of SQL, as well as any built-in or user defined functions. Another benefit of this is that all supported CLI's work directly with SPARQL, with no modifications. For example, one may write a PHP web page querying the triple store using the PHP to ODBC bridge. The SPARQL text simply has to be prefixed with the SPARQL keyword to distinguish it from SQL. A SPARQL end point for HTTP is equally available. We have further Virtuoso drivers implemented the popular Jena, Sesame and Redland RDF frameworks. Thus application written in these can transparently use Virtuoso as the storage and query processor.

Internally, SPARQL is translated into SQL at the time of parsing the query. If all triples are in one table, the translation is straightforward, with union becoming a SQL union and optional becoming a left outer join. Since outer joins can be nested to arbitrary depths inside derived tables in Virtuoso SQL, no special problems are encountered. The translator optimizes the data transferred between parts of the queries, so that variables needed only inside a derived table are not copied outside of it. If cardinalities are correctly predicted, the resulting execution plans are sensible. SPARQL features like construct and describe are implemented as user defined aggregates.

SQL Cost Model and RDF Queries

When all triples are stored in a single table, correct join order and join type decisions are difficult to make given only the table and column cardinalities for the RDF triple or quad table. Histograms for ranges of P, G, O, and S are also not useful. Our solution for this problem is to go look at the data itself when compiling the query. Since the SQL compiler is in the same process as the index hosting the data, this can be done whenever one or more leading key parts of an index are constants known at compile time. For example, in the previous example, of people knowing both John and Mary, the G, P and O are known for two triples. A single lookup in log(n) time retrieves the first part of the bitmap for

```
((G = <my-friends>) and (P = sioc:knows) and
  (O = <http://people.com/people#John>) )
```

The entire bitmap may span multiple pages in the index tree but reading the first bitts and knowing how many sibling leaves are referenced from upper levels of the tree with the same P, G, O allows calculating a ballpark cardinality for the P, G, O combination. The same estimate can be made either for the whole index, with no key part known, using a few random samples or any number of leading key parts given. While primarily motivated by RDF, the same technique works equally well with any relational index.

Basic RDF Inferencing

Much of basic T box inferencing such as subclasses and subproperties can be accomplished by query rewrite. We have integrated this capability directly in the Virtuoso SQL execution engine. With a query like

```
select ?person where { ?person a lubm:Professor }
```

we add an extra query graph node that will iterate over the subclasses of class lubm:Professor and retrieve all persons that have any of these as rdf:type. When asking for the class of an IRI, we also return any superclasses. Thus the behavior is indistinguishable from having all the implied classes explicitly stored in the database.

For A box reasoning, Virtuoso has special support for owl:sameAs. When either an O or S is compared with equality with an IRI, the IRI is expanded into the transitive closure of its owl:sameAs synonyms and each of these is tried in turn. Thus, when owl:sameAs expansion is enabled, the SQL query graph is transparently expanded to have an extra node joining each S or O to all synonyms of the given value. Thus,

```
select ?lat where { <Berlin> has_latitude ?lat }
```

will give the latitude of Berlin even if <Berlin> has no direct latitude but geo:Berlin does have a latitude and is declared to be synonym of <Berlin>.

The owl:sameAs predicate of classes and properties can be handled in the T box through the same mechanism as subclasses and subproperties.

Virtuoso has SPARQL extensions for subqueries, including a transitive subquery feature. For example the pattern `<john> foaf:knows ?person option (transitive)` will bind `?person` to everybody `<john>` knows plus everybody they know and so on up to full transitive closure. There are further options for limiting the depth and returning the path leading to each binding and so forth. The query

```
select ?p2 where {
  {  select ?p1 ?p2 where {
       ?p1 foaf:knows ?p2 . ?p2 foaf:knows ?p1 }
    } option transitive (in (?p1) out (?p2)) .
filter (?p1 = <john>) }
```

would only consider reciprocal `foaf:knows` relations. Thus the step in the transitivity can be complex. If both ends of a transitive relation are given, then the feature can be used for obtaining the paths that connect the two ends. More examples are at [6].

Data Manipulation

Virtuoso supports the SPARUL SPARQL extension, compatible with JENA [14]. Updates can be run either transactionally or with automatic commit after each modified triple. The latter mode is good for large batch updates since rollback information does not have to be kept and locking is minimal.

Full Text

All or selected string valued objects can be full text indexed. Queries like

```
select ?person from <people> where {
    ?person a person ; has_resume ?r .
    ?r bif:contains 'SQL and "semantic web"' }
```

will use the text index for resolving the pseudo-predicate `bif:contains`.

Business Intelligence Extensions

For SPARQL to compete with SQL for analytics, extensions such as returning expressions, subqueries in group patterns and in expressions, explicit grouping and the like are needed. The extended language is referred to as "SPAQL-BI".

Basic SQL style aggregation is supported through queries like

```
select ?product sum (?value) from <sales> where {
    <ACME> has_order ?o .    ?o has_line ?ol .
    ?ol has_product ?product ; has_value ?value }
```

This returns the total value of orders by ACME grouped by product.

RDF Sponge

The Virtuoso SPARQL protocol end point can retrieve external resources for querying. Having retrieved an initial resource, it can automatically follow selected IRI's for retrieving additional resources. Several modes are possible: follow only selected links, such as `sioc:see_also` or try dereferencing any intermediate query results, for example. Resources thus retrieved are kept in their private graphs or they can be merged into a common graph. When they are kept in private graphs, HTTP caching headers are observed for caching, the local copy of a retrieved remote graph is usually kept for some limited time. The sponge procedure is extensible so it can extract RDF data from non-RDF resources with pluggable RDF-izers called cartridges. Over 30 such cartridges exist to date, covering GRDDL, RDFA, microformats, many XML formats such as XBRL and more. This provides a common tool for traversing sets of interlinked documents such as personal FOAFs that refer to each other.

4   Clustering and Scalability

For the entire history of RDF and the Semantic Web, one of the dominant themes of the discourse has been scalability. The data web can be said to be one of the frontiers of databasing, as data volumes are easily very large and since there is generally no application-specific table layout and index structure, things take more space than with the corresponding relational representation. This section discusses the work done in scale-out clustering in Virtuoso. At the time of writing, Virtuoso has a cluster edition that runs on shared nothing clusters of commodity servers. This is being used for hosting large parts of the linked open data cloud.

As we move in the direction of parallelism, dynamics of performance change significantly: when moving from a single CPU to multiple CPU's or cores, the cost of resource contention between threads jumps significantly. Thus, if special care is not taken, a thread blocking to wait for another is so expensive that any gains from parallelism may be entirely lost. A single wait may cost whole microseconds. When we move from one multithreaded process to multiple multithreaded server processes connected by a network, the network latency becomes the dominant cost factor. Within a single machine, a message round trip with empty message and no processing costs about 50 microseconds, including thread switching at both ends. With a 1Gbit Ethernet added to the mix, the cost goes to about 150 microseconds, assuming no contention on the network.

These basic facts dictate the architecture of any DBMS for server clusters. The issues of query optimization are largely the same as for single servers but the execution engine has entirely different priorities.

The cost of finding a single quad from 100 million is about 5 microseconds. This is a tiny fraction of the overhead of doing any operation involving any interprocess communication. For this reason, it is vital to group as many operations as possible within a single message.

Clustered databases usually use some partitioning scheme, where the values of one or more key columns dictate which server will store the row. Also non-partitioned cluster systems such as Oracle RAC exist. With Virtuoso, we decided to use hash partitioning according to the subject or object of a triple. Thus, a single triple is indexed many ways and each index may be partitioned differently, there is no need for all the entries of a single quad to be on the same server. In relational applications, Virtuoso allows specifying partitioning index by index. We do not use the graph or predicate of a quad for partitioning since these may have very uneven distributions.

4.1 Query Execution Model

When the network latency is the main cost factor, having a maximally asynchronous and non-blocking message flow between the processes participating in a query is necessary. A query is addressed to an arbitrary node of the cluster. This node is called the *query coordinator* and it is responsible for dispatching the query to the relevant cluster nodes and assembling the response.

The basic query is a set of nested loops. Take for example

```
select * where {
    <john> foaf:knows ?person .
    ?person foaf:mbox ?mbox ; foaf:nick ?nick }
```

This can be seen as a pipeline of 3 stages. The first produces all the friends of <john>. The second takes the set of friends and adds the foaf:mbox for each. The third adds the foaf:nick to the binding. The results from the 3rd stage can be returned to the client. This joins from subject to object. Suppose the index from subject to object is partitioned by subject, which is quite natural. For the first, we know the subject, so we know which partition has the friends of <john>. We ask for them and get them in a single message exchange, unless there are megabytes worth of them in which case we would ask for the next batch when near the end of the first batch. Each of these is a subject in its turn, thus for each friend we know which partition has the foaf:mbox. We group all messages headed for each partition together and send them and again gather the results. The same process is repeated for the foaf:nick.

This is the naive way of evaluating the query. Even this produces fair parallelism through bundling messages in sufficiently large batches. We firstly note that the subject for the foaf:mbox and foaf:nick patterns is the same, hence they are always in the same partition since the index is partitioned by subject. Thus we get the two in a single operation: we send to each partition that has a friend of <john> the query fragment

```
{ ?person foaf:mbox ?mbox ; foaf:nick ?nick }
```

If there were other conditions such as filter (?nick != "Alice") we could bundle these in as well. This removes a whole message round trip and nearly halves the network traffic for the query.

Next we see that the partition that evaluates

```
{ <john> foaf:knows ?person }
```

does not have to return the set of friends to the query coordinator but can by it-self dispatch these to the appropriate partition for each. This eliminates yet another round trip. Now the query runs in 3 message steps: 1. ask for friends of <john> 2. each partition that can have one of the friends gets all the friends in its range and gets their nicks and mboxes and 3. all completed bindings are returned to the coordinator.

This last optimization is applicable when the results do not have to be returned in any given order. Adding an *order by* at the end of the query takes care of this. The cost of the final *order by* is negligible compared to the latency and data transfer savings.

When the query involves aggregation or grouping, the aggregation takes place one each involved partition separately and is collected to the coordinator at the end in one message round trip.

Consider

```
base <http://myopenlink.net/dataspace/>
select ?o ?distance
   ((select count (*) where {?o foaf:knows ?xx}))
where {
     { select ?s ?o where { ?s foaf:knows ?o }
       } option (transitive, t_in(?s), t_out(?o),
       t_min (1), t_max (4), t_distinct,
       t_step ('step_no') as ?distance) .
     filter (?s = <person/kidehen#this>)
   } order by ?distance desc 3 limit 50
```

This starts with <http://myopenlink...kidehen#this> and gets all the distinct subjects related to this by 1 to 3 consecutive foaf:knows steps. For each such person, the people this person foaf:knows are counted. The results are returned sorted by distance and descending count of friends.

This is 3 round trips for the transitive foaf:knows up to 3 deep. Each step must return results to the coordinator for handling the distinctness. Then there is one round trip for the friend counts of each person, since

```
select count (*) where {?o foaf:knows ?xx}
```

can be evaluated within one partition for each ?o. Thus the count subquery in the selection is also parallelized. The rest is local processing on the coordinator.

Since the coordinator is an arbitrary node of the cluster, it will itself handle the bindings that fall into its partition in addition to overall query coordinating.

4.2   Performance

With the Billion Triples Challenge data set, we have 25 million foaf:knows triples. Of these, 92K are such that for ?x foaf:knows ?y there is a ?y foaf:knows ?x in some graph. The query is

```
select count (*) where {
  ?x foaf:knows ?y . ?y foaf:knows ?x }
```

This runs in 7.7 seconds on two dual 4 core Xeon machines, for a total of 3.4 million random triple lookups per second. The database is partitioned in 12, 6 partitions per machine. We have 11.5 of the 16 cores busy for the query, where 12 cores would be the maximum. The interconnect traffic is only 19 MB/s, meaning that we are using a fraction of the total interconnect bandwidth of dual 1Gb Ethernets.

In this situation, each partition directly sends the (?x, ?y) pairs to the partition that holds the possibly existing (?y, ?x) pair. If instead we pass these through a single coordinator node, the execution time jumps to 35 seconds and the interconnect traffic to 39 MB/s.

Experience shows that series of simple joins, single triple optionals or existence tests followed by aggregation or sorting will scale near linearly with the addition of hardware. More complex query structures require passing data through the coordinator at least part of the time, for example between steps of a transitive subquery or for distincts and complex existence subqueries. Even then, between 4 to 6 cores can be busy for a single query. Disk performance always increases linearly with clustering, since even the most naive message pattern will deliver tasks over an order of magnitude faster than a disk bound process can handle them. Since each node of the cluster caches its partition of the data and nothing else, any added nodes linearly add to the main memory, which is the most determining resource in any DBMS that is primarily doing random access.

5   Mapping Relational Data into RDF for SPARQL Access

RDF and ontologies form the final completing piece of the enterprise data integration puzzle. Many disparate legacy systems may be projected onto a common ontology using different rules, providing instant content for the semantic web. One example of this is OpenLink's ongoing project of mapping popular Web 2.0 applications such as Wordpress, Mediawiki, PHP BB and others onto SIOC through Virtuoso's RDF Views system.

Most data integration done with RDF to date is based on extracting triples from different relational databases and importing these into a triple store. However, when the data volumes are very large or the data is rapidly changing, this becomes impractical. Also, RDBMS's are generally more efficient than triple stores for analytics queries. Maintaining a separate RDF warehouse is extra work. For these reasons, exposing RDB assets as RDF without extract-transform-load (ETL) is desirable.

On the other hand, if the number of distinct data sources is very large, if there is high cost of access or if complex inference and postprocessing of the data is needed, then a degree of RDF warehousing is appropriate.

The problem domain is well recognized, with work by D2RQ [3], SPASQL [10], DBLP [5] among others. Virtuoso differs from these primarily in that it combines the mapping with native triple storage and may offer better distributed SQL query optimization through its long history as a SQL federated database.

In Virtuoso, an RDF mapping schema consists of declarations of one or more quad storages. The default quad storage declares that the system table RDF_QUAD consists of four columns (G, S, P and O) that contain fields of stored triples, using special formats that are suitable for arbitrary RDF nodes and literals. The storage can be extended as follows:

An IRI class defines that an SQL value or a tuple of SQL values can be converted into an IRI in a certain way, e.g., an IRI of a user account can be built from the user ID, a permalink of a blog post consists of host name, user name and post ID etc. A conversion of this sort may be declared as bijection so an IRI can be parsed into original SQL values. The compiler knows that a join on two IRIs calculated by same IRI class can be replaced with join on raw SQL values that can efficiently use native indexes of relational tables. It is also possible to declare one IRI class A as subClassOf other class B so the optimizer may simplify joins between values made by A and B if A is bijection.

Most of IRI classes are defined by format strings similar to one used in standard C sprintf function. Complex transformations may be specified by user-defined functions. In any case the definition may optionally provide a list of sprintf-style formats such that that any IRI made by the IRI class always matches one of these formats. SPARQL optimizer pays attention to formats of created IRIs to eliminate joins between IRIs created by totally disjoint IRI classes. For two given sprintf format strings the SPARQL optimizer can find a common subformat of these two or try to prove that no one IRI may match both formats.

```
prefix : <http://www.openlinksw.com/schemas/oplsioc#>
create iri class :user-iri "http://myhost/users/%s"
  ( in login_name varchar not null ) .
create iri class :blog-home "http://myhost/%s/home"
  ( in blog_home varchar not null ) .
create iri class :permalink "http://myhost/%s/%d"
  ( in blog_home varchar not null,
    in post_id integer not null ) .
make :user_iri subclass of :grantee_iri .
make :group_iri subclass of :grantee_iri .
```

IRI classes describe how to format SQL values but do not specify the origin of those values. This part of mapping declaration starts from a set of table aliases, similar to FROM and WHERE clauses of an SQL SELECT statement.

The mapping consists of quad patterns which declare how a quad (triple + graph) can be constructed from relational data. The pattern has typically a constant for the graph and the predicate and columns or groups of columns for the subject and object. The quad pattern may contain additional SQL search conditions to further restrict the scope. When a SPARQL query is compiled, each triple pattern is matched against the quad map patterns and the relevant ones are selected. There is sophisticated logic for pruning out joins that do not make sense.

```
from SYS_USERS as user from SYS_BLOGS as blog
where (^{blog.}^.OWNER_ID = ^{user.}^.U_ID)
```

A quad map value describes how to compose one of four fields of an RDF quad. It may be an RDF literal constant, an IRI constant or an IRI class with a list of columns of table aliases where SQL values come from. A special case of a value class is the identity class, which is simply marked by table alias and a column name.

Four quad map values (for G, S, P and O) form quad map pattern that specify how the column values of table aliases are combined into an RDF quad. The quad map pattern can also specify restrictions on column values that can be mapped. E.g., the following pattern will map a join of SYS_USERS and SYS_BLOGS into quads with :homepage predicate.

```
graph <http://myhost/users>
subject :user-iri (user.U_ID)
predicate :homepage
object :blog-home (blog.HOMEPAGE)
where (not ^{user.}^.U_ACCOUNT_DISABLED) .
```

Quad map patterns may be organized into trees. A quad map pattern may act as a root of a subtree if it specifies only some quad map values but not all four; other patterns of subtree specify the rest. A typical use case is a root pattern that specifies only the graph value whereas every subordinate pattern specifies S, P and O and inherits G from root, as below:

```
graph <http://myhost/users> option (exclusive) {
    :user-iri (user.U_ID)
      rdf:type foaf:Person ;
      foaf:name user.U_FULL_NAME ;
      foaf:mbox user.U_E_MAIL ;
      foaf:homepage :blog-home (blog.HOMEPAGE) . }
```

This grouping is not only a syntax sugar. In this example, exclusive option of the root pattern permits the SPARQL optimizer to assume that the RDF graph contains only triples mapped by four subordinates.

A tree of a quad map pattern and all its subordinates is called "RDF view" if the "root" pattern of the tree is not a subordinate of any other quad map pattern.

Quad map patterns can be named; these names are used to alter mapping rules without destroying and re-creating the whole mapping schema.

The top-level items of the data mapping metadata are quad storages. A quad storage is a named list of RDF views. A SPARQL query will be executed using only quad patterns of views of the specified quad storage.

Declarations of IRI classes, value classes and quad patterns are shared between all quad storages of an RDF mapping schema but any quad storage contains only a subset of all available quad patterns. Two quad storages are always defined: a default that is used if no storage is specified in the SPARQL query and a storage that refers to single table of physical quads.

The RDF mapping schema is stored as triples in a dedicated graph in the RDF_QUAD table so it can be queried via SPARQL or exported for debug/backup purposes.

Virtuoso supports SPARQL Business Intelligence extensions for RDF views as well as for plain triples. Application developers can use SPARQL for sophisticated data mining on large heterogeneous data sets and the related overhead is affordable. The following SQL query is Q18 from the industry standard decision support benchmark TPC H:

```
select
  c_name, c_custkey, o_orderkey, o_orderdate,
  o_totalprice, sum(l_quantity)
from lineitem, orders, customer
where
  o_orderkey in (
      select l_orderkey from lineitem
      group by l_orderkey
      having sum(l_quantity) > 250 )
  and c_custkey = o_custkey
  and o_orderkey = l_orderkey
group by
  c_name, c_custkey, o_orderkey, o_orderdate,
  o_totalprice
order by o_totalprice desc, o_orderdate
```

This retrieves details of large orders, where large is defined as an order where the quantity of the order lines adds up to over 250. This query can not be written in pure SPARQL but Virtuoso offers the needed extensions:

```
select ?cust+>foaf:name ?cust+>tpcd:custkey
  ?ord+>tpcd:orderkey ?ord+>tpcd:orderdate
  ?ord+>tpcd:ordertotalprice
  sum(?li+>tpcd:linequantity)
from <http://example.com/tpcd>
where {
    ?cust a tpcd:customer ; foaf:name ?c_name .
    ?ord a tpcd:order ; tpcd:has_customer ?cust .
    ?li a tpcd:lineitem ; tpcd:has_order ?ord .
      { select ?sum_order
          sum (?li2+>tpcd:linequantity) as ?sum_q
        where {
            ?li2 a tpcd:lineitem ;
              tpcd:has_order ?sum_order . } } .
    filter (?sum_order = ?ord and ?sum_q > 250)
  }
order by
  desc (?ord+>tpcd:ordertotalprice)
  ?ord+>tpcd:orderdate
```

The `?x+>property` notation is a shorthand for `?x property ?value` that does not require `?value` variable to be named and can be used inside an expression. The group by operation is implicit, grouping by the non-aggregates when the selection contains a mix of aggregates and non aggregates.

Both versions take the same time if executed on a 2GHz Xeon-based box running Virtuoso 5.0.9 with canonical TPC H "scale 1" data set. The database takes 1.52 Gb of disk space and it entirely fits into 2.02 Gb of RAM-resident disk buffers. Under such circumstances these queries produce 6621 result rows in 9425 milliseconds.

Translation from SPARQL to SQL costs a small fraction of the time required by the SQL optimizer for finding the best join order. The SQL generated from SPARQL is usually adequate and resembles the equivalent hand-written SQL. All SPARQL expressions resolve to a single SQL statement. If this statement refers to local tables or tables that are all located on the same remote database, the statement is passed as a single statement to the database holding the data. This database does not have to be Virtuoso, since Virtuoso can attach tables from any other RDBMS through its SQL federation feature.

Nevertheless the SPARQL to SQL mapping overhead may become important on a database that handles numerous trivial queries and the database is all in memory; in this case even tens of microseconds per query form a noticeable fraction of total execution time. The overhead is also important for databases with hundreds to thousands of RDF views and where queries match many views. This can happen if the mapping integrates many sources for the same entities. For example, in the Open-Link Data Spaces application suite, there are blog posts, wiki articles, news items etc that are all mapped to the `sioc:Post` RDF type. Saying that something is a `sioc:Post` will pull in a union of these three tables if there is no extra information discriminating which kind of post is meant.

In pathological cases, one can end up with SQL statements of thousands of lines for a line of SPARQL. For example `{?s ?p ?o . ?o ?p ?o2}` would be a union of all columns of all tables joined to another such union. The mapping can prune out the pairs which obviously do not join but still the statement is impractical. Thus using variables in the predicate position is discouraged and specifying RDF types for variables is encouraged.

Using parameterized queries or stored procedures eliminates any mapping overheads. Unfortunately, the SPARQL Web Service Protocol does not provide an interoperable way of passing parameters.

6   Applications and Benchmarks

As of this writing, December 2008, the native Virtuoso triple store is available as a part of the Virtuoso open source and commercial offerings. The RDF Views system is part of the offering but access to remote relational data is limited to the commercial version.

Virtuoso has been used for hosting many of the data sets in the Linking Open Data Project [4], including DBpedia [1], Musicbrainz [21], Geonames [22],

Freebase [18], PingTheSemanticWeb [23] and others. The largest databases are in the single billions of triples. Also the Neurocommons and Bio2RDF data sets are hosted on Virtuoso.

Presently, we are setting up a copy of the entire LOD cloud in a single clustered database and are planning to make this available as a ready made data set that is offered as a collection of paid machine instances on Amazon EC2. In this way, anybody can rent their private copy of the world's linked data.

Web 2.0 Applications

We can presently host many popular web 2.0 applications in Virtuoso, with Virtuoso serving as the DBMS and also optionally as the PHP web server.

We have presently mapped PHP BB, Mediawiki and Drupal into SIOC with RDF Views.

OpenLink Data Spaces (ODS)

ODS is a web applications suite consisting of a blog, wiki, social network, news reader and other components. All the data managed by these applications is available for SPARQL querying as SIOC instance data. This is done through maintaining a copy of the relevant data as physical triples as well as through accessing the relational tables themselves via RDF Views.

Berlin SPARQL Benchmark

Virtuoso was ranked the best performing triple store in the recent Berlin SPARQL benchmark. This compared representation as RDF triples, mapping of the equivalent relational data to RDF and pure relational solutions. In the relational section of this benchmark, Virtuoso also outperformed MySQL by a wide margin [7].

RDF load rates have been measured with the LUBM and US Census data sets. The rate is about 40K triples per second on a single server and 100K triples with a cluster of 2 servers. Rates vary in function of the index scheme used, the presence of text indexing, the composition of the data set etc.

7   Future Directions

Clustering

Future cluster work consists of adding parallel backward and forward chaining inference into the query engine. As Virtuoso has a highly parallel query platform, it is natural to exploit this for more complex operations. We can see backward chaining rules as a special case of a transitive subquery — each iteration makes more goals satisfied and/or completes variable bindings. The existing parallelization will work and since rule bodies will match data that is partitioned, the rule should be sent for matching to where the data resides.

On the forward chaining side, the whole database can be seen as a sort of RETE network. When a fact is added, it is matched to the database according to forward chaining rule heads. Each rule head is like a collection of stored queries that are evaluated with bindings from the incoming data. If there is a result, the rule body is instantiated, facts are added and the process repeats.

In this way, most RDF reasoning can be supported at the database level, with all the parallelism and scalability benefits this entails.

Federated query processing over multiple heterogeneous RDF end points faces many of the same problems as query evaluation on a cluster. The difference is that latencies are over two orders of magnitude longer and there is less flexibility in designing the message flow. We are planning to apply the cluster execution model to federated queries against arbitrary SPARQL end points.

A more traditional line of work is implementing a columnar representation for relational tables and further experimentation with compression. These would make Virtuoso a strong contender in the relational business intelligence arena. We are presently running the traditional relational benchmarks with Virtuoso.

Updating Relational Data by SPARUL Statements

In order to have task oriented RDF data representation, such as property tables, one needs to adapt the SPARQL update and data load logic to supporting these.

In many cases, an RDF view contains quad map patterns that map all columns of some table into triples in such a way that sets of triples made from different columns are "obviously" pairwise disjoint and invoked IRI classes are bijections. E.g., quad map patterns for RDF property tables usually satisfy these restrictions because different columns are for different predicates and column values are used unchanged as object literals. We are presently extending the SPARUL compiler and run-time in order to make such RDF views updatable [9].

The translation of a given RDF graph into SQL data manipulation statement begins with extracting all SQL values from all calculatable fields of triples and partitioning the graph into groups of triples, one group per one distinct extracted primary key of some source table. Some triples may become members of more than one group, e.g., a triple may specify relation between two table rows. After integrity check, every group is converted into one insert or delete statement.

The partitioning of N triples requires $O(N \ln N)$ operations and keeps data in memory so it's bad for big dump/restore operations but pretty efficient for transactions of limited size, like individual bookkeeping records, personal FOAF files etc.

8 Conclusion

With Virtuoso Cluster being operational at the time of this writing, we see that the greatest part of the RDF scalability issues is overcome. It remains the case that RDF, for all its flexibility, takes more space and is not as efficient as a task oriented relational representation. Advances in technology tend to benefit both RDF and relational models. For example, the clustering section applies 1:1 to relational workloads as well. Thus, for RDF to be equivalent to relational, it must accept some of the same restrictions, e.g. no graph, no variables in predicate position, strict enforcement of single value for cardinality one properties. If these are accepted, then mapping as discussed above can be used for application specific RDF representations that are essentially identical to the corresponding relational or the relational can be mapped to RDF, which are almost the same thing. The only difference is that a task specific RDF layout can still be typed at run time and use RDF data types like IRI's and typed literals. With these concessions, RDF is on a par with relational representations but does pay by embracing the same limitations.

How far one goes in the direction of application specific logical schema, indexes, materialized joins and the like is a function of the application. With Virtuoso Cluster, this is not a necessity at scales of billions and tens of billions of triples.

When relational databases replaced network databases, the argument in their favor was that one did not have to limit the set of possible queries when designing the database. Now, with linked data, the argument is that one does not have to restrict what data can be joined with when designing the database. Both represent a qualitative step in the direction of increased flexibility.

For a new technology to take hold, it must address a new class of problem: For the RDBMS, it was making the enterprise line of business applications around the database. For RDF, it is turning the Internet into a database. For the latter task, some of the flexibility for which RDF pays in 1:1 comparison against relational is necessary.

With Virtuoso, we address both sides of the matter: The generic storage of large volumes of RDF as well as exposing existing RDB's to the data web via SPARQL. Virtuoso does this with remarkable flexibility of scale, with a desktop version starting with a memory footprint of about 25MB, small enough for mobile, going up to clusters with tens and hundreds of gigabytes of memory and terabytes of disk at the high end.

The SPARQL standardization process will have to catch up with the extensions on the field, notably Jena and Virtuoso, which both implement similar extensions. To this effect, a SPARQL 2.0 working group will begin in 2009. We expect little difficulty since the need for most extensions is self-evident.

Further details on the SQL to RDF mapping and triple storage performance issues are found in separate papers on the http://virtuoso.openlinksw.com site. The Virtuoso blog http://virtuoso.openlinksw.com/blog is the most up-to-date information resource on the product.

References

1. Auer, S., Lehmann, J.: What have Innsbruck and Leipzig in common? In: Franconi, E., Kifer, M., May, W. (eds.) ESWC 2007. LNCS, vol. 4519, pp. 503–517. Springer, Heidelberg (2007)
2. Beckett, D.: Redland RDF Application Framework, http://librdf.org/

3. Bizer, C., Cyganiak, R., Garbers, J., Maresch, O.: D2RQ: Treating Non-RDF Databases as Virtual RDF Graphs, http://sites.wiwiss.fu-berlin.de/suhl/bizer/D2RQ/
4. Bizer, C., Heath, T., Ayers, D., Raimond, Y.: Interlinking Open Data on the Web. In: 4th European Semantic Web Conference, http://www.eswc2007.org/pdf/demo-pdf/LinkingOpenData.pdf
5. Chen, H., Wang, Y., Wang, H., et al.: Towards a Semantic Web of Relational Databases: a Practical Semantic Toolkit and an In-Use Case from Traditional Chinese Medicine, http://iswc2006.semanticweb.org/items/Chen2006kx.pdf
6. Erling O.: ISWC 2008: Billion Triples Challenge, http://www.openlinksw.com/dataspace/oerling/weblog/Orri%20Erling's%20Blog/1478
7. Erling, O.: Virtuoso Vs. MySQL: Setting the Berlin Record Straight, http://www.openlinksw.com/dataspace/oerling/weblog/Orri%20Erling's%20Blog/1484
8. Guo, Y., Pan, Z., Heflin, J.: LUBM: A Benchmark for OWL Knowledge Base Systems. Journal of Web Semantics 3(2), 158–182 (2005), http://www.websemanticsjournal.org/ps/pub/2005-16
9. Mikhailov, I.: Updating Relational Data Via SPARUL (Updatable RDF Views), http://esw.w3.org/topic/UpdatingRelationalDataViaSPARUL
10. Prudhommeaux, E.: SPASQL: SPARQL Support in MySQL, http://xtech06.usefulinc.com/schedule/paper/156
11. Ruttenberg, A.: Harnessing the Semantic Web to Answer Scientific Questions. In: 16th International World Wide Web Conference, http://www.w3.org/2007/Talks/www2007-AnsweringScientificQuestions-Ruttenberg.pdf
12. 3store, an RDF triple store, http://sourceforge.net/projects/threestore
13. Intellidimension RDF Gateway, http://www.intellidimension.com
14. Jena Semantic Web Framework, http://jena.sourceforge.net/
15. Northrop Grumman Corporation: Kowari Metastore, http://www.kowari.org/
16. Oracle Semantic Technologies Center, http://www.oracle.com/technology/tech/semantic_technologies/index.html
17. Semantically-Interlinked Online Communities, http://sioc-project.org/
18. Getting Started with Freebase, http://www.freebase.com/view/guid/9202a8c04000641f8000000005b82619
19. Wikipedia3: A Conversion of the English Wikipedia into RDF, http://labs.systemone.at/wikipedia3
20. Extensible Business Reporting Language (XBRL) 2.1, http://www.xbrl.org/Specification/XBRL-RECOMMENDATION-2003-12-31+Corrected-Errata-2006-12-18.rtf
21. About MusicBrainz, http://musicbrainz.org/doc/AboutMusicBrainz
22. About Geonames, http://www.geonames.org/about.html
23. Ping The Semantic Web, http://pingthesemanticweb.com/about.php
24. Oracle Real Application Clusters, http://www.oracle.com/database/rac_home.html

Semantic Task Management Framework: Bridging Information and Work

Ernie Ong, Uwe V. Riss, Olaf Grebner, and Ying Du

Abstract. Despite the growing importance of knowledge work in todays organizations, its support by means of ICT tools is still rather limited. Recent trends in semantic technologies provide novel approaches for an effective solution to these challenges in terms of semantic-based task management. However, task management involves the complex interplay of information and work activities. Thus a semantic task management framework is needed which supports an adaptable semantic foundation, to meet the challenges of knowledge work, via a set of task services on the desktop. To this end, we propose the Nepomuk Semantic Task Management Framework (STMF) as platform for a task-oriented ecosystem for desktop applications.

1 Introduction

In a world of rapid change, knowledge work (KW) plays a decisive role of growing importance in the success of knowledge intensive enterprises. The reality of

Ernie Ong
SAP Research, TEIC Building, University of Ulster, Shore Road; Newtownabbey, BT37
0QB, UK. Current address: The British Library, 96 Euston Road, London NW1 2DB, UK
e-mail: `ernest.ong@bl.uk`

Uwe V. Riss
SAP Research, Vincenz-Priessnitz-Str. 1, Germany
e-mail: `uwe.riss@sap.com`

Olaf Grebner
SAP Research, Vincenz-Priessnitz-Str. 1, Germany
e-mail: `olaf.grebner@sap.com`

Ying Du
SAP Research, TEIC Building, University of Ulster, Shore Road; Newtownabbey,
BT37 0QB, UK
e-mail: `ying.du@sap.com`

S. Schaffert et al. (Eds.): Networked Knowledge - Networked Media, SCI 221, pp. 25–43.
springerlink.com   © Springer-Verlag Berlin Heidelberg 2009

globalization of networked enterprises and economies places additional emphasis on this frontier. Consequently, the need for effective support in KW grows increasingly urgent. However, KW is quite a recalcitrant domain with respect to ICT support since it is characterised by highly variable activities of highly skilled knowledge workers (KWers) operating both autonomously and collaboratively [7]. This condition brings about two core aspects (1) supporting the management of **knowledge artifacts**, and (2) supporting the coordination of **work activities** or task management (TM) in short.

So far the support for KW by ICT tools is still rather limited. The most frequently applied tool in this respect is email although it shows a large number of drawbacks [25]. For example, it lacks appropriate support for information delivery and tracking possibilities, as well as for work organization. These observations suggest an apparent potential for efficient collaborative task management.

In the past several attempts have been undertaken to provide such support on the basis of process-aware information systems [8]. However, so far these approaches show significant shortcomings in terms of flexibility as required for KW. This results in lacking acceptance among KWers [20, 13]. We can put this down to the fact that workflow-like process structures are too rigid and their integration with information management systems is problematic due to variety of possible work situations. Often these rather resemble search activities than well defined processes.

Recently emerging trends in semantic technologies make new approaches for an effective solution to the challenges possible, to better support KW [22]. However, task management involves the complex interplay of information and work activities [17]. Consequently, support for TM within existing work processes and tools is just as crucial [10]. To this end, an effective task management framework is needed which is based on (and supports) a rich and adaptable semantic foundation, to meet the ill-defined challenges KWers face, via a set of task services which can be leveraged from within existing desktop applications.

This is the motivation for the Nepomuk Semantic Task Management Framework (STMF). To meet the challenges of KW, the successful framework must address the following challenges:

1. **Modelling:** support flexible semantic models of information artifacts and work activities in different social layers (personal vs. organizational) and in different modelling layers (application vs. domain). Here, the STMF needs an expressive and extensible model of all KW artifacts from desktop information objects and Internet resources to enterprise directories. This is the aim of the Task Model Ontology (TMO). In particular, the TMO must provide efficient access to task information and activity description. This, of course, is the subject studied by knowledge organization (KO) [24] and suggests that the TMO must support the modelling of optimized access paths to such task information.
2. **Knowledge:** capture and reuse of explicit and implicit knowledge to support knowledge work. To this end, STMF should provide opportunities for managing informational and process-oriented knowledge within common productivity applications. Seamless annotation of semantic metadata in existing work processes and tools is crucial.

3. **Infrastructure:** support a task-oriented ecosystem for all desktop applications in a networked environment. This stems from our perspective of tasks as a generic concept that is pervasive across applications and user activities on the desktop, and represents a conceptual hub for organizing information and work activities. STMF should additionally narrow the gap between semantic technologies and conventional development technologies to foster widespread adoption.

Addressing these challenges the STMF is designed as a task management component on top of the fundamental semantic layer provided by the Nepomuk middleware [12]. The STMF provides an interface to desktop applications which require a better integrated task model and specific task services. Moreover, we do not see task management as an application on the desktop among others but as another fundamental layer for applications that provides task services for desktop applications and coordinates all task related activities across all desktop applications.

In the following we first describe the Nepomuk approach and its integration in the Nepomuk Social Semantic Desktop (SSD) as the basis for our approach before we come to the description of the STMF.

2 General Approach

In this section we will explain the motivation for the introduction of the STMF. In particular we will explain the reasons for a specific task management layer between the desktop applications and the fundamental semantic layer. In short, the rationale is that the task management (TM) and the semantic infrastructure, as it has been developed in Nepomuk, supplement each other in central aspects. They represent complementary views of the knowledge artifacts KWers work with. In the following we describe the synergies that result from such integration. In order to find the synergies, however, we first have to look at the limitation of today's task management systems as well as semantic technologies [10].

To start with the analysis of TM we can refer to a study of Bellotti et al. [2] who have investigated the tools which KWers use to record and organize their to-do items and to track task execution. A central result of this and other studies is that most tasks are contained in emails or compiled on paper or print-outs. Only a minority of users applied dedicated TM systems. Another key finding of the study was that the effort of formally managing tasks is usually too high compared to the benefits that the KWer can expect in return. Therefore even writing tasks down on paper is considered as preferable compared to using TM systems. One reason for this is the missing integration of TM systems with email clients and other applications and the support in relating tasks to desktop knowledge artifacts. The deficiencies have mainly prevented an extensive usage of TM tools so far.

If we look at semantic technologies, the situation is similar. Although it is a widespread opinion that semantic technologies possess a high potential for improving KW, there are still considerable obstacles that prevent widespread adoption of such technologies. For example, Colucci and co-workers [5] have asserted that the computational complexity is often challenging. This even holds for rather simple operations. The interaction with semantic-based systems is largely tedious and users often do not possess the required skills. Finally KWers often do not realize the benefit that they could obtain from the additional effort of annotation since suitable applications that make use of semantic capabilities are still missing.

Another problem occurs in information retrieval. KWers spend a considerable amount of time looking for knowledge artifacts. By this term we mean all digital objects that are suited to increase the knowledge of KWers. However, these knowledge artifacts usually appear in one or more work contexts so that often a unique location of the respective artifact is not possible. On the other hand, KWers can often remember in which work activity they have last dealt with a specific artifact so that the work activity appears as an excellent knowledge hub. This requires a task model as formalization of work activity, which can be defined by a task ontology. This ontology can be seamlessly embedded in the ontologies used to describe knowledge artifacts (cf. section 3.1). The STMF uses this integration to translate task related user activities into metadata [10]. In this way the work processes provide the glue between knowledge artifacts and the work context which helps to interpret these. Here we use the term work context to describe all knowledge artifacts, persons, topics, sub-activities etc. that have been involved in this activity. Semantic technologies such as those developed in Nepomuk provide the basis for this integration. For example, the integration allows KWers to navigate through the entire semantic network starting from a suitable task.

The STMF approach aims at overcoming the deficiencies of both sides by employing the mutual strengths. To leverage the synergies we have to work out a way how semantic technologies can support TM and vice versa. The essential improvements of TM and semantic framework encompass the following issues:

1. **Leveraging Task Management**

    a. Semantic network providing support in handling knowledge artifacts
    b. Establishment a desktop-wide task management layer
    c. Providing a platform for application developers to include TM services

2. **Leveraging Semantic Technologies**

    a. Automatic annotation of knowledge artifacts
    b. Ensuring a consistent usage of ontologies
    c. Social aspects of TM and exchange of metadata

In the following subsections we will further focus on these opportunities of TM and Nepomuk Semantic Web Services and show how they are addressed by the STMF.

2.1 Leverage Semantic Information for Task Management

Semantic technologies can realize the information integration required for TM. The basis of this integration consists in the fact that almost all tasks are related to knowledge artifacts which can be stored at various places, e.g., on the desktop. It is often tedious for the KWer to bring all required knowledge artifacts together even if they have worked with them frequently. The reason is that the places where the objects are stored are often selected according to criteria that are not task related, e.g., if all presentations might be stored in one folder. Therefore it is often difficult for KWers to remember the place where they have stored specific information.

One of the central aspects of the STMF is the enrichment of task data by assigning resources that are used in the task. This assignment provides users with easier access to the data that they need for task execution. This is exactly the information that is transferred to the semantic network by the STMF.

This means that KWers can later see in which tasks a person or a document was involved and this information helps them to better understand the roles of these objects. The KWers can directly navigate to the respective task and might find other KWers and knowledge artifacts involved. They can replace a person or the content of a document in new work activities since the role of such objects in the context of the task is clear.

The Nepomuk system provides a compilation of topics - personal semantic concepts defined by the user - that can be extended by topics resulting form task execution. These are simply added to the existing topics and can be used in the same way and exchanged with task co-participants. In the same way new persons that are added to a task are automatically incorporated in the KWer's contact list.

Moreover, task management provides the Nepomuk system with information with whom and when a KWer collaborated and in which order specific activities took place. It also gives information about used resources. This is information that generic Nepomuk metadata annotations cannot generally provide. In this respect we can make use of the Nepomuk context management which provides low level event information, e.g., when a specific document was opened, but cannot reliably assign these events to tasks. Here additional information from the STMF is required.

TM provides a more activity-oriented view since often the mere contents of a document, for example, does not make clear which purpose it was used for. The TM logs can inform about the utilization of a resource since it provides information about what, when and how the resource was applied in the task. A service that supports such a temporal description is the Task Journal Service [19].

Since the STMF is seamlessly integrated in the Social Semantic Desktop (SSD) it can make use of SSD services that help KWers to find required resources and assign them to tasks. This is supported by the integration of the Task Management Ontology (TMO) in the ontologies that describe a conceptualization of the KWers desktop data and their personal mental models.

2.2 Establish a Desktop-Wide Task Management Layer

According to Boardman [3] we can distinguish production and support activities. While the first describe those activities that directly contribute to the KWers work goals, e.g., development environment and text editors, the former are required to organize work so that it can be performed more efficiently, e.g., employing time or task management. Usually both aspects are clearly separated, i.e., we have applications for support activities and applications for production work. However, this separation requires KWers to switch between support and production applications additionally the other frequently occurring interrupts in their work process. Such interrupts lead to a decrease in the KWers productivity and is one of their motivations to avoid TM applications, since a piece of paper can be generally used in parallel to other desktop activities.

The STMF approach consists in the provision of a desktop-wide TM framework offering a set of task services that enable a tight connection of the STMF to desktop applications. These services allow for the incorporation of TM functionalities in the production applications. Prototypically such services have been implemented in applications such as the Mozilla Firefox browser and the Microsoft Outlook client. The services are called from these applications via specific application plug-ins. These plug-ins enable KWers to directly assign websites or emails to tasks or to create new tasks that are assigned to these objects [10].

Figure 1 shows how these applications plug-ins work together with the STMF. On the right hand side of Figure 1 there is the Kasimir TM sidebar, a TM prototype that has been developed in the Nepomuk project [10]. It mainly provides a to-do list with the existing tasks of a user, showing task-subtask relations. For the selected task various views of varying detail are offered, e.g., a detailed resource view, a context view, a Task Journal view, and a Task Pattern view. Kasimir represents a traditional task management application that is enriched by information from the Nepomuk RDF repository.

Fig. 1 Application Interactions based on the STMF

On the left hand side of Figure 1 we see the Mozilla Firefox browser and Microsoft Outlook with their TM extensions which are unobtrusively integrated. Thus KWers obtain the opportunity to efficient task handling directly within the context of productive applications. For example, while browsing the internet or intranet or scanning emails the KWers can work with tasks. In this way the KWer can immediately continue the browsing (production activity). The same holds for the email client.

2.3   Platform for Application Developers

A tight integration with desktop applications, however, requires that desktop application developers can efficiently develop plug-ins, which are based on STMF services. One advantage that application developers take of the STMF is the stable interface that it provides. Furthermore, the STMF does so in a manner that is insulated from changes to the underlying ontologies. In particular application developers are not required to directly work with the semantic infrastructure consisting of several ontologies but they can access tasks directly via SOAP web services, for example. Moreover, the STMF performs additional consistency checks that are generic for the TM, e.g., they check that a requested task state transition is valid for the current task state. In this way, the STMF layers task-specific semantics on top of the basic Nepomuk services and, in so doing, the STMF ensures that the task data stored in the RDF Store and the operations on the task data make sense. The services provided by the STMF focus on the TM requirements and disburden the application developer from explicitly dealing with the generic semantic infrastructure.

For the user the integrating framework suggests a uniform access to task management functionality all over the desktop, even if this is not a mandatory consequence. A common framework might even advise application developers to follow common user interaction principles which make it easier for users to deal with TM functions. Since the STMF only provides an API, task user interfaces may be adapted in a contextual way so that the particular needs of a KWer in the specific work environment can be addressed optimally. For example, we might know that in a bibliographic environment a user mainly attaches to be read tasks to found documents so that specific functionality for such tasks can be offered and transformed into metadata. This reduces the definition effort for the KWer.

2.4   Enriching the Semantic Desktop by Task Management

The Nepomuk infrastructure provides a large number of services that help KWers to deal with the semantic data such as desktop crawlers, personal information model support [23], local and distributed search and others. For example, the crawlers support the KWer in including resources in the semantic network. However, such automatic services can only provide rather low-level semantic annotations, e.g., relations of an email to its sender and recipients. Higher level semantic information must directly come from the KWer but it is not required that the KWer manually define them.

It is particularly the TM that can provide such high-level semantic information resulting from the KWers work activities. For example, if a KWer assigns a document or a person to a task by adding it to the respective task context this implicitly means the task and the knowledge artifacts are related in a way that a crawler cannot provide. Transforming such operations into relations means that TM activities enhance the semantic network. This provides the tangible benefit of reducing the effort for users to manually annotate resource. Moreover, the relation between knowledge artifacts, such as documents and tasks, augments the knowledge artifacts with context information. This helps users to better understand the contents of these knowledge artifacts and their work. This is often a problem when using search engines which only provide rather limited access to the context of an information object.

The relationships that results from TM activities such as the assignment of a document or person to a task are immediately reflected in the semantic repository. This means that such information is immediately accessible, e.g., when the KWer browses through this network.

2.5   Ensuring Consistency of Metadata

Ontology engineering is far from being trivial and this particularly holds for KWers. The particularities of handling metadata are often not obvious. For example, the KWer might not be able to clearly decide whether a specific document is related to a task or whether it is a topic of this task. Even if the difference is clearly defined somewhere it might not be obvious to the KWer. This can lead to inconsistent usage of metadata that spoils consistent reasoning and also makes the semantic navigation more complex.

The STMF translates the handling of the ontologies into the handling of tasks. Since thereby a specific application domain is given the meaning of attributes and relations can be determined more precisely. This means that user activities, e.g., assigning a document or person to a task, can be unambiguously related to metadata which are automatically created based on the execution of these activities. The user is not required to directly deal with the ontology and the definition of metadata.

Consequently metadata are defined uniformly and inconsistencies are avoided due to the STMF that interprets all task operations performed by the KWer in the same way. Actually the KWers do not even realize that they are working with a semantic framework. Nevertheless they profit from the benefits that the semantic representation provides.

2.6   Social Aspects of Task Management

Finally, the STMF supports the social aspects of semantic information and task management. In this respect it makes specific use of the email client. For example, the STMF supports the usage of email for task handling and transfer of metadata. Today emails are extensively used for task delegation and tracking but TM functionality that is adapted to this is mainly missing. At the same time the STMF manifests the relation to the email in the TM system and augments the sent task information by metadata. In this way the delegation of tasks implicitly results in an enlargement of the delegates personal semantic network in which the received resources will be semantically included. Of course, privacy issues have to be considered but the resulting opportunities are nevertheless auspicious, in particular, since they support the networking between KWers. To deal with the resulting demands the STMF will incorporate security aspects to support social TM scenarios.

The Nepomuk SSDs are organized in a peer-to-peer network. The communication between these peers is realized by a Network Communication layer that provides a basis for collaborative TM. This particularly supports the collaboration within organizations whereas the external communication uses email as the medium for metadata exchange. In this way richer task information can be exchanged which increases the value of the TM system.

3   Semantic Task Management Framework

The design and implementation of Nepomuk STMF and its underlying task model [11] called Task Model Ontology (TMO) rely on the Nepomuk semantic foundation layer, i.e., the set of services and ontologies provided by it. The TMO addresses the need for a semantic model of TM comprising a description of information artifacts and work activities. In this way the STMF provides uniform and pervasive access to task data and services across applications and user activities on the desktop built on the task unit as a conceptual hub. Despite the central role of STMF for the Nepomuk task management it is to be remarked that access to TM data is not restricted to the STMF. Direct access to the semantic task description in RDF format is also available for applications and an user interface for such direct access has been provided [4].

3.1   Task Model Ontology (TMO)

The central idea of the STMF as described in Section 2 is a seamless integration of TM in the semantic infrastructure. To this end it was necessary to describe the task structure by means of a proper Task Model Ontology (TMO) as part of the existing Nepomuk ontologies and particularly as task specific supplement of the Personal Information Model Ontology (PIMO) [23].

The TMO is structured in two layers: (1) A set of classes and resources which describe task-oriented information and work activities, and (2) an underlying set of Nepomuk classes which support the elaboration or concretization of more generic concepts in PIMO. The embedding in the Nepomuk ontologies guarantees that task information can be employed throughout the entire SSD.

3.2   STMF Services

The STMF offers TM functionality in the form of a set of services, so-called Task Management Services, that can be used by all desktop applications via respective plug-ins. These services can be grouped by their functionality in terms of the provide functionality. Figure 3 shows the task services and their classification.

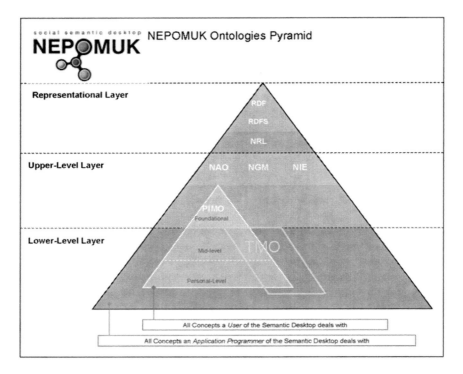

Fig. 2 TMO in the Nepomuk Ontologies Pyramid

The STMF offers services for Core Task Management, Task Experience Management, and Social Task Management as described in Figure 3. The Core Task Management includes Personal Task Management and Task Information Management. The former category includes the main services to organize the personal to-do items while the latter support the task context by enabling the KWer to attach various kinds of information objects to a task. The Social Task Management consists of the categories of (proper) Social Task Management and Social Network Management. The former provides collaborative task services in a wider sense whereas the latter services offers functionality to manage persons and their relations. Finally, the Task Experience Management offers two categories of services, Task Contingency and Task Structure Services. The former category includes services that help KWers to understand individual event that occurs in a task while the latter categories provide services that support the reuse of repetitive task structures. In the following sections, we present these 6 sub-categories in more detail and describe the functionality offered by the services of each of them.

The Personal Task Management offers services for handling tasks and task lists that support KWers in maintaining their task. These services provide methods for **basic task handling** such as task create, read, update and delete. It includes the handling of specific task attributes such as priority, due date, task state. The **task**

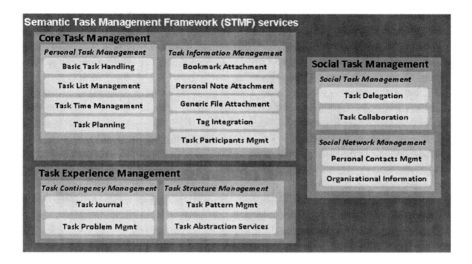

Fig. 3 Overview of STMF Services

list management offers functionality to organize task lists according to different criteria. There are filter and sorting mechanisms. Filter criteria are for example involved persons, involved documents, assigned tags or due dates. It also supports task-subtask relations. The **task time management** enables KWers to monitor and plan the time spent on specific tasks. There are methods to analyze existing sets of tasks as well as for planning future tasks. There are also methods for keeping detailed track records of spent. Similar for planning, task service methods allow for defining and retrieving the target effort. The **task planning** methods allow to (re-)structure tasks. For example, this allows reorganize the task hierarchy.

Task Information Management enables KWers to describe the context of tasks in terms of relevant information. To this end they can attach selected information objects to the task. There information objects include **bookmarks, personal notes,** all kind of **desktop files, tags,** and **persons** involved in tasks. For the selection of these objects the services can make use of the semantic network provided by the Nepomuk infrastructure. This allows auto-completion and recommendation on the UI level. A particular role plays the **task participants management**. It does not only allow KWers to assign persons to a task but it also enable them to give them specific roles in the task as for example task owner or involved. These roles are also used to indicate to which person a subtask has been delegated or who is the delegatee.

Social Task Management supports the collaboration between KWers in terms of TM. Thus, the STMF enables different kinds of social task interaction as for example **task delegation**. The service also encompasses the exchange of metadata that belongs to these information objects including attributes and relations. Beside task delegation there are services to support **task collaboration** in which KWers share a common task information space. In this respect the distinction of private and public information is supported. Delegation protocols help to control the processes of delegation and metadata transfer in order to realize task synchronization. **Personal contacts management** enables KWers to exchange personal contacts and related information. Finally it is possible to make use of **organizational information** retrieved from organizational repositories.

The Task Contingency Management makes information collected during task execution available. In this way it works with information that is specific for individual tasks. This concerns **task journal** in which the STMF register task events such as the point of time when a particular person was involved in a task or a subtask was delegated. It also includes **task problem handling** of specific situations that occurred during the task execution for later reuse. Aspects of Task Contingency Management have been described in [19].

In contrast the Task Structure Management does not concern singular events but repetitive task structures. This concerns **task patterns** that describe this reoccurring task feature that provide guidance how to perform new tasks on the basis of completed tasks as well as services that recommend information objects to be used in task execution based on previous tasks. Services that help KWers to find suitable information objects on the basis of completed tasks are called **abstraction services**. These services enable the knowledge transfer from personal to collective where the knowledge reuse and organizational learning is possible [16].

3.3   Core STMF Architecture

The core STMF architecture and its environment is depicted in Figure 4. Security related aspects of the STMF architecture will be described in the next section. It shows the Nepomuk Middleware that is organized into Core and Extension Services. Core Services provide the foundational functionality on which the Extension Services are built. Conceptually, Extension Services could be used to provide domain- and application-specific support for domain- and application-specific ontologies within the Nepomuk semantic middleware such as that which the STMF provides for the TMO. All internal communication within the Middleware is based on Java-OSGI whereas applications external to the Middleware rely on platform and language agnostic technologies based on HTTP such as SOAP web services when interacting with a Nepomuk service. The STMF provides Extension Services that use the following Nepomuk Core Services:

- **Nepomuk Desktop Bus:** This acts as service and application registry for semantic Nepomuk services. It also enables the communication between the STMF and the Core Services.
- **Data Wrapper:** The Aperture Data Wrapper crawls the desktop for Desktop Objects such as emails, documents and spreadsheets, and adds their semantic data to the RDF Store.
- **Local Storage:** The RDF Store based on Sesame2 provides the semantic data base for all semantic data ranging from tasks and other concepts like persons to the Nepomuk ontologies such as TMO and PIMO.

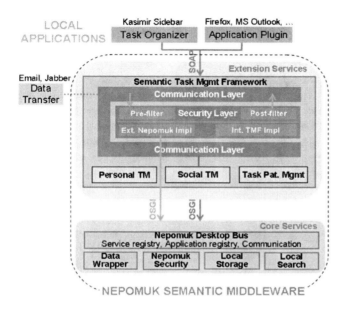

Fig. 4 STMF Architecture within the Nepomuk Semantic Middleware

- **Local Search:** This provides access to semantic data in the RDF Store via SeRQL and SPARQL queries.

From an architectural perspective, the STMF services are platform and language independent. This is realized by the provision of SOAP web services. From an interface perspective, the STMF services are exposed via two API sets. The first and lower-level API (task RDF API) focuses on data access and comprises an RDF interface that aims at exposing task data to semantics-aware applications capable of exploiting the semantic data. This interface provides client applications with direct access to the task data in the RDF Store with both SeRQL and SPARQL query support.

Since most conventional applications are not capable of processing semantic data and do not use semantic technologies such as RDF and SPARQL but are based on more conventional object-oriented technologies, the STMF provides a transformation (adapter) layer which converts RDF data to the object paradigm thus enabling the easy integration of task management within such applications. To this end the STMF also provides a second and higher-level API (task service API) that provides task management specific services on top of the Nepomuk semantic middleware, thus enabling both data access and the task management services described above. Internally, the task service API uses the task RDF API to realize the data access.

The STMF defines a Communication Layer and accompanying Data Transfer adaptors to manage the transmission of task-related messages between Nepomuk desktops, e.g., for task delegation and synchronization of semantic data and information objects. The actual implementation of the Data Transfer adaptors can be realized in various ways, e.g., via email or other transport mechanisms such as Jabber/XMPP. In the current implementation, the STMF provides adaptors via the standard email protocols (STMP/POP) and Microsoft Outlook using COM technology. In addition to data transfer, the Outlook adaptor also provides full access to the Outlook application model. This can be exploited to access and manipulate Outlook objects including email, address book entries and calendar entries from within the STMF thus enabling bi-directional synchronisation between semantic and Outlook application repositories. The end result is a much closer integration between popular desktop productivity and information management tools and semantic task management.

Communication scenarios, which deal with data and information object exchange between task participants, do however pose one key challenge: security. For example, the initiation of task communication such as task delegation requests presents vulnerabilities where information may be exposed to other users. The Security Layer in the STMF manages the task-specific security requirements in such scenarios.

The STMF implements security at two levels corresponding to the exchange of data and information objects described above. At the metadata level, the STMF limits access to task data in two ways. First, private data is automatically omitted or post-filtered so that external users are neither aware of nor able to access the data which the owner considers private. This, in turn, is supported directly in the TMO which provides privacy attributes for tasks, task attachments and task journal entries. And second, the STMF enables the user to exchange non-private task data with task participants. This ensures that the user is responsible for explicitly selecting the data to be supplied to the addressee. In short, these mechanisms realise the principle of least privilege in regard to task data in communication scenarios.

The Security Layer is also responsible for mediating access to information objects referenced in tasks. Once again, it adheres to the principle of least privilege by pre-filtering and rejecting unauthorised requests for information objects. Examples of these include

1. **Requester-integrity verification:** The requester must be a task participant.
2. **Private exclusion:** Only information objects still marked public are allowed.
3. **Explicit authorisation:** The owner must explicitly authorise access to the requested information object.
4. **Trusted return address:** The owner, and not the requester, is responsible for determining the return address for the information objects.

From a design perspective, the STMF applies aspect-oriented principles to separate security concerns from the rest of the STMF. For example, pre-filters are designed to prevent the completion of a STMF operation, i.e., a document request, in the event a precondition is not satisfied, e.g., the person requesting access to a document is not a task participant. This is realised in a manner that is transparent to core STMF functionality by encapsulating the Security Layer within the Communication Layer. This provides a clear separation of concerns between main STMF operation, communication and security within the STMF.

The Internal (Security) TMF Implementation component addresses only task-specific security requirements. On the other hand, the External Nepomuk (Security) Implementation component cooperates with the Nepomuk Security component to address more generic security aspects which are common across the Nepomuk infrastructure, e.g., role-based access control, encryption and digital signatures. This provides a clear separation between task-specific and generic security aspects.

3.4   Layered Refinements on the TMO and STMF

The STMF is an implementation of a set of domain-specific services to support task management within the Nepomuk semantic framework. To this end, the STMF has been designed with extensibility to its underlying data model and its functionality to support orthogonal domain- and new application-specific concepts and services. Orthogonal concepts can be accommodated by composing the TMO while the new services can be introduced by extending the STMF.

The TMO task data model can be extended with orthogonal organizational, domain and application specific ontologies which describe the modelling needs in specific situations. Such ontologies can be aligned with the TMO via ontology composition which aims to harmonise and align two or more ontologies.

From a technical perspective, since the STMF uses data access objects (DAO) to mediate access to the task data in the RDF Store, extensions to the task data model can be supported by introducing new DAOs to encapsulate access to the extended data model. Due to the availability of different RDF data access frameworks, abstract DAO factories are used to instantiate concrete DAO classes and the respective mappings between the RDF data model and the corresponding Transfer or Value Object (VO) in the object realm.

At present, the STMF uses the RDF Reactor framework in its DAO layer to provide object-oriented Java proxies to the underlying RDF data. Consequently, changes to the underlying RDF framework or the use of a different data management technology can be implemented without any impact to the non-storage classes in the STMF, e.g., by adding new concrete DAO classes. This design aims at providing configurability within the STMF while maintaining stability both within and without, e.g., for applications using the STMF. This is crucial since the STMF goes beyond supporting task data but furthermore mediates access to other data in the RDF Store and other Nepomuk services. It is to be remarked that the use of DAO, DAO factories and VO provide a flexible framework for integrating business objects from enterprise systems.

Whereas extensions to the STMF data model are realized via ontology composition, refinements and extensions to the STMF functionality are realized via service composition. This may be used to provide new domain- and application-specific services based on the STMF. The current STMF is designed as a façade (design pattern) that uses service composition to mediate access to the underlying Nepomuk services. Extensions to the STMF can also be realized in a similar fashion where core STMF functionality is delegated to the existing base STMF. New functionality, on the other hand, can be intercepted and handled separately. From an architectural perspective, an STMF extension can be realized as an additional Extension Service within the Nepomuk Semantic Middleware or as a separate web service. In this way, multiple monotonic variants of the STMF can co-exist on the same desktop. Very importantly, this adheres to the Open-Closed Principle of object-oriented software construction.

The STMF together with the TMO therefore provide strong technical and semantic foundation, respectively, on which to build and customise task management services according to the needs of different application and organisational situations.

4 Related Work

One of the core insights of the present paper is the strong relation between Personal Information Management and Task Management. In fact, every task execution requires and produces information (which is primarily on the personal level). In this respect the current approach is similar to the one that is realized in the OntoPIM approach [14]. In a similar way as the Nepomuk project, OntoPIM fosters the idea of a Personal Ontology reflecting the users perspective of their work domain. To obtain the Personal Ontology, OntoPIM extracts information from the information objects the users are working with, e.g., emails, via an inference engine. In this way OntoPIM supports users in performing tasks. For example it proposes new tasks as successors of current tasks on the basis of existing task logs, e.g., suggesting a FindFlight task after a FindHotel task.

The STMF is aiming at the same goal. However, the central idea of STMF is to go beyond automatic inference and derive relevant information directly from the users' work activities, i.e., not from task logs. Moreover, it aims at providing relevant task information based on social experience. Here the STMF focuses on social interaction, i.e., exchange of metadata between KWers, and the idea of abstract task patterns [18].

One of the first TM products that has followed the idea of information integration was Caramba [9], supporting TM for virtual teams by enabling links to information objects, tasks, and resources. However, it does so on a non-semantic basis.

Another related approach is the Haystack system [1] that goes already back to 1999. It is also rooted in a semantic network technologies based on an RDF infrastructure and includes tasks as a central concept as well. In this respect it is more comparable to the Nepomuk infrastructure. Haystack also shares in the insight that task handling determines a significant percentage of the users working time. To support tasks the Haystack system provides a task pane that gives access to task relevant objects. On the other hand, Haystack follows the traditional view of regarding task management as an application among others and not as a service layer that can be accessed by various applications. Thus the only closer integration of tasks is realized for email. Nevertheless, from a general perspective Haystack follows a similar cross-domain approach as Nepomuk.

The approach that is closest to the STMF is the Unified Activity Management (UAM) project at IBM Research [15]. The WAX system that results from this approach provides a Web service framework that applies a semantic representation of activities (or tasks) in a similar way as the present approach [6]. In the same way as the STMF it focuses on collaborative task handling, support of unstructured information and a plug-in approach. Moreover, we share with UAM the belief that the formalization of tasks opens a wide range of opportunities for better support of KW. The main difference between UAM and STMF is the integration in the SSD.

The key difference between the STMF and other approaches as UAM mainly consists in the fact that the STMF is essentially embedded in the SSD and extensively utilizes this integration. In this way the STMF does not only provide information to the SSD but also supports the task management by making information objects from the SSD available to the task management and the knowledge artefacts related to tasks are not only available for TM applications but to all desktop applications that are connected to the STMF. A further benefit here is that extensions to the standard SSD semantic model with personal, domain or organisational ontologies are also well integrated into the STMF. The SSD therefore provides not only services on which to realize the STMF but also a solid modelling foundation on which to enrich task descriptions.

5   Conclusions and Future Work

Despite its growing importance, the support for task management for knowledge workers by desktop ICT tools is still limited. Thus, task support tools are clearly separated from the tools KWers mainly use in their daily work activities. This results in additional cognitive and administrative overhead. The goal of the STMF is to realize a task-oriented operating environment for the desktop that provides KWers with more effective support in a manner that can be fully integrated with tools they already use. To this end, it addresses two key challenges, namely providing a uniform task model across all applications and user activities, and realizing a pervasive set of task services thereby elevating tasks and task services to first class citizens across the desktop.

However, the STMF initiative is far from complete. In the short-term, we plan to realize a task-oriented messaging bus to support multi-directional events between the STMF and any STMF-aware services and applications. This leads us closer to the ideal of a universal task-oriented operating environment on the desktop by providing a communication layer for supporting complex interactions between desktop applications, events, and enterprise systems from which such events may arise.

In order to incorporate security aspects, the STMF will provide security-based filtering, e.g., of private information, so that external users are neither aware of nor able to access the information that the owner considers as private. The security concept will also include the access to semantic relations between resources, i.e., indirectly related resources.

A central aspect of our further development is to use the STMF to provide more effective support for experience management and reuse via task patterns [19]. Experience management in the field of knowledge work requires a tight integration of process and knowledge management. The STMF provides an ideal platform to bring both aspects together. Moreover, the web service approach of the SSD offers interesting integration opportunities for business process management and the STMF [21]. These advantages can even be increased by the integration of external ontologies and the corresponding metadata and the introduction of multi-faceted context management within the STMF. The former leverages the potential of Nepomuk to integrate the TMO with personal, domain and organisational ontologies. This provides a richer means by which multi-faceted task context can be described. The multi-faceted nature of task context is necessary to provide an effective basis for understanding the different aspects of the information and work process needs embedded within tasks. This in turn forms the foundation for task pattern abstraction based on the needs of the KWer. The realization of the STMF within the Nepomuk SSD is therefore highly valuable, not just from a technical perspective but also as a means to gain clearer insights into the needs and preferences of the KWer.

Acknowledgements. This work has been partially funded by the European Commission as part of the Nepomuk IP (grant no. Grant 027705) within the 6th Framework Programme of IST.

References

1. Adar, E., Karger, D.R., Stein, L.A.: Haystack: Per-user information environments. In: CIKM, pp. 413–422. ACM, New York (1999)
2. Bellotti, V., Dalal, B., Good, N., Flynn, P., Bobrow, D.G., Ducheneaut, N.: What a to-do: studies of task management towards the design of a personal task list manager. In: Dykstra-Erickson, E., Tscheligi, M. (eds.) CHI, pp. 735–742. ACM, New York (2004)
3. Boardman, R.: Improving tool support for personal information management. Ph.D. thesis, Department of Electrical and Electronic Engineering Imperial College, University of London (2004)
4. Brunzel, M., Mueller, R.M.: Handling of task hierarchies on the nepomuk social semantic desktop. In: Sebillo, M., Vitiello, G., Schaefer, G. (eds.) VISUAL 2008. LNCS, vol. 5188, pp. 315–318. Springer, Heidelberg (2008)
5. Colucci, S., Noia, T.D., Sciascio, E.D., Donini, F.M., Ragone, A., Rizzi, R.: A semantic-based fully visual application for matchmaking and query refinement in b2c e-marketplaces. In: Fox, M.S., Spencer, B. (eds.) ICEC. ACM International Conference Proceeding Series, vol. 156, pp. 174–184. ACM, New York (2006)
6. Cozzi, A., Farrell, S., Lau, T., Smith, B.A., Drews, C., Lin, J., Stachel, B., Moran, T.P.: Activity management as a web service. IBM Systems Journal 45(4), 695–712 (2006)
7. DeFillippi, R., Arthur, M.B., Lindsay, V.J.: Knowledge at Work. Blackwell, Oxford (2006)
8. Dumas, M., van der Aalst, W.M., ter Hofstede, A.H.: Process-aware information systems: bridging people and software through process technology. John Wiley & Sons, Inc, New York (2005)
9. Dustdar, S.: Caramba - a process-aware collaboration system supporting ad hoc and collaborative processes in virtual teams. Distributed and Parallel Databases 15(1), 45–66 (2004)
10. Grebner, O., Ong, E., Riss, U.V.: Kasimir - work process embedded task management leveraging the semantic desktop. In: Bichler, M., Hess, T., Krcmar, H., Lechner, U., Matthes, F., Picot, A., Speitkamp, B., Wolf, P. (eds.) Multikonferenz Wirtschaftsinformatik, GITO-Verlag, Berlin (2008)

11. Grebner, O., Ong, E., Riss, U.V., Brunzel, M., Bernardi, A., Roth-Berghofer, T.: Task management model - Nepomuk deliverable D3.1. Tech. rep. (2006), `http://nepomuk.semanticdesktop.org/xwiki/bin/download/Main1/D%3%2D1/`

12. Groza, T., Handschuh, S., Moeller, K., Grimnes, G., Sauermann, L., Minack, E., Mesnage, C., Jazayeri, M., Reif, G., Gudjonsdottir, R.: The nepomuk project - on the way to the social semantic desktop. In: Pellegrini, T., Schaffert, S. (eds.) Proceedings of I-Semantics 2007, pp. 201–211. JUCS (2007)

13. Holz, H., Maus, H., Bernardi, A., Rostanin, O.: From lightweight, proactive information delivery to business process-oriented knowledge management. Journal of Universal Knowledge Management 2, 101–127 (2005)

14. Lepouras, G., Dix, A., Katifori, A., Catarci, T., Habegger, B., Poggi, A., Ioannidis, Y.: Ontopim: From personal information management to task information management. In: Personal Information Management, SIGIR 2006 workshop, Seattle, Washington, August 10-11 (2006)

15. Moran, T.P., Cozzi, A., Farrell, S.P.: Unified activity management: Supporting people in e-business. Communications of the ACM 48(12), 67–70 (2005)

16. Ong, E., Grebner, O., Riss, U.V.: Pattern-based task management: Pattern lifecycle and knowledge management - benefits and open issues. In: Gronau, N. (ed.) 4th Conference on Professional Knowledge Management - Experiences and Visions, Potsdam, Germany, March 28 - 30, GITO, Berlin (2007)

17. Riss, U.V.: Knowledge, action, and context: Impact on knowledge management. In: Althoff, K.-D., Dengel, A.R., Bergmann, R., Nick, M., Roth-Berghofer, T.R. (eds.) WM 2005. LNCS (LNAI), vol. 3782, pp. 598–608. Springer, Heidelberg (2005)

18. Riss, U.V., Cress, U., Kimmerle, J., Martin, S.: Knowledge transfer by sharing task templates: two approaches and their psychological requirements. Knowledge Management Research and Practice 5, 287–296 (2007)

19. Riss, U.V., Grebner, O., Du, Y.: Task journals as means to describe temporal task aspects for reuse in task patterns. In: ECKM 2008 Proceedings of the 9th European Conference on Knowledge Management, pp. 721–729 (2008)

20. Riss, U.V., Rickayzen, A., Maus, H., van der Aalst, W.: Challenges for business process and task management. Journal of Universal Knowledge Management 2, 77–100 (2005)

21. Riss, U.V., Weber, I., Grebner, O.: Business process modeling, task management, and the semantic link. In: Hinkelmann, K. (ed.) AAAI Spring Symposium on AI Meets Business Rules and Process Management, Stanford Univ., pp. 99–104. American Association for Artificial Intelligence, Menlo Park (2008)

22. Sauermann, L., Bernardi, A., Dengel, A.: Overview and Outlook on the Semantic Desktop. In: Decker, S., Park, J., Quan, D., Sauermann, L. (eds.) Proceedings of the 1st Workshop on The Semantic Desktop at the ISWC 2005 Conference, CEUR Workshop Proceedings, vol. 175, pp. 1–19. CEUR-WS (2005)

23. Sauermann, L., Elst, L., Dengel, A.: Pimo - a framework for representing personal information models. In: Pellegrini, T., Schaffert, S. (eds.) Proceedings of Iqand I-fInternational Conferenceson New Media Technology and Semantic Systems as part of TRIPLE-I 2007, pp. 270–277. J.UCS, Know-Center, Austria (2007)

24. Sigel, A.: Towards knowledge organization with Topic Maps (2000)

25. Whittaker, S., Sidner, C.L.: Email overload: Exploring personal information management of email. In: CHI, pp. 276–283 (1996)

AUTOMS-F: A Framework for the Synthesis of Ontology Mapping Methods

Alexandros G. Valarakos, Vassilis Spiliopoulos, and George A. Vouros

Abstract. Effective information integration is still one of today's emerging research goals. The explosive growth of heterogeneous information sources makes the task harder and more challenging. Although ontologies promise an effective solution towards information management and coordination, it would be a surprise if two independent parties have constructed the same ontology to manage information for the same domain. Hence, to integrate information effectively, ontology mapping methods are invaluable. This paper presents the AUTOMS-F framework, which aims to facilitate the development of synthesized methods for the efficient and effective automatic mapping of ontologies. AUTOMS-F is highly extendable and customizable, providing facilities for supporting the rapid prototyping of synthesized mapping methods, adapting some well established programming design patterns. The paper presents the AUTOMS mapping method as an evaluated case of AUTOMS-F's potential.

1 Introduction

During the last years the world is faced with the information overload phenomenon: Information is growing exponentially, is being provided in various forms and is stored in decentralized systems that range from inter-/intra-organization systems to those operating over the World Wide Web. Meanwhile, the need for transparent and bidirectional communication between these decentralized systems is more vital than ever before, as the exploitation of the available information is required for the right decision at the right time. To effectively deal with information heterogeneity, state-of-the-art approaches utilize ontologies. Ontologies formalize a conceptualization of

Alexandros G. Valarakos, Vassilis Spiliopoulos, and George A. Vouros
AI Lab, Information and Communication Systems Engineering Department,
University of the Aegean, Samos 83 200, Greece
e-mail: `alexv,vspiliop,georgev@aegean.gr`

S. Schaffert et al. (Eds.): Networked Knowledge - Networked Media, SCI 221, pp. 45–59.
springerlink.com © Springer-Verlag Berlin Heidelberg 2009

a certain domain by defining specific elements (concepts and properties) and the relations among them. Ontologies provide the key technology for the fulfilment of the Semantic Web vision, where - in contrast to what is happening today - provided information will not be mainly targeted to humans, but will be machine understandable and exploitable, as well: Innovative Semantic Web applications are expected to be able to deal effectively with the information overload phenomenon and manage available information successfully.

In spite of the fact that ontologies provide a formal and unambiguous representation of domain conceptualizations, it would be a surprise if two independent parties would have constructed the same ontology to manage information even for the same domain. This is true, because ontologies are mainly developed in a decentralized fashion and are freely provided in the World Wide Web for being used in numerous applications. This heterogeneity introduces ambiguity on the appropriateness of information and restrains interoperability between different information sources. Simple examples of ontologies heterogeneity include ontologies which use different lexicalizations for the same ontology elements: For example car and vehicle may denote the same class of entities. More complicated situations appear in cases where ontologies formalize different conceptualizations of the same domain, comprising different elements, and being structured (in terms of ontology elements relations) in different ways.

True interoperability, data integration and effective management of information will be admittedly achieved through reaching an agreement, by producing a single and well-agreed ontology or by coordinating source ontologies so that each party uses its own ontology, but refers to the information of the other party, by exploiting concept and relation mappings between the two ontologies. Ontology Mapping is of increasing importance towards this goal. Specifically, given two ontologies O_1 and O_2, mapping one of them to the other involves computing pairs of elements with highly similar intended meaning.

Towards this goal, state-of-the-art ontology mapping systems exploit synthesized mapping methods, each one targeting different kinds of ontological features, by utilizing different similarity strategies. All these efforts have as common goal the optimum synthesis of individual (atomic) mapping methods, in order to maximize their efficiency. In the context of the Ontology Alignment Evaluation Initiative (OAEI) [5], for instance, all participating systems (especially the best performing ones), heavily focus on the effective and efficient synthesis of individual mapping methods. As a result, the investigation of the optimum synthesis of individual mapping methods is of paramount importance. Therefore, for the proper investigation of the best performing synthesis of atomic methods and for the production of ontology mapping systems that achieve the effectiveness needed in real-world applications, solid, generic, expandable and configurable ontology mapping frameworks must exist, facilitating the development and evaluation of synthesized methods.

AUTOMS-F (**AUT**omated **O**ntology **M**apping through **S**ynthesis - **F**ramework) is a Java application programming interface (API) that aims to facilitate the development of integrated tools for the automatic mapping of domain ontologies. The main concern of AUTOMS-F is the provision of facilities for the advanced, flexible and rapid synthesis of several ontology mapping methods. As already stated, the ultimate goal is to provide synthesized approaches realized as integrated tools that produce better results and performance measures than each of the synthesized individual mapping methods alone. The framework has been used for the implementation of the AUTOMS mapping method [3] which is described as a case study in the fourth section of this article.

The paper is structured as follows: Section 2 presents the ontologies mapping problem, the requirements and the assumptions made towards implementing AUTOMS-F. Section 3 describes AUTOMS-F in detail. Section 4 presents AUTOMS, a specific mapping tool implemented using AUTOMS-F as a case study of using the proposed framework. Section 5 presents related work, and section 6 concludes the paper, sketching our future plans.

2   Problem Statement and Requirements

A mapping between two ontologies is expressed by a one-to-one function between (matching) ontology elements (i.e., ontology concepts and properties). Therefore, establishing a mapping [8] between ontology elements involves the computation of pairs of elements whose meaning is assessed to be similar. Similarity in meaning can be computed using a number of metrics that exploit ontology elements features. It is important to note that the mapping process does not modify the involved ontologies: It produces, as output, a set of mapping pairs together with their computed similarity (match) measure.

The majority of the mapping methods can be described by the generic mapping process [9] depicted in Fig. 1. The discrete steps of this process are as follows:

1. *F*eature Engineering: Ontologies are transformed into an internal representation. This step selects a fragment of the ontology to be processed.
2. *S*earch Step Selection: Element pairs from the two input ontologies are being selected, with the one element belonging to the first ontology and the other to the second. Depending on the mapping method, all element pairs or only a subset of them may be considered. The set of pairs constitute the search space of the method.
3. *S*imilarity Computation: This step computes the similarity of the previously selected pairs. Many different similarity metrics may be utilized by a single method.
4. *S*imilarity Aggregation: In this step all similarity metrics, which may exploit different ontological features, are aggregated into a single one.

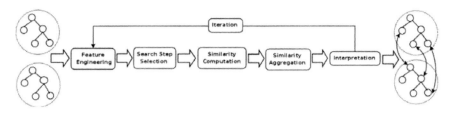

Fig. 1 The commonly accepted discrete steps of the generic mapping process

5. *I*nterpretation: This step concludes to a set of matching pairs by exploiting the aggregated similarities computed in the previous step (e.g., a trivial case is the use of threshold value(s)).
6. *I*teration: The whole process may be repeated several times, by propagating and updating the assessed similarities, taking into account the structure of the input ontologies.

Any framework that aims to facilitate the development of ontology mapping methods must support the development of the generic steps exposed in Fig. 1. AUTOMS-F, aiming to the provision of a generic framework for the development of mapping methods, in accordance to the steps proposed, poses a number of requirements:

1. According to the *F*eature Engineering step, a mapping method may utilize only a subset of the available information provided by the input ontologies. Different mapping methods should be able to use different sets of features.
2. The manipulation of the input ontologies must abstract from their specific representation formalism. Thus, ontologies in various representation formalisms, such as xml dialects, plain texts, rdfs, owl etc., must be handled.
3. According to the *S*earch Step Selection step, a method may examine only a subset of the candidate matching pairs, while different methods should be able to select different subsets of pairs, under well-defined conditions.
4. Moreover, a method may be applied to the candidate matching pairs produced by other methods.
5. According to the *S*imilarity Computation step, different mapping methods may need to compute different similarity measures for the assessment of matching pairs.
6. Also, a mapping method must be able to re-examine the results of other methods, supporting the development of more effective (in terms of correct mappings) mapping methods.
7. According to the *S*imilarity Aggregation step, the synthesis of different mapping methods and the aggregation of their corresponding similarity measures must be robust, expandable and easily supported by the framework.
8. According to the *I*nterpretation step, the matching pairs may be produced based on the aggregated similarity values assessed, and after the

application of a selection policy, aiming at choosing the best matching element pairs of the input ontologies.

Concerning the requirements of the framework's Application Programming Interface (API) the following are required:

1. *S*implicity: The API should be the result of an abstract specification of the ontology mapping process, and should be independent of the particular implementation of the constituent mapping methods and their specific configurations. Moreover, it must support the development of easily configurable and extensible systems, reducing effectively the time and cost of development.
2. *F*lexibility: It must cleanly separate the implementation of the above mentioned distinct steps of the mapping process, resulting in an easily configurable and extensible API, supporting reusability and thus, reducing the development cost and time.

3 AUTOMS-F: Architecture and Implementation

AUTOMS-F is an open source toolkit implemented using the Java programming language. It provides a basic framework for developing customized and synthesized ontology mapping methods. The framework is accessible by a comprehensive API.

In this section, we firstly present the conceptualization of AUTOMS-F, exposing its main components. Then, we present the AUTOMS-F components in accordance to the steps of the generic mapping process presented in section 2. Secondly, we specify key programming issues concerning the implementation of AUTOMS-F, towards the rapid and effective development of synthesized ontology mapping methods.

3.1 *Framework's Conceptualization*

AUTOMS-F, aiming at the satisfaction of the requirements stated in section 2, is broken into operation-specific component parts. The main types of components defined in AUTOMS-F and which are further detailed in the paragraphs that follow, are: 1) The *M*apping Method, 2) the *M*apping Task, 3) the *M*apping Association Tree, 4) the *P*arser, 5) the *C*oncept Property Selector, 6) the *A*ggregation Operator, 7) the *S*imilarity Method, 8) the *P*air Selector and 9) the *R*esult Renderer. These types of components are sufficient for describing an ontology mapping task according to the presented mapping process. They constitute the backbone of the framework and their specific implementation leads to different specifications of the ontology mapping process. Their manipulation/implementation is achieved through the AUTOMS-F's API, resulting to individual mapping methods.

3.1.1   Mapping Method and Mapping Task

The mapping method is the central component of AUTOMS-F. This component aggregates all the necessary information that is exploited in the various
steps of the mapping process: a) The elements of the input ontologies selected to participate in the candidate matching pairs, b) the metric used for
assessing the similarity between the elements in the candidate matching pair,
c) the logic used for combining the results of the various mapping methods,
resulting in a new set of assessed matching pairs, d) the logic used for selecting valid matching pairs form the resulting ones, and e) the representation
format that will be used for visualizing the valid matching pairs.

A mapping method can be associated with other mapping methods. When
a mapping method is associated with at least another mapping method or
another association of mapping methods, then this association constitutes a
mapping task (or synthesized mapping method). A task specifies the synthesis of different (atomic or synthesized) methods.

Tasks, due to their recursive definition specify a hierarchical tree of arbitrary complexity, which is named the Mapping Association Tree (MAT).
Fig. 2 depicts an example of MAT that consists of 2 mapping tasks (T_1 and
T_2), each with 2 mapping methods (m_1, m_2 and m_4, m_5, respectively), and 2
mapping methods (m_3 and m_6) that are siblings to these tasks. A mapping
task is depicted by a rectangular, whereas a mapping method is depicted in
oval. The specific configuration of a method or task is shown by the corresponding symbols attached to it, e.g., P_1 for parser, etc (these are further
explained in the next subsections).

The root mapping method is always a mapping task (TR) since it is always associated with other methods. The MAT defines a hierarchical structure that among others specifies the execution order of mapping methods. A

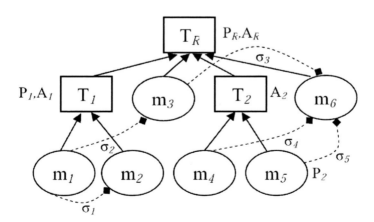

Fig. 2 An example of a Mapping Association Tree: Tasks (T), methods (m),
parsers (P), aggregator operators (A) and concept-property selectors (σ)

left-to-right depth-first execution order of the $methods$ and $tasks$ in the MAT has been adopted. Hence, according to Fig. 2, the method m_2 follows the execution of method m_1. The execution order of the $methods$ and $tasks$ in MAT is: m_1, m_2, T_1, m_3, T_2, m_4, m_5, m_6, TR. Moreover, this hierarchical structure implies inheritance relations that are exploited for usability and performance reasons, as it will be shown in the next subsections.

Similarity measures are not specified in $tasks$, since their role is to combine/manipulate the results produced by the subsequent $methods$ and $tasks$ (those that are rooted by this $task$ in the MAT). The root $task$ (TR) has a default manipulation $method$ which unifies the results produced by its subsequent $methods$ and $tasks$. This is in contrast to the other $tasks$, which can be associated with different combination/manipulation $methods$.

3.1.2 The Parser

The $parser$ is responsible for collecting the candidate matching pairs of ontologies elements involved in the mapping process. This collection is an $(n \times m)$ similarity matrix, where n and m are the number of elements of the target and source ontology, respectively. AUTOMS-F's internal representation distinguishes ontology elements in concepts (C) and properties (P). Candidate matching pairs between concepts and properties of the two input ontologies come from the cartesian product of their respective sets. Hence, the candidate matching pairs of concepts is the $C_1 \times C_2 = (c_{11}, c_{21}), (c_{11}, c_{22}), \ldots, (c_{12}, c_{21}), \ldots, (c_{1n}, c_{2m})$, where C_1 is the set of concepts in the first ontology, and C_2 is the set of concepts in the second ontology. Therefore, c_{1i} and c_{2j} are concepts from the first and second ontology, respectively.

A $parser$ is assigned to a $mapping$ task or $method$. According to the MAT structure a $parser$ is inherited to subsequent $tasks$ (i.e., $tasks$ lower in the hierarchy) and $methods$ (i.e., $methods$ lower in the hierarchy) that have not been associated to any $parser$. Supporting the $parsers$ inheritance property, and for consistency preservation reasons, we assume that $tasks$ or $methods$ in the MAT use $parsers$ that collect pairs of ontological elements that are supersets of the sets collected by subsequent $tasks$ or $methods$ parsers. Different $parsers$ can be defined at any level of the tree. Because of this, different $methods$ may exploit different collections of element pairs: Generally, the similarity matrix of a method contributes to the computation of the similarity matrix of the root $task$, which always contains the super-set collection of ontological element pairs.

In the MAT example (Fig. 2) a $parser$ (PR) has been assigned to the root $task$ (TR) which is inherited to its subsequent $methods$ and $tasks$, given that no $parser$ is specified for them. Thus, the $parser$ is inherited to the $methods$ m_3, m_4 and to the task T_2, in contrast to the methods m_1 and m_2 that inherit the $parser$ (P_1) that is assigned to the task T_2. Finally, method m_5 is associated to the $parser$ P_5.

It is possible for a *m*ethod to have two different *p*arsers attached: one for collecting elements of the first input ontology and one for the second. This feature is useful in cases where the two input ontologies are represented in different formalisms: A situation usually appearing in integrating legacy systems (schema oriented databases) with ontology-based applications.

Whenever there is not an one-to-one correspondence between the internal representation of AUTOMS-F (which is an ontology based one: concepts and properties along with all of their features) and the input ontologies/schemata, a transformation method is employed for defining correspondences between the appropriate elements of the ontologies/schemata with the elements in AUTOMS-F internal representation. For, example, the user may explicitly define which xml tag (e.g., tag <description>) of the input ontology/schema corresponds to which ontology element (e.g., <rdfs:comment> element) of the internal representation of the AUTOMS-F. The framework provides the necessary infrastructure for extending and adapting this behavior as needed by the specific needs of the input ontologies/schemata and their implementation.

3.1.3   Similarity Method

A *s*imilarity method is assigned to every *m*apping method and it specifies the way a match between the candidate matching pairs is being computed. Every *s*imilarity method results to an $(n_1 \times m_2)$ similarity matrix, where n_1 and m_2 are the number of the elements (concepts or properties) of the two input ontologies, respectively. For each element a different similarity matrix is produced. The value of each matrix entry specifies the similarity of the specific pair of elements (assessed matching pair) to which the entry corresponds. The candidate matching pairs, to which the similarity method is applied, are produced by the *p*arser of the corresponding *m*apping method or by a *s*elector component (the concept-property selector component is explained in the next subsection) applied to the candidate matching pairs computed by the *p*arser of another mapping method.

Also, since the internal representation of the AUTOMS-F is entirely based on Jena's ontology model, every similarity method has access to a copiousness of features regarding the selected elements of the input ontologies and the ontologies themselves. The set of the available features, which is the minimum and complete set concerning the manipulation of an ontology, is provided by Jena's ontology model. For example, a *s*imilarity method can directly access the local name of a concept and the property names of the concepts that constitute its vicinity.

A sophisticated *s*imilarity assessment method exploits information beyond the one found in the candidate matching pairs. Therefore, a *s*imilarity method has direct access to the involved ontologies. The framework's API supports all available settings of Jena's [1] ontology models, supporting the creation of advanced *s*imilarity methods. In order to facilitate synthesis of *m*apping methods, every *s*imilarity method of a *m*apping method has direct access to a previously-executed *m*apping method's similarity matrix.

3.1.4   Concept-Property Selector

A concept-property selector is assigned to a mapping method for producing candidate matching pairs on which the method's similarity method will be applied. Since a concept-property selector and a parser have the same effect (producing candidate matching pairs for the similarity method), when both of them exist in a mapping method the concept-property selectors override the parser. It must be noticed that, in contrast to a parser, the concept-property selector of a mapping method makes the combination of mapping methods results that do not belong to the same branch in the MAT, feasible. This feature makes possible the implementation of rules, such as, in similarity method m_2 exploit as candidate matching pairs only those that have not been assessed as such by the mapping method m_1.

Also, the way matching pairs are being selected by the concept-property selectors preserve the consistency of the results produced: Indeed, the candidate mapping pair set produced by a concept-property selector in a method mi should be at least a subset of the set produced by their super methods or tasks.

The selection of the candidate matching pairs is based on a similarity matrix of a previously executed mapping method and the models of the involved ontologies. In Fig. 2 the dashed lines represent the concept-property selectors. The square at the one end of the line denotes the method to which the selector is assigned, whereas the other end of the line denotes the mapping method that provides the similarity matrix. As it is depicted in Fig. 2, σ_1 is assigned to the method m_2 using m_1's similarity matrix, σ_2 is assigned to the method m_3 using m_1's similarity matrix, σ_3 is assigned to the method m_6 using m_3's similarity matrix, and σ_3, σ_4 and σ_5 are assigned to the method m_6 using m_3's, m_4's and m_5's similarity matrices, respectively.

3.1.5   Aggregation Operator

An aggregation operator is assigned to every mapping task in a MAT and it is responsible for specifying the way similarity matrices of direct subsequent methods or tasks are being combined. Hence, as can be seen in Fig. 2, the aggregator of the task T_1 combines the similarity matrices of the methods m_1 and m_2. The aggregation operator of the task T_2 combines the similarities matrices attached to methods m_4 and m_5, whereas the aggregation operator of the root task (TR) combines the similarities matrices of the methods m_3 and m_6 and the tasks T_1 and T_2. The root task (TR) is being related to a default aggregation operator which selects the best similarity value amongst the values produced by the mapping methods and tasks in the MAT, for every assessed mapping pair. Also, the models of the ontologies involved are accessible by aggregation operators so as to facilitate advanced aggregation techniques and tests.

3.1.6 Pairs Selector

Every *m*apping method and *t*ask is assigned a *p*air selector. A *p*air selector defines the criteria for selecting the best matching pairs from the matching pairs assessed by the *s*imilarity method. For example, a *p*air selector may define that the best matching pair is the one with the highest similarity value (resulting in one-to-one matching pairs) or define that the n% of the candidate matching pairs with the highest similarity value, are the best mapping pairs (resulting in one-to-many mapping pairs). A common strategy found in state-of-the-art mapping systems is the application of the *p*air selector only in the aggregated similarity matrix of the root *t*ask. The selected mapping pairs are passed to the *r*esult renderer component in order to be visualized. Also, the models of the involved ontologies are accessible by this component, enabling the development of advanced selection techniques, beyond the ones based on threshold values.

3.1.7 Result Renderer

A *r*esult renderer is responsible for the presentation of the mapping pairs. Every *m*apping method and *t*ask in the *MAT* is assigned with a *r*esult renderer. This facilitates the separate evaluation of each *m*ethod and *t*ask, leading to better decisions concerning their individual vs. synthesized deployment. Furthermore, this component is responsible for the storage of results.

3.2 Synthesizing Mapping methods

To the best of our knowledge, all the available frameworks adapt a sequential synthesis of mapping processes following the sequential execution order of the mapping processes. This results in a linear synthesis of atomic mapping methods. AUTOMS-F's mapping method adequately represents more complex synthesis patterns, such as the one presented in section 2: In contrast to other existing frameworks, AUTOMS-F facilitates a non-linear synthesis of the mapping methods and tasks, introducing the notion of *M*AT in combination with *s*electors and *a*ggregation operators. According to the above subsections, the synthesis of *m*apping methods in AUTOMS-F is supported in three ways:

1. By allowing a *m*apping method to have direct access to the similarity matrix computed by another method or task,
2. By combing the similarity matrices of *m*apping methods or *t*asks using specific *a*ggregation operators, and
3. By selecting candidate matching pairs from other methods or tasks, by exploiting *c*oncept-property selectors. These pairs are being used as input to the mapping methods.

The first and the third way facilitate a non-linear synthesis of *m*apping methods.

3.3 Implementation Issues

AUTOMS-F has been developed using the Jena Java Framework [1]. It has been implemented in Java for ensuring platform independency. A great concern during its development was the easy extensibility of the framework API, hence well-established programming design patterns [2] for ensuring usability, reuse, extensibility and abstraction were employed.

Fig. 3 depicts a UML diagram of the main classes of the framework according to its conceptualization (section 3.1). The MappingMethodImpl class is linked through an aggregation relation with itself and it aggregates at least one MappingMethodImpl class. Also, the same class is linked with exactly one of the following abstract classes: SimilarityMethod, PairFilter, Parser, Operator and ResultRenderer. The MappingMethodImpl class stores a list with the methods-tasks to which it is linked using the mappingMethodList attribute. This method is responsible for doing the necessary initializations (initialize operation) and for performing the mapping operations (the match operation of the MappingMethodImpl class). The SimilarityMethod class supports various manipulations of the similarity matrices to support the synthesis of methods. Due to space restrictions we present only some of the attributes and operations of the system. All the classes, except the MappingMethodImpl class, constitute hot spots for the framework, hence these are the classes that can be further extended.

The Strategy pattern - behavioural design pattern - is used in the MappingMethodImpl class to support the creation of different mapping methods. The template method pattern in the SimilarityMethod class - a behavioural pattern - is used for the computation of the similarity of a pair of ontology elements. Thus, instantiating the framework, one can define - override - the methods that measure the similarity between a pair of ontological elements and leave the construction of the similarity matrix to the SimilarityMethod

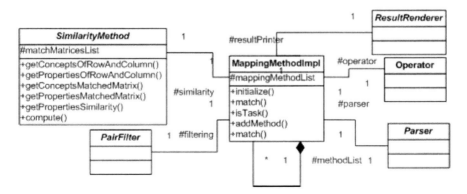

Fig. 3 UML diagram of the main AUTOMS-F classes, attributes and operations

class. Also, the composition pattern - a structural pattern - is exploited for the specification of the MAT.

AUTOMS-F, as it exploits Jena's model loader, can handle ontologies that are implemented in RDF, RDFS, OWL and DAML+OIL formalisms. The ontologies can be read from the local disk or be accessed through their URLs. An ontology element can be any of Jena's ontology class (OntClass) or property (OntProperty) objects. Hence, a method can retrieve any information about an ontology element, i.e. label, super-concepts, class properties etc.

AUTOMS-F contains samples of all the extensible classes resulting in a default $mapping$ method. More advanced $mapping$ methods can be developed by extending the SimilarityMethod class and overriding the methods that measure the similarity between ontology elements, i.e., concepts and properties. However, one may integrate a method into the framework by extending the SimilarityMethod class and overriding the compute operation that executes the similarity method. This means that the new class computes the mapping and the similarity matrix defined in the extended class. In this way, special attention is given to the manipulation of the ontology elements pairs, since the manipulation of the candidate matching pairs is left to the specific implementation of the method. Also, for the selection of the ontology elements we recommend the unified use of the framework's-based defined $parser$: This ensures consistency between the produced candidate matching pairs.

4 A Case Study: The AUTOMS Ontology Mapping Tool

AUTOMS-F has been used for developing the AUTOMS ontology mapping tool. AUTOMS synthesizes 6 mapping methods [3]: The lexical, the semantic, the simple structural, the properties-based, the instances-based and the iterative structural methods. Fig. 4 depicts the association tree of AUTOMS and the position of the mapping methods in it. The lexical and semantic methods are executed first. Then, the structural matching method follows by exploiting the results of the previously run methods, whose results have been aggregated by task T_2. Afterwards, AUTOMS executes the properties-based and instances-based mapping methods, and finally, the iterative structural matching method is being executed by exploiting results from the other $methods$ in its level, as well as from the $task$ that aggregates results from lower levels. AUTOMS uses the same $parser$ and $aggregation$ operator in any of its $tasks$. The $parser$ is defined in the TR $task$ and the $aggregation$ operator of each $task$ selects the best values of each assessed matching pair from the similarity matrices of its constituent $methods$ and $tasks$.

The requirements of AUTOMS have been satisfied by the flexibility and extensibility provided by the framework. The learning curve of the framework was rather short. In some cases, AUTOMS developers needed to extend the framework for capturing OAEI contest's requirements [5]; however this did

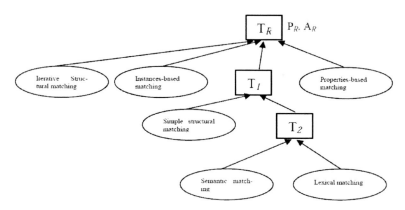

Fig. 4 AUTOMS's *M*apping Association Tree and its particular configurations

not affect the development of AUTOMS and AUTOMS-F proved to be a very robust and flexible framework. AUTOMS-F developers have been provided with an optimized version of their API, resulting in quite short (comparing to other tools of the OAEI contest) AUTOMS execution times. Scalability was a weak point at the time AUTOMS-F was used to develop AUTOMS, since very large ontologies (30MB) provided by the OAEI organizers could not be loaded and parsed. AUTOMS was evaluated in the OAEI 2006 contest among 10 other systems, achieving very good results as far as its efficiency and effectiveness are concerned.

5   Related Work

To the extent of our knowledge the works that are related to AUTOMS-F are the following: The Alignment API [4] and the COMA++ system [10, 11]. The Alignment API has been used for the evaluation of the ontology mapping methods that participated in the Ontology Alignment Evaluation Initiative workshop [5]. It has been implemented using Java and provides an API for incorporating, evaluating and presenting the results of different mapping algorithms.

AUTOMS-F and the Alignment API are based on different semantic-web technologies. AUTOMS-F uses the Jena Java framework whereas the Alignment API uses the OWL API [6]. Moreover, the Alignment API executes mapping methods in a pipeline, in contrast to AUTOMS-F which defines an execution structure of the mapping methods - the *M*apping Association Tree - facilitating the effective synthesis of different mapping methods, as well as their parallel execution. The Alignment API supports the combination of two methods by means of the fixed operators [7] compose, join, inverse and meet, which combine the result matrices of the constituent methods. However, these operators have not been fully implemented as far as the version 2.5 is concerned. On the other hand, using AUTOMS-F one has the flexibility to define his/her own *a*ggregation operators, combining more than two matrices. Also, AUTOMS-F supports the use of different *p*arsers, on each of the involved ontologies, for collecting their elements. Different *p*arsers can be applied in the context of a specific *m*ethod or *t*ask. Furthermore, AUTOMS-F incorporates *s*electors, which are built in components: This makes their exploitation very easy and straightforward. These facilities are not provided by the Alignment API. At the current version, AUTOMS-F does not provide any evaluation utilities, something that Alignment API does. In general, AUTOMS-F provides more hot spots than the Alignment API, thus making itself more extensible and customizable. Alignment API is under LGPL license. The new version of AUTOMS-F will be available soon in www.icsd.aegean.gr/ai-lab under GPL licence.

The COMA++ system, although it provides the necessary interfaces for intergrading arbitrary mapping methods and taking advantage of its matching-pairs visualization features, it is not an extendible API framework. More precisely, it is not possible for the user to define its own *s*imilarity aggregation or *i*nterpretation (pair selection in AUTOMS-F) policies as presented in Fig. 1. For this purpose, a predefined list must be exploited. Finally, in comparison to AUTOMS-F it does not support advanced synthesis utilities such as the *M*apping Association Tree and the notion of *s*electors. To sum up, in contrast to AUTOMS-F and the Alignment API, the main focus of COMA++ implementation is not to provide the infrastructure for facilitating the building of mapping tools, but the development of a mapping tool.

6   Concluding Remarks and Future Work

AUTOMS-F addresses the ontology mapping problem providing advanced methods synthesis facilities. The framework provides solutions to integrating and combining different *m*apping methods that aim to solve the ontology mapping problem. AUTOMS-F successfully meets all the requirements specified in section 2 and smoothly implements all the steps of the generic mapping process (presented also in section 2) except the step concerning the iteration of the mapping process which constitutes future research work. AUTOMS [3] is an evaluated case of the framework's potential. Although the full automation of ontology mapping is still a challenge, AUTOMS-F provides a robust framework for synthesizing different mapping methods, increasing the benefits of deploying state of the art mapping technology.

We plan to extend AUTOMS-F in several ways. Firstly, execution threads will be added to the *m*ethods of a *t*ask at each task level, in order to decrease the execution time of systems that combine many different methods. Secondly, we will introduce a consistency checking method that will ensure consistency between the resulted matching pairs. Thirdly, we will investigate a way to introduce iterative execution of the *m*apping methods and *t*asks in the framework preserving their synthesis capability, hence satisfying all the steps of the generic mapping process shown in Fig. 1. Fourthly, we will investigate the scalability issue: mapping between large ontologies and last but not least, we plan to add evaluation utilities for appropriate assessing and comparison of the implemented *m*apping methods and *t*asks.

Acknowledgements. This work is part of the 03ED781 research project, implemented within the framework of the Reinforcement Programme of Human Research Manpower (PENED) and co-financed by National and Community Funds (25% from the Greek Ministry of Development-General Secretariat of Research and Technology and 75% from E.U.-European Social Fund). The authors are grateful to K. Kotis for his valuable comments concerning the evaluation of the AUTOMS mapping tool.

References

1. Jena - A Semantic Web Framework for Java, `http://jena.sourceforge.net/` (accessed December 24, 2007)
2. Brandon, G.: The Joy of Patterns: Using Patterns for Enterprise Development. Addison-Wesley Pub.Co., Reading (2001)
3. Kotis, K., Valarakos, A., Vouros, G.: AUTOMS: Automating Ontology Mapping through Synthesis of Methods, OAEI (Ontology Alignment Evaluation Initiative), 2006 contest, Ontology Matching International Workshop, Atlanta USA (2006)
4. Alignment API and Alignment Server, `http://alignapi.gforge.inria.fr/` (accessed December 24, 2007)
5. Ontology Alignment Evaluation Initiative (2006), `http://oaei.ontologymatching.org/2006/newindex.html` (accessed December 24, 2007)
6. Bechhofer, S., Volz, R., Lord, P.: Cooking the semantic web with the OWL API. In: Fensel, D., Sycara, K., Mylopoulos, J. (eds.) ISWC 2003. LNCS, vol. 2870, pp. 659–675. Springer, Heidelberg (2003)
7. The Alignment API documentation (2006), `http://gforge.inria.fr/docman/view.php/117/251/align.pdf` (accessed August 24, 2007)
8. INTEROP - Network of Excellence, State of the art and state of the practice including initial possible research orientations. Deliverable d8.1, NoE INTEROP, IST Project n. 508011 (2004)
9. Ehrig, M., Staab, S.: QOM - Quick Ontology Mapping. GI Jahrestagung (1), 356–361 (2004)
10. Aumueller, D., Do, H., Massmann, S., Rahm, E.: Schema and ontology matching with COMA++. In: Proceedings of SIGMOD, Demonstration (2005)
11. Do, H., Rahm, E.: COMA - A System for Flexible Combination of Schema Matching Approaches. In: Proceedings of VLDB (2002)

Developing Semantic Web Applications with the OntoWiki Framework

Norman Heino, Sebastian Dietzold, Michael Martin, and Sören Auer

Abstract. In this paper, we introduce the OntoWiki Application Framework for developing Semantic Web applications with a strong emphasis on collaboration. After presenting OntoWiki as our main show case for the framework, we give both an architectural overview and a detailed view on the included components. We conclude this paper with a presentation of different use cases where the framework was strongly involved.

Introduction

Web application development usually begins with clarifying requirements, goals and usage scenarios. Today more than ever, requirements of modern Web applications lead to a strong need for semantic technologies. Depending on the context of the Web application, expectations for integrating those technologies vary greatly. They can range from advantages in search handling, categorization and content management to flexibility in handling different data schemes. Particularly, semantic technologies have the potential to facilitate data exchange between Web applications and allow them to be used together in unforeseeable ways.

In this paper we present a framework for developing Semantic Web applications. The OntoWiki Application Framework has been developed with applications in mind that have a strong emphasis on collaboration. As the OntoWiki Application Framework is a successor of OntoWiki, a visual Semantic Wiki [2], we will first introduce OntoWiki. The role of OntoWiki is, however, not limited

Norman Heino, Sebastian Dietzold, Michael Martin, and Sören Auer
Institute of Computer Science
University of Leipzig
Johannisgasse 26
04103 Leipzig
e-mail: {heino,dietzold,martin,auer}@informatik.uni-leipzig.de

S. Schaffert et al. (Eds.): Networked Knowledge - Networked Media, SCI 221, pp. 61–77.
springerlink.com   © Springer-Verlag Berlin Heidelberg 2009

to a show case for the OntoWiki Application Framework, but it is intended as a generic management backend for Semantic Web applications.

The paper is structured as follows. In section 1 we will give a short introduction to the central goals of OntoWiki. Section 2 will present the OntoWiki Application Framework including its components and features. Subsequently, in section 3 we demonstrate OntoWiki usage in three different application scenarios. The article concludes with lessons learned from the use cases and future developments.

1   OntoWiki – A Visual Semantic Wiki

In this section we present OntoWiki. It is a tool which, briefly speaking, supports presentation and knowledge engineering in a Web environment. We will sketch central issues which have resulted in the development of the tool, explain why it is called OntoWiki and outline the problems while using conventional Wiki systems. Subsequently, we explain the major goals of OntoWiki, some general use cases and describe existing views and workflows.

1.1   OntoWiki – Not a Classical Wiki

The driving force behind OntoWiki development was the need of a Web tool for rapid and simple knowledge acquisition in a collaborative way. Therefore, technologies were required for presenting information in a human-readable and machine-interpretable fashion. The tool presented is called OntoWiki, since it is inspired by classical Wiki systems. Its design, however, is independent and complementary to conventional Wiki technologies. The approach taken with OntoWiki differs from previously emerged strategies to integrate Wiki systems and the Semantic Web (cf. [4, 3, 8, 10, 12]). In these works it is proposed to integrate RDF triples into text-based Wiki systems by means of a special syntax. It is a straightforward combination of existing Wiki systems and the Semantic Web knowledge representation paradigms. Yet, we see the following obstacles:

Usability: The main advantage of Wiki systems is their unbeatable usability. Adding more and more syntactic possibilities counteracts ease of use for editors.

Redundancy: To allow the answering of real-time queries to the knowledge base, statements have to be additionally kept in a triple store. This introduces a redundancy, which complicates the implementation.

Evolution: As a result of storing information in both Wiki texts and triple store, supporting evolution of knowledge is difficult.

In contrast to other semantic Wiki approaches, in OntoWiki text editing and knowledge engineering (i. e. working with structured knowledge bases) are not mixed. Instead, OntoWiki directly applies the Wiki paradigm of "making it easy to correct mistakes, rather than making it hard to make them" [9] to collaborative management of structured knowledge. This paradigm is achieved by interpreting knowledge bases as *information maps* where every node is represented visually and interlinked to related resources. Furthermore, it is possible to enhance the knowledge schema gradually as well as the related instance data agreeing on it. As a result, the following requirements have been determined for OntoWiki:

Intuitive display and editing   of instance data should be provided in generic ways, yet enabling means for domain-specific presentation of knowledge.

Semantic views   allow the generation of different views and aggregations of the knowledge base.

Versioning and evolution   provides the opportunity to track, review and roll-back changes selectively.

Semantic search   facilitates easy-to-use full-text searches on all literal data, search results can be filtered and sorted (using semantic relations).

Community support   enables discussions about small information chunks. Users are encouraged to vote about distinct facts or prospective changes.

Online statistics   interactively measures the popularity of content and activity of users.

Semantic syndication   supports the distribution of information and their integration into desktop applications.

OntoWiki enables the easy creation of highly structured content by distributed communities. The following points summarize some limitations and weaknesses of OntoWiki and thus characterize the application domain:

Environment: OntoWiki is a Web application and presumes all collaborators to work in a Web environment, possibly distributed.

Usage Scenario: OntoWiki focuses on knowledge engineering projects where a single, precise usage scenario is either initially (yet) unknown or not (easily) definable.

Reasoning: Application of reasoning services was (initially) not the primary focus.

1.2   Generic and Domain-Specific Views

OntoWiki can be used as a tool for presenting, authoring and managing knowledge bases adhering to the RDF data model. As such, it provides generic methods and views, independent of the domain concerned. Two coarse-grained generic views included in OntoWiki are the resource view and the list view. While the former is generally used for displaying all known information about a resource, the latter can present a set of resources, typically instances of a certain concept. That concept not necessarily has to be explicitly defined as `rdfs:Class` or `owl:Class` in the knowledge base. Via its facet-based browsing, OntoWiki allows the construction of complex concept definitions, with a pre-defined class as a starting point by means of property value restrictions. These two views are sufficient for browsing and editing all information contained in a knowledge base in a generic way.

For domain-specific use cases, OntoWiki provides an easy-to-use extension interface that enables the integration of custom components. By providing such a custom view, it is even possible to hide completely the fact that an RDF knowledge base is worked on. This permits OntoWiki to be used as a data-entry frontend for users with a less profound knowledge of Semantic Web technologies.

1.3 Workflow

With the use of RDFS [5] and OWL [11] as ontology languages, resource definition is divisible into different layers: a terminology box for conceptual information (i. e. classes and properties) and an assertion box for entities using

Fig. 1 The list and details view in OntoWiki

the concepts defined (i. e. instances). There are characteristics of RDF which, for end users, are not easy to comprehend (e. g. *classes* can be defined as *instances* of `owl:Class`). OntoWiki's user interface, therefore, provides elements for these two layers, simultaneously increasing usability and improving a user's comprehension for the structure of the data.

After starting and logging in into OntoWiki with registered user credentials, it is possible to select one of the existing ontologies. The user is then presented with general information about the ontology (i. e. all statements expressed about the knowledge base as a resource) and a list of defined classes, as part of the conceptual layer.

By selecting one of these classes, the user receives a list of resources that are instances of it. In figure 1 the class `Student` has been selected and yields a list of students being either instance of `Student` directly or of its subclass `PhDStudent`; OntoWiki applies basic `rdfs:subClassOf` reasoning automatically. After selecting an instance from the list – or alternatively creating a new one – it is possible to manage (i. e. insert, edit and update) information in the details view, which is depicted in figure 1 as well.

OntoWiki focuses primarily on the assertion layer, but also provides ways to manage resources on the conceptual layer. By enabling the visualization of schema elements, called *System Classes* in the OntoWiki nomenclature, conceptional resources can be managed in a similar fashion as instance data. One of the missing features for schema management is a knowledge base consistency check, which will be included as part of the upcoming reasoning support in the near future.

2 The OntoWiki Application Framework

In the previous section we have shown how OntoWiki can be used as a Semantic Wiki. In order to render its functionality, OntoWiki relies on several APIs that are also available to third-party developers. Usage of these programming interfaces enables them to extend, customize and tailor OntoWiki in several ways. In this section we describe the OntoWiki Application Framework that builds the foundation for OntoWiki and related applications. To get an idea as to what can be achieved with the framework, we refer to the use cases described in section 3.

2.1 Architecture Overview

As depicted in figure 2, the OntoWiki Application Framework consists of three separate layers. The persistence layer consists of the Erfurt API which provides an interface to different RDF stores. In addition to the Erfurt API, the application layer is built by a) the underlying Zend Framework[1] and b) an

---

[1] `http://framework.zend.com/`

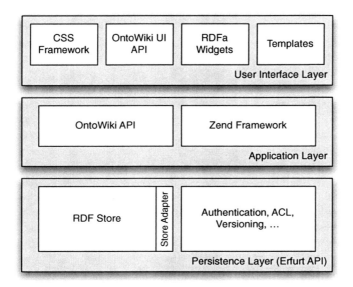

Fig. 2 The OntoWiki Application Framework with its three layers: persistence layer, application layer, user interface layer

API for OntoWiki extension development. With the exception of templates, the user interface layer is primarily active on the client side, providing the CSS framework, a JavaScript UI API, RDFa widgets and HTML templates generated on the Web-server side.

2.2 *Persistence Layer*

Persistent data storage as well as associated functionality such as versioning and access control are provided by the Erfurt API. This API consists of the components described in the subsequent paragraphs.

2.2.1 Authentication and Access Control Components

For Semantic Web applications it might be useful to have a means of authenticating users against an RDF store, instead of a database table. Erfurt therefore includes an authentication component that provides an API for user management.

Although Leuf and Cunningham define openness to everyone as one of the key concepts for a Wiki software [9], we think that especially in enterprise scenarios it might be useful to have access control at read and write level. Therefore, Erfurt allows fine-grained access control for both – groups of users and individuals. Access control rules can be defined in OntoWiki itself by modifying the system configuration model. It provides a class for models as well as actions, whose instances are objects of access control statements. Since in OntoWiki, each registered user has his/her URI which can be used as subject of a rule statement, an example access control statement, which grants the admin user the right to register new users, would be as follows:

```
<http://localhost/OntoWiki/Config/Admin>
  <http://ns.ontowiki.net/SysOnt/grantAccess>
    <http://ns.ontowiki.net/SysOnt/registerNewUser>.
```

The above statement is, of course, unnecessary in OntoWiki since the admin user is granted any action by default.

2.2.2   Caching Component

Erfurt supports several caching mechanisms based on `Zend_Cache`. Almost any entity from objects to function return-values can be stored for faster retrieval. `Zend_Cache` allows the usage of several cache backends of which database and file backends are the most important. Developers are encouraged to make use of Erfurt's caching facilities as it will greatly improve user experience.

2.2.3   Event Dispatcher

The Erfurt event dispatcher builds the foundation of Erfurt's and of OntoWiki's plug-in system. Since the dispatcher implements the Observer pattern, extensions can register code for execution when certain events occur. The registrants can be either classes or objects. In both cases, a method with the same name as the event, must exist and will be executed once the event is triggered. Events can be triggered by using the event dispatchers `trigger` method. For a detailed description of OntoWiki's plug-in architecture, see section 2.3.

2.2.4   RDF/RDFS/OWL API

These classes provide a resource-, property- or model-centric view on the triples in an RDF store, taking into account additional inbuilt semantics that are provided by different layers of the Semantic Web stack.

Once the heart of the Erfurt API (then named pOWL [1]), they provided an easy-to-use interface which unfortunately led to extensibility and scalability problems. Thus, the current state of Erfurt contains only the most important classes with a reduced method set.

Functionality currently provided by these classes includes the following:

- adding/removing statements,
- updating models with a statement diff,
- namespace handling,
- URI handling,

- transitive closure calculation and
- `owl:imports` handling.

For performance reasons, complex retrieval tasks are not covered by the API and should be done through SPARQL [6] in combination with domain-specific MVC models instead (see section 2.3 on how this is done in OntoWiki core components).

2.2.5 Store Component

Triple storage and retrieval are provided by Erfurt's storage component. Erfurt allows for easy integration of RDF stores via adapters that mediate between the store's communication protocols and Erfurt's PHP API. The API provided by adapters is not directly exposed to framework clients. Instead, a lightweight intermediate layer (`Erfurt_Store`) is used to assure that access control rules are adhered to and that versioning information is kept along with changes to the RDF store. This architectural decision has two implications:

- Versioning and ACL enforcement is completely transparent to store adapters.
- The store architecture is open to extension, for instance different import and export formats can thus be supported by the API.

Erfurt comes with store adapters for MySQL and OpenLink Virtuoso [7]. The list of supported stores will be expanded in future versions of the framework. As a matter of fact, work is currently being done on Redland and Oracle adapters.

2.2.6 Versioning Component

As its name implies, this component is responsible for keeping versioning information on an RDF store. Versioning is handled on statement level, i. e. actions that are recorded are *statement-added*, *statement-removed* and *statement-changed*. The usual entry point is a resource URI which yields all changes that have been made to statements about that specific resource. In addition, Erfurt's versioning component provides other entry points such as user URI or model URI, where all changes are returned that have been made by a specific user or have been made to statements in a specific model, respectively.

2.3 Application Layer

OntoWiki as a Web application is based on the Zend Framework which lays out the basic architecture and is primarily responsible for request handling. In the following paragraphs we cover custom OntoWiki classes and aspects of the Zend Framework that need to be considered when developing Semantic Web applications with the OntoWiki Application Framework.

2.3.1 OntoWiki Request Lifecycle

The single entry point to the application is the `index.php` file which sets up the basic environment and starts the `OntoWiki_Application` singleton. The latter initializes the OntoWiki application itself and serves as a global registry for objects and simple values. Thereafter, the Zend Framework takes over control and dispatches the request to an appropriate controller with an action that handles the request. The content is then rendered into templates, as described in the Templates paragraph of section 2.4.

2.3.2 OntoWiki MVC Models

One of OntoWiki's most outstanding features is that it automatically displays human-readable representations of resources instead of URI strings. The naming or title properties it uses are configurable both on a global level and per model. SPARQL queries that test all naming properties can be quite complex. OntoWiki therefore provides a model base class that builds SPARQL query fragments and fetches the correct naming property value from an Erfurt store result set.

Fig. 3 Screenshot of OntoWiki with OntoWiki Application Framework components: 1) menu, 2) toolbar, 3) navigation, 4) module window and 5) message

2.3.3 Menus

Menus in OntoWiki (see 1 in figure 3) consist of instances of `OntoWiki_Menu`. Entries are set by using the `setEntry` instance method that takes two arguments: the name of the menu entry and the content, which can be a string, another instance of `OntoWiki_Menu` or a menu separator stated by `OntoWiki_Menu::SEPARATOR`. An optional third parameter denotes whether entries of the same name should be replaced or not.

2.3.4 Toolbar

To ensure a consistent user interface throughout all views, the toolbar is centrally managed. In each request there exists an instance of `OntoWiki_Toolbar` to which buttons and separators can be appended or prepended. An example toolbar is depicted under 2 in figure 3. Table 1 shows default buttons that are available.

Table 1 Toolbar buttons available in OntoWiki.

Constant	Name	CSS class	Function
CANCEL	Cancel		Cancel an operation
SAVE	Save		Save current changes
EDIT	Edit	edit-enable	Enter editing mode
ADD	Add		Add a new entity
EDITADD			Add a new entity by editing another
DELETE	Delete		Delete the current selection
SUBMIT	Submit	submit	Save changes
RESET	Reset	reset	Reset changes

The name or CSS class of default buttons can be overwritten by providing the `appendButton` or `prependButton` method with a configuration array as the second parameter. If the configuration array is the only parameter, a custom button will be generated (in that case an image URL should be provided, as well).

2.3.5 Navigation

Without any customization, OntoWiki's main navigation is displayed as a tab bar in the upper part of the main window (see 3 in figure 3). Components can register one or more actions with the navigation. A component's default action is registered automatically by the component manager. Disabling the navigation is possible by calling `OntoWiki_Navigation::disableNavigation()`.

2.3.6 Extension Architecture

The OntoWiki Application Framework differentiates between three kinds of extensions:

Plug-ins   are the most basic, yet most flexible types of extensions. They consist of arbitrary code that is executed on certain events. Plug-ins need to be registered for events in the `plugin.ini` config file that has to be placed in the same folder as the plug-in class.

Modules   display little windows that provide additional user interface elements with which the user can affect the main window's content. Since some modules are highly dynamic extensions, they can be configured both statically and dynamically. Static configuration works in the same way as with other extensions; a module.ini file is placed in the module's root directory. In addition, a module class needs to extend `OntoWiki_Module` and can redefine several of its methods in order to allow for dynamic customization. If present, return values will overwrite static configuration settings in the `module.ini` file.

Components   are pluggable MVC controllers to which requests are dispatched. Usually but not necessarily, components provide the main window's content and, in that case, can register with the navigation to be accessible by the user. In other cases components can function as controllers that serve asynchronous requests. Components are statically configured by a `component.ini` file within the component's folder.

2.3.7   Localization

`Zend_Translate` along with CSV files are used to translate user interface strings. Extensions can provide their own translation files. If done so, the folder containing the translations must be set in the configuration file. Translatable strings are printed using the the _ member function of `OntoWiki_View`. Alternatively, the translate object that can be requested from `OntoWiki_Application` provides a `translate` method.

2.3.8   URLs and URI Parameters

By convention, the URL parameter that identifies a resource is named r. If this parameter contains only a URI's local part or a cURI[2], OntoWiki automatically expands it into a full URI by using namespace prefixes from imported knowledge base files.

For constructing URLs, usage of `OntoWiki_Url` is recommended. This class initializes itself with the currently active URL but all parameters including controller and action can be replaced. Apart from name and value, the `setParam` method accepts a third optional parameter that, if set to `true`, enables automatic URI compacting by replacing namespaces with their prefixes or no prefix at all for the currently active model. This behaviour allows for user-friendly short URLs with almost no extra effort for the developer.

---

[2] `http://www.w3.org/TR/rdfa-syntax/#s_curieprocessing`

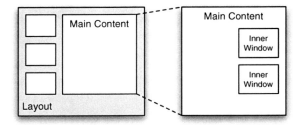

Fig. 4 OntoWiki template hierarchy. The main content is produced by the controller action. Layout content is applied automatically by the template system.

2.3.9 Messages

User notifications are represented by `OntoWiki_Message` which is instantiated by passing a message text and a type constant. Recognized types are `SUCCESS`, `INFO`, `WARNING` and `ERROR`. The main `OntoWiki_Application` object keeps a message stack that is automatically displayed in the upper part of the page. In figure 3, a message is depicted under 5. Elements can be added to this stack via `OntoWiki_Message` member functions `appendMessage` or `prependMessage`.

2.4 User Interface Layer

2.4.1 Templates

Content is rendered in OntoWiki through templates, as suggested by the `Zend_View` template system. The controller action serving the request renders its output in a template. In doing so, it has control over inner windows within the main content and can explicitly include modules (see figure 4). In order to build a complete page the main content is inserted into a layout template that defines the position of main content and side windows.

2.4.2 User Interface API and CSS Framework

While much of the user interface dynamism found in OntoWiki is made available via a JavaScript API, its look and feel results from a sophisticated CSS framework that makes it almost unnecessary to provide custom style sheets. Since the API itself relies heavily on jQuery[3], large parts of it are implemented as jQuery plug-ins. This design allows, not only style, but also behaviour be used automatically on HTML elements that carry the respective CSS classes.

---

[3] `http://jquery.com/`

2.4.3 RDFa Widgets

By exposing RDFa, structured data is available in rendered HTML code. A set of JavaScript-based widgets that make use of statements extracted from RDFa provides editing functionality to be directly invoked from the client side (i. e. inside the user's web browser). Since complete statements are available to those widgets and they can even fetch additional metadata, e. g. `rdf:range` or `rdf:datatype` constraints, it is possible to provide the user with well-suited edit forms. Changed statements are then sent back asynchronously, so no HTML page refresh is required after performing an edit action.

3 Use Cases

In this section we introduce exemplarily three projects that make use of the OntoWiki Application Framework to different extents.

3.1 SoftWiki – Requirements Engineering the Wiki Way

SoftWiki is a specialized Wiki application for end-user-centered requirements engineering. The aim of the SoftWiki application is to support the collaboration of all stakeholders in software development processes in particular with respect to software requirements. Potentially very large and spatially distributed user groups shall be enabled to collect, semantically enrich, classify and aggregate software requirements. Thus, SoftWiki is a prime example for such knowledge-rich applications which can be developed with the OntoWiki Application Framework.

Requirements for the SoftWiki application can be outlined as follows:

- The main entity in SoftWiki is a requirement with its attributes and relations between requirements.
- Users should be enabled to create and manage requirements as well as relations between them in an easy way with respect to two different management schemes, namely topic hierarchies and tag clouds.
- Users should be supported in their collaboration, for instance, if they want to discuss and vote on particular requirements.

Based on these requirements, SoftWiki was developed as a plug-in for OntoWiki instead of developing a new Semantic Web application from scratch. SoftWiki uses the majority of the backend functionality including versioning, access control and authentication. In addition to the OntoWiki-provided backend, SoftWiki implements a dynamic user interface, built upon asynchronously loaded GUI components from the OntoWiki base system.

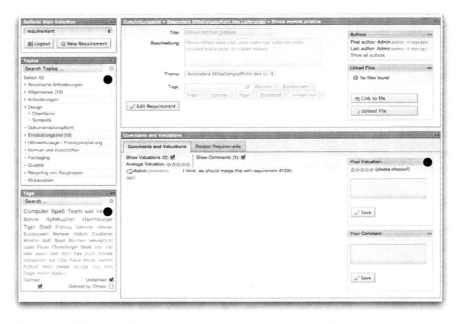

Fig. 5 SoftWiki detail view for a specific requirement. All marked windows are reused GUI components.

The list of GUI components includes

- a tag cloud which is based on the tag ontology from Ayers et al.[4],
- a generic hierarchy browser to visualize container hierarchy (this component is used for a class tree but SoftWiki uses it for a hierarchy of SKOS concepts which are used to manage requirements) and
- a generic vote and discussion component which enables users to comment on a specific resource (in SoftWiki these resources are limited to requirements.).

Fig. 5 shows a screenshot of a detailed view for a specific requirement in SoftWiki, where components reused from generic OntoWiki application are marked with a black dot. These components are integrated by Ajax calls from the SoftWiki application. Every component is a specific action and can be used from inside or outside OntoWiki.

Reusing and integrating OntoWiki core functionality was very important in the SoftWiki development. However, the main development effort was done on the SoftWiki plug-in controller which implements specialized views for listing and editing requirements. This controller uses the Erfurt API to store and query statements on a project's requirements and generates the custom views using the CSS framework. Both new and reused components are connected

---

[4] http://www.holygoat.co.uk/owl/redwood/0.1/tags/

by JavaScript functionality which handles all user input and requests the specific GUI elements.

3.2 Caucasian Spiders Database

The Caucasian Spiders database [5] is a faunistic knowledge base on the spiders of the Caucasus. It consists of several components:

- a biological taxonomy of spiders,
- concepts for records, locations and publications,
- instance data about individual spiders with localities and
- geographical information.

The knowledge base has a size of about 240k triples. It is browsable in OntoWiki with only minor tweaks to internal inference algorithms[6]. The project uses OntoWiki as both a knowledge-base editor for data entry and a browser for displaying instance data. OntoWiki supports these use cases with generic user interface components like class tree, facet-based browsing and Map component just to name a few. Significant customizations of the user interface were not required.

The project might, however, benefit from a custom hierarchy tree that can be based on arbitrary properties. This would allow tree-based browsing of the biological taxonomy instead of the `rdfs:subClassOf` hierarchy, which is more natural to biologists. Such a component has been developed for another project (see section 3.1) and will be integrated into OntoWiki in one of the upcoming versions.

3.3 Professor Catalogue of the University of Leipzig

In the course of the 600th anniversary of the University of Leipzig a database of Leipzig professors from the nineteenth and twentieth century[7] has been built in the department of history.

The database has been realized using Semantic Web technologies allowing it to be queried in various ways. One could for instance try to find the names of professors who were taught by Nobel Prize winners. However useful that query might be, it shows the flexibility that is gained by building upon RDF and related technologies.

Since OntoWiki can work with any RDF knowledge base, it was a natural choice as a generic data wiki for collaboratively building the database and

---

[5] `http://caucasus-spiders.info/`

[6] E. g. disabling inference that resource A is a class if there exists at least one resource a_i that has A as its `rdf:type` (called *implicit class* in OntoWiki terminology).

[7] `http://www.uni-leipzig.de/unigeschichte/professorenkatalog/`

entering instance data. The result was a knowledge base with about 60 schema elements and 800 entries.

4 Conclusion

In this paper we presented the OntoWiki Application Framework which can be used a basis for developing Semantic Web applications in different environments. We described the usage of the framework and gave example use-cases that were implemented with the OntoWiki Application Framework.

There is, of course, room for future improvements and refinements of the OntoWiki Application Framework. One often-requested feature is reasoning support which is currently integrated as a component. Work is also done on scalability problems that occurred when using very large datasets or complex queries. Scalability becomes paramount with statement-based access control, which heavily decelerated the use of the OntoWiki Application Framework. Further improvements could also include integration with different Semantic Web endpoints like DBpedia[8], Sindice[9] or Linked Data providers.

References

1. Auer, S.: Powl – A Web Based Platform for Collaborative Semantic Web Development. In: Proceedings of the 1st Workshop on Scripting for the Semantic Web at the ESWC, CEUR Workshop Proceedings, Heraklion, Greece (May 30, 2005)
2. Auer, S., Dietzold, S., Riechert, T.: OntoWiki - A Tool for Social, Semantic Collaboration. In: Cruz, I., Decker, S., Allemang, D., Preist, C., Schwabe, D., Mika, P., Uschold, M., Aroyo, L.M. (eds.) ISWC 2006. LNCS, vol. 4273, pp. 736–749. Springer, Heidelberg (2006)
3. Aumüller, D.: Semantic Authoring and Retrieval within a Wiki (WikSAR). In: Demo Session at the Second European Semantic Web Conference (ESWC 2005) (May 2005), http://wiksar.sf.net
4. Aumüller, D.: SHAWN: Structure Helps a Wiki Nvigate. In: Proceedings of the BTW-Workshop "WebDB Meets IR" (2005)
5. Brickley, D., Guha, R.V.: RDF Vocabulary Description Language 1.0: RDF Schema. W3C recommendation, W3C (February 2004), http://www.w3.org/TR/2004/REC-rdf-schema-20040210/
6. Clark, K.G., Feigenbaum, L., Torres, E.: SPARQL Protocol for RDF. W3c recommendation, W3C (January 2008), http://www.w3.org/TR/rdf-sparql-protocol/
7. Erling, O., Mikhailov, I.: RDF Support in the Virtuoso DBMS. In: Auer, S., Bizer, C., Müller, C., Zhdanova, A.V. (eds.) The Social Semantic Web 2007, Proceedings of the 1st Conference on Social Semantic Web (CSSW), Leipzig, Germany, September 26-28, 2007. LNI, vol. 113, pp. 59–68 (2007)

---

[8] http://dbpedia.org/
[9] http://sindice.com/

8. Krötzsch, M., Vrandecic, D., Völkel, M.: Wikipedia and the Semantic Web - The Missing Links. In: Voss, J., Lih, A. (eds.) Proceedings of Wikimania 2005, Frankfurt, Germany (2005)
9. Leuf, B., Cunningham, W.: The Wiki Way: Collaboration and Sharing on the Internet. Addison-Wesley Professional, Reading (2001)
10. Oren, E.: SemperWiki: A Semantic Personal Wiki. In: Decker, S., Park, J., Quan, D., Sauermann, L. (eds.) Proc. of Semantic Desktop Workshop at the ISWC, Galway, Ireland, November 6, vol. 175 (2005)
11. Patel-Schneider, P.F., Hayes, P., Horrocks, I.: OWL Web Ontology Language - Semantics and Abstract Syntax. W3c:rec, W3C (February 10, 2004), http://www.w3.org/TR/owl-semantics/
12. Souzis, A.: Building a Semantic Wiki. IEEE Intelligent Systems 20(5), 87–91 (2005)

Conceptual Foundations for a Service-oriented Knowledge and Learning Architecture: Supporting Content, Process and Ontology Maturing

Andreas Schmidt, Knut Hinkelmann, Tobias Ley, Stefanie Lindstaedt,
Ronald Maier, and Uwe Riss

Abstract. Effective learning support in organizations requires a flexible and person-alized toolset that brings together the individual and the organizational perspective on learning. Such toolsets need a service-oriented infrastructure of reusable knowledge and learning services as an enabler. This contribution focuses on conceptual foundations for such an infrastructure as it is being developed within the MATURE IP and builds on the knowledge maturing process model on the one hand, and the seeding-evolutionary growth-reseeding model on the other hand. These theories are used to derive maturing services, for which initial examples are presented.

Andreas Schmidt
FZI Research Center for Information Technologies, Haid-und-Neu-Straße 10-14,
76131 Karlsruhe, Germany
e-mail: andreas.schmidt@fzi.de

Knut Hinkelmann
University of Applied Sciences Northwestern Switzerland, Riggenbachstrasse 16,
4600 Olten, Switzerland
e-mail: knut.hinkelmann@fhnw.ch

Tobias Ley
Know-Center, Inffeldgasse 21a, 8010 Graz, Austria
e-mail: tley@know-center.at

Stefanie Lindstaedt
Know-Center, Inffeldgasse 21a, 8010 Graz, Austria
e-mail: slind@know-center.at

Ronald Maier
University of Innsbruck, Universitätsstrasse 15, 6020 Innsbruck, Austria
e-mail: ronald.maier@uibk.ac.at

Uwe Riss
SAP AG, CEC Karlsruhe, Vincenz-Priessnitz-Strasse 1, 76131 Karlsruhe, Germany
e-mail: uwe.riss@sap.com

S. Schaffert et al. (Eds.): Networked Knowledge - Networked Media, SCI 221, pp. 79–94.
springerlink.com   © Springer-Verlag Berlin Heidelberg 2009

1   Introduction

In a world of constant change, enterprises need to become increasingly agile in order to compete successfully. They need to adapt to changes, deliver new or improved product and service offers. To do so, they need to leverage their employees' creativity and hands-on experience, and improve the sharing of knowledge within the enterprise (and often also across its borders). To support these activities, we need to move away from systems conceived and operated in a top-down way (like traditional learning or knowledge management systems). These systems are slow to adapt to new developments, and hardly adapt to the personal needs of individuals and their situations. As a consequence, they lack user acceptance and don't live up to the initial expectations.

To avoid that, we need a balance of bottom-up and top-down development of systems supporting learning, knowledge handling and innovation in businesses and organisations. Web 2.0-style engagement of individuals in sharing and other social activities shows that we clearly need to take into account the motivational aspects of knowledge workers. Motivational theories like the self-determination theory of Deci and Ryan [23] emphasize the important needs of experiencing competence, autonomy, and relatedness – which cannot be achieved in the context of top-down systems. To realize that, *personal learning environments* [2] have been proposed, consisting of work-integrated, personalized tools for communicating, collaborating, structuring, reflecting, and awareness building. The individual learner should be able to easily combine these tools according to his own needs and preferences and readily interoperate with others' personal learning environments to account for the social nature of learning processes.

One of the challenges the MATURE IP (http://mature-ip.eu) is facing, is embedding the paradigm of personal learning environments into organizations. To that end, we need a new form of organizational guidance, realized through a complementary organizational learning environment. Such an environment has a two-fold purpose: (1) It is supposed to give the individual the possibility to view their contributions in an organisational context and encourage participation toward organization goals. (2) It should give the organization the opportunity to analyze bottom-up activities within the sum of individual PLEs. The results of these analyses should promote the consolidation of such activities towards organizational goals, enable the breeding of strategically important communities, and help enriching existing knowledge resources so that they can be readily reused as learning objects.

Such environments need to be flexible, and personalized, which calls for an infrastructure providing reusable knowledge services that can be easily recombined. But the notion of service also goes beyond components; it usually assumes that the granularity of functionality as well as packaging is motivated by usage patterns (e.g., by personal and organizational learning environments) and not purely technical (software engineering) considerations. Engineering of such knowledge and learning architectures thus requires a thorough understanding of individual and organizational learning and its effective support.

In this paper, we present an approach to conceptualizing knowledge services based on the knowledge maturing model [14]. This model helps to understand the flow of knowledge and its barriers within and across organizations from a macroscopical point of view. We extend this by differentiating between knowledge assets of varying degrees of maturity (section 2). We then derive intervention strategies from the SER model (section 3) that form the basis for maturing (support) services (section 4) and give examples for such services.

2 Knowledge Maturing

The knowledge maturing model views learning activities as embedded into, interwoven with, and even indistinguishable from everyday work processes. Learning is understood as a social and collaborative activity, in which individual learning processes are interdependent and dynamically interlinked with each other: the output of one learning process is input to the next. If we have a look at this phenomenon from a macroscopic perspective, we can observe that knowledge is continuously repackaged, enriched, shared, reconstructed, translated and integrated etc. across different interlinked individual learning processes. During this process knowledge becomes less contextualized, more explicitly linked, easier to communicate, in short: it matures. The knowledge maturing process model structures this process into five phases (based on experiences from several practical cases as well as a comprehensive empirical study, [25], [14]):

- **Expressing ideas.** New ideas are developed by individuals from personal experiences or in highly informal discussions. The knowledge is subjective and deeply embedded within the context of the originator. The vocabulary is vague and often restricted to the person expressing the idea.
- **Distributing in communities.** This phase accomplishes the development of common terminology shared among community members, e.g. in discussion forum entries, blog postings or wikis.
- **Formalizing.** Artefacts created in the preceding two phases are inherently unstructured and still highly subjective and embedded in the context of the community. In this phase, purpose-driven structured documents are created, e.g. project reports or design documents or process models in which knowledge is 'desubjectified' and the context is made explicit.
- **Ad-hoc learning.** Documents produced in the preceding phase are not well suited as learning material because no didactical considerations were taken into account. Now the topic is refined to improve comprehensibility in order to ease its consumption or re-use. The material is ideally prepared in a pedagogically sound way, enabling broader dissemination, e.g. service instructions or manuals.
- **Standardization.** The ultimate maturity phase puts together individual learning objects to cover a broader subject area. Thus, the subject area becomes teachable to novices. Tests and certificates confirm that participants of formal training achieved a certain degree of proficiency.

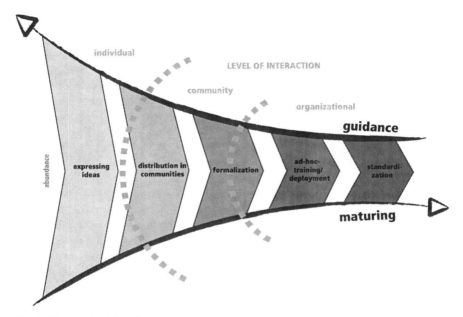

Fig. 1 Knowledge Maturing Process model

This maturing process is most intuitively recognized in the case of 'content objects' (knowledge represented in the form of documents, drawings, etc.). However, it also applies to other types of knowledge representations vital for operating and developing any kind of organisation: namely processes and semantics [21]:

- **Contents** provide a static picture of the world and are probably the best managed type of knowledge asset. The term knowledge asset points towards a value-oriented perspective on knowledge elements (business value) suggesting the importance of knowledge for the functioning of an organisation's business processes. It can take the form of notes, contributions and threads, protocols, lessons learnt, learning objects, courses, etc.
- **Processes.** This type of knowledge asset is more related to the dynamic aspect of the organisation. Large organisations already support this by developing business process models and workflows. Taking into account that organisational learning processes are much more agile and the costs of modelling approaches are considerable, a more suitable approach is to enable recording and sharing of individual work practices. Processes can take the form of e.g. individual task lists and routines, task patterns, good practices, best practices, work flows or standard operating procedures.
- **Semantics.** This type of knowledge asset is probably the least visible within organizations. Semantics connect the different assets and supports the individual learning processes by providing the basis for mutual understanding. Without semantic integration, grassroot approaches encouraging people to contribute their individual views, experiences and insights would get stuck in misinterpretations

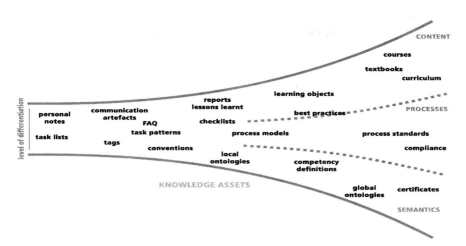

Fig. 2 Knowledge Maturing Process model

and lengthy negotiation processes. These knowledge assets can take the form of tag clouds and emerging folksonomies, folder structures, competence models, local or global enterprise ontologies.

These three knowledge asset types – and thus the three strands of maturing – are closely interwoven and they depend on each other in various respects. Contents and processes require semantics to become communicable. Therefore, semantics is the fundament for every community-based approach and fosters collaboration between individual knowledge workers. Without process integration, semantics and contents are not directly applicable to work procedures so that additional transformation efforts by the knowledge workers are required. More mature content allows a worker to deal with the high complexity and variability of knowledge-intensive processes and adapt to unpredictable situations [5]. Finally, contents are required to explicate semantics and processes so that these are comprehensible to knowledge workers with different backgrounds. While semantics and processes focus on the actual doing, contents aim at understanding and reflection.

Figure 2 depicts the described situation schematically. Knowledge asset types are not well differentiated in the early maturing phases; notes can contain content, process, and semantic aspects, sometimes all at the same time. Only with a deepened understanding, this differentiation can take place. This corresponds with a decrease in abundance: while there are many notes and communication artefacts at the beginning of the maturing process, formal training materials are rather scarce at its end. It also shows that the maturing process is accompanied by a process of organisational guidance that supports the identification of significant emerging topics and their transformation to more mature forms of knowledge. As the process of guidance already indicates, the development should not be misunderstood as a continuous linear process. On the contrary, maturing is made up of a complex pattern of individual steps. Not all knowledge assets are developed up to the ultimate maturity phase, some of them end up in a stalemate or are discarded; others are combined with other assets at various maturity levels, or split up into more differentiated assets. What we observe is an evolution of knowledge assets.

3   Seeding – Evolutionary Growth – Reseeding

In order to describe the individual steps of the maturing process in more detail, we applied Fischer's Seeding, Evolutionary growth, and Reseeding (SER) model [6]. The SER model was originally developed to describe and help to understand the evolution of complex software environments. Instead of viewing a software environment as the final product of the software development process which led to its existence, the SER model views the software system as the starting point (seed) for a complex, socially driven, evolutionary further 'development' process. In this process, users interact with the environment, its units, its structures and its tools - and thus develop them further. New units are built during these interactions, new tools are developed (by adaptation or end-user programming capabilities), and a variety of relationships or structures are discovered and expressed. The provided tools afford the creation of new and the combination of existing units, structures, and tools, the more the users have the opportunity to express their creativity and to satisfy their needs. Community activity leads to evolutionary, undirected (and often confusing) growth of the original software system. Fischer observed that typically such an evolutionary growth phase is followed by what he calls a reseeding phase: At some point in time, the environment becomes too complex to be managed. Many new units and tools have evolved and structures have become frizzled. Restructuring and redesign of the environment is initiated by some triggering event (e.g., design breakdown). This reseeding can happen in a form of consolidation and negotiation processes in which the variety of units, structures, and tools are pruned. In traditional software systems, this reseeding has to be accomplished by programmers, since the end-users will not be able to do so themselves. Fischer argues that in order to build and maintain useful software systems, we need to provide the end-user not only with tools which support evolutionary growth activities (e.g., combine, specialize) but also with tools which enable her to participate in the reseeding phase (e.g., visualization of structures, negotiation).

In order to reflect on applying the SER model to the knowledge maturing process consider for example the maturity phase 'distributing in communities'. First, a community 'space' is seeded with an initial idea or topic. This involves creating an initial knowledge structure together with its knowledge units and their capabilities and characteristics. This community environment needs to be equipped with tools for combination, analysis, and change of the structures and the units themselves in order to enable evolutionary growth. Such tools enable the users to combine knowledge units to build (increasingly complex) knowledge structures and to change the knowledge units themselves according to their needs. Analysis tools enable the community to monitor and guide its activities. If the development of the topic reaches a certain level, the decision whether to take the topic to the maturity phase "formalizing" has to be made. If the development of the topic stagnates, reseeding might be an option. This includes pruning the current knowledge base, introducing new ideas, knowledge elements or people into the community or changing the topic.

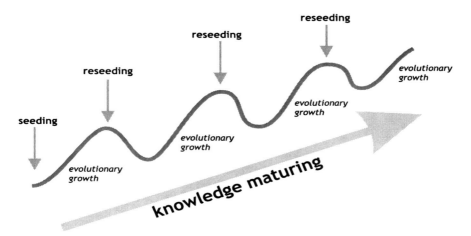

Fig. 3 The SER model and knowledge maturing

It is tempting to equate a SER cycle with a knowledge maturing phase. However, this conceptualization of knowledge maturing evokes the false impression that maturing is a collection of discrete steps which will happen in strict order. By applying the SER model, we not only stress that evolutionary growth and reseeding are important recurring phases of the maturing process, but that they are really inseparably interlinked and interwoven. That is, a user might engage in growth activities at one moment involving one knowledge asset type (content, semantics, process; compare fig. 2) while the same user might engage in reseeding activities in parallel. This interplay of growth and reseeding activities invokes the association to the interplay of assimilation and accommodation processes during knowledge construction in informal learning [20]. Here, a person integrates new knowledge into her own mental model of the topic by either adding the knowledge into already existing knowledge structures or this new piece of knowledge causes her to restructure her mental model in order to accommodate it.

Based on these insights, we treat maturing as an organizationally guided learning process which interweaves informal learning processes of many individuals - first on a group or community level, then on an organizational level. Since these individuals utilize different types of knowledge representations (content, semantics, process) to document the gained insights, tools are needed to do so with low effort and to identify relationships between them. Our future research will specifically focus on identifying the factors which influence assimilation versus accommodation activities and the barriers people experience when doing so.

When analyzing tools supporting knowledge work, we find a variety of (mostly) independent tools separated along two dimensions: (1) types of knowledge assets

(content, semantics, process) and (2) level of interaction (organization, commu-
nity/group, individual). The first dimension corresponds to different ways of knowledge construction and the second to the breadth of knowledge sharing. The separation
of these tools reflects existing gaps in support of maturing processes (fig. 4).

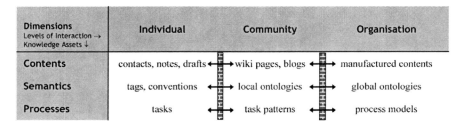

Fig. 4 Separation of systems

4 Maturing Services

In the following, we will use the concept of *maturing services* to refer to integrated
support for the maturing process. That is, maturing services will bridge the separation along both dimensions of knowledge construction and knowledge sharing as
outlined in the previous section. They are needed not only to help knowledge workers to handle these different knowledge assets, but also to entice them in sharing and
negotiating among them. Generally, a service consists of contract, interface and implementation. It has distinctive functional meaning typically reflecting a high-level
business concept covering data and business logic [11]. A service is an abstract
resource that represents a capability of performing tasks that form a coherent functionality from the point of view of providers entities and requesters entities. Service
descriptions provide information about:

- *service capability:* conceptual purpose and expected result,
- *service interface:* the service's signature, i.e. input, output, error parameters and
  message types,
- *service behavior:* a detailed workflow invoking other services,
- *quality of service:* functional and non-functional quality attributes, e.g., service
  metering, costs, performance metrics and security attributes.

The service concept has gained popularity with the advent of a set of standards for
open interaction between software applications using Web services (such as WSDL,
SOAP and UDDI). Whereas the technical definition of services is supported by standards, it is the conceptual part (i.e. defining types of services that are useful) that
is currently lacking. Knowledge management (KM) services or knowledge services
are a subset of services, both basic and composed, whose functionality supports
high-level KM instruments as part of on-demand KM initiatives, e.g., find expert,
submit experience, publish skill profile, revisit learning resource or join community-
of-interest [13]. These services might cater to the special needs of one or a small

number of organizational units, e.g., a process, work group, department, subsidiary, factory or outlet in order to provide solutions to defined business problems. KM services describe aspects of KM instruments supported by heterogeneous application systems.

For example, a complex KM service "search for experts" might be composed of the basic KM services (1) expert search, (2) keyword search, (3) author search, (4) employee search and (5) check availability. The (1) expert search service delivers a list of IDs, e.g., personnel numbers, for experts matching the input parameter of an area of expertise. The (3) author search service requires a list of keywords describing the area of expertise. Thus, the complex KM service search for experts also comprises an integration service for the task of finding keywords that describe the area of expertise, here called (2) keyword search. The keywords are assigned to areas of expertise either in a simple database solution or in a more advanced semantic integration system based on an ontology. With the help of an inference engine, these relationships together with rules in the ontology can be used to determine a list of keywords. The (3) author search service then returns a list of IDs of matching authors or active contributors to the CMS. An (4) employee search service takes the personnel numbers found in the expert search and the author search and returns contact details, e.g., telephone number, email address, instant messaging address. Finally, the (5) check availability service delivers the current status of the experts and a decision on their availability.

We conceptualize maturing services as complex services that are in turn composed of basic services either already offered in heterogeneous systems as part of an enterprise application landscape, implemented additionally to enrich the services offered in an organization or invoked over the Web from a provider of maturing services. In the following, we introduce three types of maturing service which we will consider in the future:

- **Seeding services** enable the user to set up and initialize knowledge units and structures within a community. Seeding services also include functionalities to use the instantiated structures.
- **Growth services** allow users to add new knowledge units (e.g., documents or users), to adapt their characteristics (e.g., the users' competencies), to provide comments and to change the system behaviour. Growth services are based on a form of using the Web often cited as Web 2.0 in which users can produce their own content (user-generated content) and which utilizes collective usage data and user feedback to improve the system's value and performance due to network effects and phenomena which have been termed "collective intelligence" or "wisdom of the crowds" [28].
- **Reseeding services** allow the user to analyse and visualize the collective activities of the community, negotiate between conceptualizations of different users and finally (and most importantly) to change the underlying structures and functionalities. These reseeding services will go beyond the services offered under the umbrella term Web2.0 by enabling users to not only add and change content, but also to change the underlying structure and functionality of the evolving knowledge system.

In the rest of this section we present several examples for maturing services which help to illustrate the ideas we have put forward in the previous sections. In the following, we will briefly describe three examples, one for each of the three knowledge asset types (contents, semantics and processes).

4.1 Semantic Wiki Services for Career Guidance (Contents)

Wikis are prime examples of tools that allow a collective construction of knowledge in a community setting. There are certainly good examples of Wikis being used as tools for creating a collective online encyclopaedia, for teaching and learning purposes, and for organizational knowledge management ([10], [19], [15]). In our perspective, Wikis are very well suited for enabling the evolutionary growth phase, especially because of the ease of editing the content and the policy that everyone can edit anything. Additionally, they make the collective construction process traceable (utilizing their history functionality) and allow for discussion processes around artefacts.

A problem with Wikis, however, is their inability to deal with more formal content or structures. The way a standard Wiki works seems to suggest that any artefact is constructed basically from scratch in a community setting, and that there is no end to this construction process. This is an unrealistic proposition in most settings and especially in an organizational setting where knowledge generation uses artefacts that fluctuate between the informal and the formal pole. In this sense, the use of Wikis illustrates one of the barriers given in fig. 4, namely that between the community and the organizational level.

An example may help to illustrate our reasoning. We are currently examining the use of knowledge in a career guidance setting. Career advisors have the task to personally consult individuals (such as pupils or graduates or their parents) on their job prospects, and advise on potential careers given their interests and the general job situation in the region. In doing so, they make use of a large body of formally documented knowledge artefacts, for instance statistics and reports on job opportunities or labour market development in certain employment sectors and regions. Additionally, they draw on a considerable amount of informal knowledge derived from their experiences with concrete cases. This knowledge in use is more or less systematically applied in their job, and it is more or less systematically shared among practitioners.

We regard these processes of generation, application and sharing of both formal and informal knowledge as a knowledge maturing process. To support the practitioners in this process, we are employing a Semantic Media Wiki [12]. Several maturing services have been designed that try to bridge the gaps in the maturing process. First of all, an integrated search mechanism enables the practitioners to draw in a large array of different kinds of existing resources from a number of relevant sources (formal reports, statistics, videos etc.) - thus seeding the Wiki with relevant material. The Wiki then renders these existing resources so that discussions and knowledge construction in the Wiki can take place in the context of the formal documents. The idea being, that these informal discussions and knowledge construction draw in

Fig. 5 Design study for markup suggestion

practitioners' knowledge in use, which documents experiences from their practice. This should enhance the evolutionary growth of the knowledge base.

We then explore some of Semantic Media Wiki functionalities to capture the context this informal knowledge has been applied to (such as the region, the target group or the employment sector). With some information extraction and classification algorithms, we are able to suggest semantic mark-up which might be applied to an article (see fig. 5). A visualization of the whole network made up of semantic categories, textual similarity measures, and links between articles provides an overview of the whole available content, and enables detection of similarities for some gardening or reseeding activities. In addition, we will be visualizing indicators for the use frequency of articles and text readability scores. This will allow the gardening activities to focus on parts of the content that are especially important (highly used), but of poor quality (low readability). Finally, the Wiki also provides a way to export a newly created article or a collection of articles as a report so as to document the current status on a higher level of maturity.

4.2 From Collaborative Tagging to Emerging Semantics (Semantics)

Tagging resources can be seen as a first step of providing semantic descriptions for these resources. The results of such activities are knowledge assets (tags) which are used on an individual level (see fig. 4). Collaborative tagging environments (such as http://www.flickr.com or http://www.del.icio.us) make it possible to share these in a community setting.

How can services be designed to facilitate the seeding and evolutionary growth in the community setting? We have basically taken two approaches to this problem: (1) improving the quality of the folksonomy by providing tagging support, and (2) supporting the creation of ontologies from folksonomies as part of the community process.

In the first approach, we use cognitive models that have been extensively used for modelling individual cognitive processes of knowledge encoding, representation and retrieval. An example here is the declarative knowledge module in ACT-R [1] which models knowledge as an associative network. We then seek to transfer these models to a distributed community setting where several actors and shared artefacts are involved. What we are aiming to do is to describe knowledge maturing in an organisation as a distributed cognitive process. This cognitive process is based on a knowledge representation that describes the knowledge of a whole community. In the example of the collaborative tagging environment, the folksonomy (shared tags) is modelled as an associative network using tag co-occurrences [27]. Tags are modelled as nodes in a network where co-occurrence with other tags determines the associations, or the weights on the edges.

We have modelled a folksonomy in this way for a flickr data set [17]. After an appropriate model has been established (and evaluated for its validity) intelligent services can be built upon it by simulating cognitive processes on a community level, such as knowledge retrieval. In the flickr example, the service we implemented was to recommend tags when users upload new pictures. This service simulates tag associations in a distributed cognitive structure. In another case, we employ spreading activation mechanisms for these processes (which are also implemented in the ACT-R architecture) [24]. First experiments have shown that this service reduces the overall number of tags people apply as they make use of existing tags. In our view this helps to emerge a shared understanding, as the system grows evolutionary.

There is recently also growing empirical research into how information from such an associative network of tag co-occurence allows the emergence of semantic relations between tags such as discovering broader or narrower terms or synonyms ([26], [9]).

The second, complementary approach aims at community tools to engineer taxonomies or ontologies in a collaborative and lightweight manner [4], building on, but also extending the tagging paradigm. Here, the collaborative tagging environment is enhanced by providing a (lightweight), collaborative ontology editor that allows for introducing *broader*, *narrower*, and synonym relationships to cover for the most common problems in folksonomies. It is assumed that an ontology evolves or matures based on community activities. For that process, we want to provide the community with supporting services that help them to consolidate part of the folksonomy into an ontology by spotting candidates for merging or heavily used tags (where it would be worth consolidating), and by facilitating the consolidation task as such (e.g., by proving argumentation support [18]). Here, analysis services particularly help in reseeding activities. First experiments have been made as part of the semantic social bookmarking application SOBOLEO [29], and in approach to collaborative building of competence models based on people tagging [3]. Evaluation results have shown the general feasibility of the approach and indicated required supporting services.

4.3   From Task Management to Process Management (Processes)

Almost all knowledge assets a user is working with are related to some work activities. For example, a travel plan might be related the organization of a business trip or a report might be related to the regular administration activities in a project. There is also a semantic dimension of this relation [7] since semantic technologies can be applied to formally describe these connections in order to use them later for information retrieving.

Generally the representation of work activities in tasks can be considered as the first step to monitor the actual processes that take place in an organization. However, isolated tasks do not allow for the analysis of collaborative processes so that the specific relations between individual tasks must be represented. The main relation in this respect is the task-subtask-relation which describes that a specific (sub)task contributes to the accomplishment of a larger task. For example, the provision of a travel plan is only one task among others contributing to the task that describes the entire business trip. The collaborative character of tasks is expressed by the fact that the executor of a subtask is not necessarily identical to the executor of the task to which this subtask belongs. Including task-subtask-relation we obtain a network of related activities conducted by various users with different dependencies that provides a detailed picture of the activities in an organization.

The individual task with the involved people and the used resources do not only describe the actual processes in an organization but are also first-class knowledge assets. They contain the information how specific work has been conducted and can help other employees to better perform their work. They can be used to derive general task patterns, i.e., descriptions how a specific type of task can be accomplished. The feedback that is provided by employees who use these patterns can directly be incorporated in this pattern resulting in a task pattern lifecycle [16]. This lifecycle represents a typical maturing process that is to be supported by additional services. For example, this concerns the identification of similar activities in order to streamline the pattern portfolio or the support in augmenting the patterns by additional information and services.

Coming to the organizational level further development is possible. Here we find automated processes such as workflows that significantly increase the productivity of an organization. However, especially in the realm of knowledge work it has been found that workflow approaches face significant problems since they do not provide the flexibility that is required here ([8], [22]). This opens opportunities for the analysis of task patterns and concrete work processes in order to identify exactly those process aspects that are suited for process automation. Usually the underlying process models are developed by conducting interviews with employees and managers on the work process. Process maturing services can provide information to which process models can be extended that correspond to realistic work activities and where people had to deviate from the given schema in order to cope with particular circumstances. In this way old processes cannot only be updated but also completely new processes can be derived from the actual work activities. The integrated process framework does not only provide opportunities for the design of new processes but can also help to bring existing process support to the individual users due to the semantic relations by which information and processes are related.

5   Conclusions

In this contribution we have presented conceptual foundations for a service-oriented infrastructure to support learning activities in organizations. These foundations consist of a combination of two models:

- The knowledge maturing process model describes how individual learning processes are interlinked within an organizational context and the different forms of knowledge and assets involved.
- The SER model describes interventions into collaborative processes to foster a goal-oriented development.

From these two theoretical approaches, we can categorize the services according to (1) the phases of maturing and barriers/transitions they address (and the types of knowledge assets) and according to (2) the type of intervention. Additionally, we can distinguish services that address content, process, and semantic knowledge assets.

Within the MATURE IP (which has started in April 2008), this categorization will be developed into a general knowledge and learning architecture. This architecture does not only contain reusable maturing services, but will also provide flexible toolsets to the end user based on the mashup paradigm, which empowers the end user to perform situation-dependent integration between different tools (and thus create *situational applications* for learning). These toolsets can be arranged into two families:

- a **Personal Learning and Maturing Environment** for supporting the individual's learning processes embedded into work processes and for fostering the individual's engagement in maturing processes.
- a **Organizational Learning and Maturing Environment** for taking the organizational perspective or intervening into individual learning processes from an organizational perspective

The maturing services will co-evolve with these environments in a participatory design approach. Within the first year, several design studies have been prepared and have been evaluated with various end users, bringing end users, experts on individual and organizational learning, and developers into an intensive and creative discussion process.

Acknowledgements. This work has been partially funded by the European Commission as part of the MATURE IP (grant no. 216346) within the 7th Framework Programme of IST.

The Know-Center is funded within the Austrian COMET Program - Competence Centers for Excellent Technologies - under the auspices of the Austrian Ministry of Transport, Innovation and Technology, the Austrian Ministry of Economics and Labor and by the State of Styria.

References

1. Anderson, J.R., Bothell, D., Byrne, M., Douglass, S., Lebiere, C., Qin, Y.: An integrated theory of the mind. Psychological Review 111(4), 1036–1060 (2004)
2. Attwell, G.: The personal learning environments - the future of elearning. eLearning Papers 2(1) (2007)
3. Braun, S., Schmidt, A.: People tagging & ontology maturing: Towards collaborative competence management. In: 8th International Conference on the Design of Cooperative Systems (COOP 2008), Carry-le-Rouet, France, May 20-23 (2008)
4. Braun, S., Schmidt, A., Walter, A., Nagypal, G., Zacharias, V.: Ontology maturing: a collaborative web 2.0 approach to ontology engineering. In: Noy, N., Alani, H., Stumme, G., Mika, P., Sure, Y., Vrandecic, D. (eds.) Proceedings of the Workshop on Social and Collaborative Construction of Structured Knowledge (CKC 2007) at the 16th International World Wide Web Conference (WWW 2007), CEUR Workshop Proceedings, Banff, Canada, May 8, vol. 273 (2007)
5. Feldkamp, D., Hinkelmann, K., Thönssen, B.: Kiss- knowledge-intensive service support: An approach for agile process management. In: Paschke, A., Biletskiy, Y. (eds.) RuleML 2007. LNCS, vol. 4824, pp. 25–38. Springer, Heidelberg (2007)
6. Fischer, G., Grudin, J., McCall, R., Ostwald, J., Redmiles, D., Reeves, B., Shipman, F.: Seeding, evolutionary growth and reseeding: The incremental development of collaborative design environments. In: Olson, G., Malone, T., Smith, J. (eds.) Coordination Theory and Collaboration Technology. Lawrence Erlbaum Associates, Mahwah (2001)
7. Grebner, O., Ong, E., Riss, U.: Kasimir - work process embedded task management leveraging the semantic desktop. In: Proceedings of the Multikonferenz Wirtschaftsinformatik. Workshop Semantic Web Technology in Business Information Systems, Munich, Germany, pp. 716–726 (2008)
8. Holz, H., Maus, H., Rostanin, O.: From lightweight, proactive information delivery to business process-oriented knowledge management. Journal of Universal Knowledge Management (2), 101–127 (2005)
9. Hotho, A., Jäschke, R., Schmitz, C., Stumme, G.: Information retrieval in folksonomies: Search and ranking. In: Sure, Y., Domingue, J. (eds.) ESWC 2006. LNCS, vol. 4011, pp. 411–426. Springer, Heidelberg (2006)
10. Jaksch, B., Kepp, S.J., Womser-Hacker, C.: Integration of a wiki for collaborative knowledge development in an e-learning context for university teaching. In: Holzinger, A. (ed.) USAB 2008. LNCS, vol. 5298, pp. 77–96. Springer, Heidelberg (2008)
11. Krafzig, D., Banke, K., Slama, D.: Enterprise SOA: Service-Oriented Architecture Best Practices. Upper Saddle River (2005)
12. Krötzsch, M., Vrandecic, D., Völkel, M., Haller, H., Studer, R.: Semantic wikipedia. Journal of Web Semantics 5, 251–261 (2007)
13. Maier, R., Hädrich, T., Peinl, R.: Enterprise Knowledge Infrastructures, 2nd edn. Springer, Heidelberg (2008)
14. Maier, R., Schmidt, A.: Characterizing knowledge maturing: A conceptual process model for integrating e-learning and knowledge management. In: Gronau, N. (ed.) 4th Conference Professional Knowledge Management - Experiences and Visions (WM 2007), Potsdam, vol. 2, pp. 325–334. GITO-Verlag, Berlin (2007)

15. Majchrzak, A., Wagner, C., Yates, D.: Corporate wiki users: results of a survey. In: Riehle, D., Noble, J. (eds.) WikiSym 2006: Proceedings of the 2006 international symposium on Wikis, pp. 99–104. ACM, New York (2006)
16. Ong, E., Grebner, O., Riss, U.: Pattern-based task management: Pattern lifecycle and knowledge management. In: Proceedings of the 4th Conference Professional Knowledge Management (WM 2007), Potsdam, pp. 357–364. Gito (2007)
17. Pammer, V., Ley, T., Lindstaedt, S.: Tagr: Unterstützung in kollaborativen Tagging-Umgebungen durch semantische und assoziative Netzwerke. In: Gaiser, B., Hampel, T., Panke, S. (eds.) Good Tags - Bad Tags: Social Tagging in der Wissensorganisation, Waxmann (2008)
18. Ravenscroft, A., Braun, S., Cook, J., Schmidt, A., Bimrose, J., Brown, A., Bradley, C.: Ontologies, dialogue and knowledge maturing: Towards a mashup and design study. In: Schmidt, A., Attwell, G., Braun, S., Lindstaedt, S., Maier, R., Ras, E. (eds.) 1st International Workshop on Learning in Enterprise 2.0 and Beyond. CEUR Workshop Proceedings, vol. 383 (2008)
19. Reinhold, S.: Wikitrails: augmenting wiki structure for collaborative, interdisciplinary learning. In: Riehle, D., Noble, J. (eds.) Proceedings of the 2006 international symposium on Wikis, pp. 47–58. ACM Press, Denmark (2006)
20. Riss, U., Cress, U., Kimmerle, J., Martin, S.: Knowledge sharing: From experiment to application. In: Proceedings of KMAC 2006, The Third Knowledge Management Aston Conference, pp. 121–133. Operational Research Society, Birmingham (2006)
21. Riss, U.V.: Knowledge, action, and context: Impact on knowledge management. In: Althoff, K.D., Dengel, A., Bergmann, R., Nick, M., Roth-Berghofer, T. (eds.) WM 2005. LNCS (LNAI), vol. 3782, pp. 598–608. Springer, Heidelberg (2005)
22. Riss, U.V., Cress, U., Kimmerle, J., Martin, S.: Knowledge transfer by sharing task templates: two approaches and their psychological requirements. Knowledge Management Research and Practice 5, 287–296 (2007)
23. Ryan, R.M., Deci, E.L.: Self-determination theory and the facilitation of intrinsic motivation, social development, and well-being. American Psychologist 55(1), 68–78 (2000)
24. Scheir, P., Ghidini, C., Lindstaedt, S.N.: Improving search on the semantic desktop using associative retrieval techniques. In: Proceedings of I-MEDIA 2007 and I-SEMANTICS 2007, Graz, Austria, pp. 221–228 (2007)
25. Schmidt, A.: Knowledge maturing and the continuity of context as a unifying concept for knowledge management and e-learning. In: Proceedings of the Fifth International Conference on Knowledge Management (I-KNOW 2005), Graz, Austria (2005)
26. Specia, L., Motta, E.: Integrating folksonomies with the semantic web. In: Franconi, E., Kifer, M., May, W. (eds.) ESWC 2007. LNCS, vol. 4519, pp. 624–639. Springer, Heidelberg (2007)
27. Steels, L.: Collaborative tagging as distributed cognition. Pragmatics & Cognition 14(2), 287–292 (2006)
28. Surowiecki, J.: The Wisdom of Crowds: Why the Many Are Smarter Than the Few and How Collective Wisdom Shapes Business, Economies, Societies and Nations. Doubleday (2004)
29. Zacharias, V., Braun, S., Schmidt, A.: Social semantic bookmarking with Soboleo. In: Murugesan, S. (ed.) Handbook of Research on Web 2.0, 3.0 and X.0: Technologies, Business, and Social Applications, IGI Global (2009)

ARS/SD: An Associative Retrieval Service for the Semantic Desktop

Peter Scheir, Chiara Ghidini, Roman Kern, Michael Granitzer,
and Stefanie N. Lindstaedt

Abstract. While it is agreed that semantic enrichment of resources would lead to better search results, at present the low coverage of resources on the web with semantic information presents a major hurdle in realizing the vision of search on the Semantic Web. To address this problem we investigate how to improve retrieval performance in a setting where resources are sparsely annotated with semantic information. We suggest employing techniques from associative information retrieval to find relevant material, which was not originally annotated with the concepts used in a query. We present an associative retrieval service for the Semantic Desktop and evaluate if the use of associative retrieval techniques increases retrieval performance.

Evaluation of new retrieval paradigms, as retrieval in the Semantic Web or on the Semantic Desktop, presents an additional challenge as no off-the-shelf test corpora for evaluation exist. Hence we give a detailed description of the

Peter Scheir
Graz University of Technology, Austria
e-mail: `peter.scheir@tugraz.at`

Chiara Ghidini
Fondazione Bruno Kessler, Italy
e-mail: `ghidini@fbk.eu`

Roman Kern
Know-Center, Inffeldgasse 21a, 8010 Graz, Austria
e-mail: `rkern@know-center.at`

Michael Granitzer
Know-Center, Inffeldgasse 21a, 8010 Graz, Austria
e-mail: `mgrani@know-center.at`

Stefanie Lindstaedt
Know-Center, Inffeldgasse 21a, 8010 Graz, Austria
e-mail: `slind@know-center.at`

S. Schaffert et al. (Eds.): Networked Knowledge - Networked Media, SCI 221, pp. 95–111.
springerlink.com   © Springer-Verlag Berlin Heidelberg 2009

approach taken to the evaluation of the information retrieval service we have
built for the Semantic Desktop.

1 Introduction

It is largely agreed that the semantic enrichment of resources provides for
more information to be used during search (see e.g. [12] or [26]). In turn, this
can lead to greatly improve the effectiveness of retrieval systems, not only for
resources on the web but also for personal desktops. However, critics [17] as
well as advocates [21] of the Semantic Web agree that only a small fraction
of resources on the current web are enriched with semantic information. The
sparse annotation of resources with semantic information presents a major
obstacle in realizing search applications for the Semantic Web or the Semantic Desktop, which operate on semantically enriched resources. To overcome
this problem, we propose the use of techniques from associative information
retrieval in order to find relevant resources, even if no semantic information
is provided for those resources.

The main idea of our approach is to perform search using spreading activation in a two layer network structure (graphically illustrated in Figure 1)
which consists of (1) a layer of concepts, used to semantically annotate a
pool of resources, and (2) a layer of resources (documents). The combination
of spreading activation in both layers, traditionally performed either to find
similar concepts or to find similar text, allows extending search to a wider
network of concepts and resources, which can lead to the retrieval of relevant
resources with no annotation.

In this paper we describe our approach towards information retrieval on
the Semantic Desktop and present a retrieval service developed during the
first year of the APOSDLE[1] project. The rest of this paper is organized as
follows: in section 2 we introduce the concept of the Semantic Desktop and
of associative information retrieval. In section 3 we describe the approach
taken to the realization of the retrieval service. In section 4 we present the
setting (APOSDLE) in which the retrieval service for the Semantic Desktop
was employed and in section 5 we focus on the evaluation of the retrieval
service. We present related work in section 6 and our conclusion in section 7.

2 Basic Concepts

The work presented in this paper provides a first implementation of an associative retrieval service for the Semantic Desktop. In this section we briefly
introduce the main ideas and goals of the Semantic Desktop and of associative
information retrieval.

---

[1] http://www.aposdle.org/ (14.04.2008)

2.1 Semantic Desktop

The Semantic Desktop [24] [9] paradigm stems from the Semantic Web movement and aims at applying technologies developed for the Semantic Web to
desktop computing. In recent years the Semantic Web movement led to the
development of new, standardized forms of knowledge representation and
technologies for coping with them such as ontology editors, triple stores or
query languages. The Semantic Desktop founds on this set of technologies and
introduces them to the desktop to ultimately provide for a closer integration
between (semantic) web and (semantic) desktop.

2.2 Associative Information Retrieval

Crestani [8] understands *associative retrieval* as a form of information retrieval which tries to find relevant information by retrieving information that
is by some means *associated* with information that is already known to be
relevant. Information items which are associated can be documents, parts of
documents, extracted terms, concepts, etc. The idea of associative retrieval
dates back to the 1960s, when researches [22], [23] in the field of information
retrieval tried to increase retrieval performance using associations between
documents or index terms, which were determined in advance.

Association of information is frequently modeled as graph, which is referred
to as *associative network* [8]. Nodes in this network represent information items
such as documents, terms or concepts. Edges represent associations between
information items and can be weighted and / or labeled, expressing the degree
and type of association between two information items, respectively.

3 An Associative Information Retrieval Service for the Semantic Desktop

The service presented here relies upon the existence of two sources of information: first a domain ontology, used to define the vocabulary (concepts)
used to annotate resources, and then the resources themselves in the form of
textual documents. On top of these two sources of information we build an
associative network consisting of two interconnected layers, one for concepts
and one for documents (see Figure 1).

Nodes in the concepts layer correspond to concepts in the domain ontology. Nodes in the document layer correspond to documents on the Semantic
Desktop. Concept nodes are associated by means of semantic similarity (cf.
section 3.1), while document nodes are associated by means of textual similarity (cf. section 3.2). The link between the two layers of the network is
provided by annotations: a concept node is associated with a document node
if the concept is used to annotate that document (cf. sections 3.3 and 3.4).

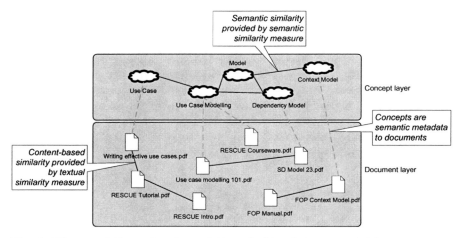

Fig. 1 The associative network consisting of of two interconnected layers

Finally, the network is searched using a spreading activation algorithm which combines spread of activation in the concept layer and spread of activation in the document layer (cf. section 3.5).

3.1 Calculating Semantic Similarity of Concepts

Concept nodes are associated in the concept layer by means of semantic similarity. For calculating the similarity of two ontological concepts a symmetric semantic similarity measure is used. The method was presented in [27] and requires two concepts belonging to the same ontology as input. It calculates the semantic similarity between these two concepts according to equation 1. This similarity measure builds on the path length to the root node from the least common subsumer (lcs) of the two concepts, which is the most specific concept they share as an ancestor. This value is scaled by the sum of the path lengths from the individual concepts to the root.

$$sim(c_1, c_2) = \frac{2 \cdot lcs(c_1, c_2)}{depth(c_1) + depth(c_2)} \tag{1}$$

With:
- c_1 ... first concept
- c_2 ... second concept
- lcs ... least common subsumer of two concepts
- $depth$ ... depth of concept in the class hierarchy

Depending on the features present in an ontology different similarity measures qualify to be applied. We chose the measure presented in [27], as a prominent feature of our ontology are taxonomic relations between concepts.

An advantage of the used measure is that it tries to address one of the typical problems of taxonomy-based approaches to similarity: relations in the taxonomy do not always represent a uniform (semantic) distance. The more specific the hierarchy becomes, the more similar a child node is to its father node in the taxonomy.

3.2 Calculating Text-Based Similarity of Documents

Document nodes are associated in the document layer by means of textual similarity. As similarity measure for text-documents we use an asymmetric measure based on the vector space model implemented in the open-source search-engine Lucene[2]. The similarity between two documents is calculated as shown in equation 2.

$$sim(d1, d2) = score(d1_{25}, d2) \tag{2}$$

With:

- $d1$ ... document vector of the first document
- $d2$ ... document vector of the second document
- $d1_{25}$ ... document vector of the first document with all term weights removed except the 25 highest terms weights

$d1_{25}$ is used as query vector for the *score*-measure of Lucene. For extracting the 25 terms with the highest weights, both the document content and the document title are taken into account. The calculation of Lucene's score is depicted in equation 3.

$$score(q, d) = coord(q, d) \cdot queryNorm(q)$$
$$\cdot \sum_{t\_in\_q} (tf(t\_in\_d) \cdot idf(t)^2 \cdot t.getBoost() \cdot norm(t, d)) \tag{3}$$

With:

- q ... query vector
- d ... document vector
- $coord(q, d) = numberOfMatchingTerms/numberOfQueryTerms$
- $numberOfMatchingTerms$ ... number of terms in document matching query
- $numberOfQueryTerms$ ... number of terms in the query
- $queryNorm(q)$ ... normalization of the query vector, Lucene default used
- $tf(t\_in\_d)$ ... term frequency of current term in document, Lucene default used

---

[2] http://lucene.apache.org/ (14.04.2008)

- $idf(t)$ ... inverse document frequency of current term in the document collection, Lucene default used
- $t.getBoost() = tf(t\_in\_q) \cdot idf(t)$
- $tf(t\_in\_q)$ ... term frequency of current term in query
- $norm(t, d) = 1/sqrt(numberOfDocumentTerms)$
- $numberOfDocumentTerms$ ... number of terms in the current document

Out of the various components that control the final score of a document matching a query, $coord(q, d)$ deserves special attention because it had shown in practice to contribute much to the final result. Thus a document that matches the set of query terms will be ranked higher than a document that only contains a smaller subset of all input query terms. Another important aspect of the scoring function is the document normalization factor, $norm(t, d)$. Documents that contain fewer terms will yield a higher score then long documents. This applies not only to the document content, but also to the document titles. Therefore the similarity of the title terms contributes more to the final score than the terms from the document body. On the other hand, the $t.getBoost()$ factor can be ignored in our case, because all query terms are weighted equally.

A detailed and a more in depth explanation of the various parameters that can be used to adapt the behavior of Lucene can be found in the Javadoc of the `org.apache.lucene.search.Similarity` class.

3.3 Semantic Annotation of Documents

The link between the two layers of the network is provided by annotations of resources with ontological concepts. As Handschuh [11] notes, different approaches to semantic annotation exist in literature. The author refers to [2] who differentiates between the following ways of semantic annotation:

- *Decoration*: Annotation of resources with a comment of the user.
- *Linking*: Annotation of resources with additional links.
- *Instance identification*: Annotation of resources with a concept. The annotated resource is an instance of the concept.
- *Instance reference*: Annotation of resources with a concept. The annotated resource references an individual in the world which is an instance of the concept.
- *Aboutness*: Annotation of resources with a concept. The annotated resource is about the concept.
- *Pertinence*: Annotation of resources with a concept. The annotated resource provides further information about the concept.

Semantic annotations in the present system are based on the *Aboutness* of resources. This means that we annotate whole documents with a set of concepts the content of the document is about. This is partly due to the usage of the current implementation inside the APOSDLE system. There annotations are used to express exactly the *aboutness* of resources and are formally described with the property *deals with* that is modeled inside the knowledge base of APOSDLE, that is used to store the semantic annotations (see section 4).

In approaches based on *Instance identification* or *Instance reference* as [6] or [14] annotation is treated on a more fine-grained level: Single words in documents are annotated with concepts stemming from the ontology.

We follow our approach for two reasons: (1) Although the complete semantics of words contained in a document are not recognized using this approach, the additional information added to the document still provides opportunities to be used at a later time in retrieving material [26], by a limited amount of human involvement. (2) We think that for the near future it makes sense to work on making the Semantic Web a reality, by focusing on bringing little semantics [13] into the current web and taking small steps. We follow this pragmatic approach and try to apply it to the Semantic Desktop in the context of our work.

3.4 Weighting the Annotations

In our (and other) approach(es) to semantic annotation, a document is either annotated with certain concepts or it is not. From a retrieval point of view this means that a document is either retrieved, if it is annotated with a concept present in the query, or it is not retrieved, if none of the concepts in the query are assigned to the document. Ranking the retrieved document set is impossible.

To allow for ranking the result set and to increase the performance of our service we weight the annotations between documents and concepts using a tf-idf-based weighting scheme. This is a standard instrument in information retrieval to improve retrieval results [19]. Our weighting approach is related to the one presented by [6], who are also weighting semantic annotations using a tf-idf-based measure.

$$weight(c, d) = tf(c, d) \cdot idf(c) = tf(c, d) \cdot \log \frac{D}{a(c)} \qquad (4)$$

With:

- c ... a concept
- d ... a document
- $tf(c, d)$ ... 1 if d is annotated with c, 0 otherwise
- $idf(c)$ ... inverse document frequency of concept c
- D ... total number of documents
- $a(c)$ ... number of documents annotated with concept c

3.5   Searching the Network

The network structure underlying the service is searched by spreading activation. Starting from a set of initially activated nodes in the network, activation spreads over the network and activates nodes associated with the initial set of nodes. Originally stemming from the field of cognitive psychology, where it serves as a model for operations in the human mind, spreading activation found its way over applications in both neural and semantic networks to information retrieval [8]. It is comparable to other retrieval techniques regarding its performance [16].

Beside systems that use spreading activation for finding similarities between text documents or search terms and text documents, approaches exist, which employ spreading activation for finding similar concepts in knowledge representations [1] [20]. The novelty of our approach lies in combining spreading activation search in a document collection with spreading activation search in a knowledge representation. The formula we use to calculate the spread of activation in our network is depicted in equation 5.

$$A(n_j) = \sum_{i=1}^{t} \frac{A(n_i) \cdot w_{i,j}}{\sum_{k=1}^{s} w_{i,k}} \tag{5}$$

With:

- $A(n_j)$ ... activation of node n_j
- $A(n_i)$ ... activation of node n_i
- t ... number of nodes adjacent to node n_j
- $w_{i,j}$ ... weight of edge between node n_i and node n_j
- s ... number of nodes adjacent to node n_i
- $w_{i,k}$ ... weight of edge between node n_i and node n_k

Search in our network is performed as follows:

1. Search starts with a set of concepts, representing the information need of the knowledge-worker. The concept nodes representing these concepts are activated.
2. *Optionally,* activation spreads from the set of initially activated concepts over the edges created by semantic similarity to other concepts nodes in the network.
3. Activation spreads from the currently activated set of concept nodes to the document nodes over the edges created by semantic annotation to find documents that deal with the concepts representing the information need.
4. *Optionally,* activation spreads from the documents nodes currently activated to document nodes that are related by means of textual similarity and are therefore associated with the document nodes.
5. Those documents corresponding to the finally activated set of document nodes are returned as search result to the user.

4   Implementation Inside the APOSDLE Project

The associative network structure and the spreading activation algorithm presented in section 3 have been implemented to support the retrieval of resources inside the first prototype of the APOSDLE system.

The goal of the current version of the APOSDLE system is to help knowledge-workers understanding the field of requirements engineering. In order to meet its goals APOSDLE uses a knowledge base in the form of a domain ontology, which described the field of requirements engineering in which the first prototype of APOSDLE operates, and a document base, which contains learning material (definitions, examples, tutorials, etc.) about requirements engineering that are partly annotated with concepts from the domain ontology.

The domain ontology consists of 70 concepts, 21 of which are used to annotate documents. The document base consists of 1016 documents, 496 documents of which are annotated with one or more concepts from the knowledge base. As we can see the scenario of APOSDLE provides a typical example of scarce annotations: only parts of the ontology are used for annotation and only parts of the documents are annotated. We see this setting corresponding to the coverage problematic presented in section 1 and employing associative retrieval techniques appropriate to finding relevant material that was not originally annotated with concepts from the domain ontology.

The service implemented in the APOSDLE project and presented in this section relies on knowledge contained in an ontology and the statistical information in a collection of documents. The service is queried with a set of concepts from the ontology and returns a set of documents. Documents in the system are (partly) annotated with ontological concepts if a document *deals with* a concept. For example, if the document is an introduction to use case models it is annotated with the corresponding concept in the ontology. In APOSDLE, the annotation process is performed manually but is supported by statistical techniques (e.g. identification of frequent words in the document collection) [18].

Concepts from the ontology are used as metadata for documents in the system. Opposed to classical metadata, the ontology specifies relations between the concepts. For example, class-subclass relationships are defined as well as arbitrary semantic relations between concepts are modeled (e. g. `UseCase isComposedOf Action`). The structure of the ontology has been used for calculating the similarity between two concepts in the ontology according to the measure presented in section 3.1. This similarity has been used to expand a query with similar concepts before retrieving documents dealing with a set of concepts. After retrieval of documents was performed, the result set was expanded by means of textual similarity as introduced in section 3.2. The implementation of a specific associative network inside the APOSDLE system has allowed developing and testing different combinations of query and result expansion that are based on the spreading activation algorithm presented in

section 3.5. The next section contains an evaluation of the performance of
different combinations and a discussion of the results obtained.

5 Evaluation

In this section we describe the evaluation that we performed. We talk about
the evaluation measures, the queries used for evaluation, how we collected
relevance judgments and about the service configuration rankings obtained.

5.1 *Semantic Web Information Retrieval and Evaluation*

At present information retrieval in the Semantic Web (on the Semantic Desktop) is an inhomogeneous field (c.f. [25]. Although a good amount of approaches does exist, different information is used for the retrieval process,
different input is accepted and different output is produced. This complicates to define generally applicable rules for the evaluation of an information
retrieval system for the Semantic Web (or the Semantic Desktop) and to
create a test collection for this application area of information retrieval.

The present approach to retrieval on the Semantic Desktop is different from
current attempts to retrieval in a desktop environment: (1) the semantic information present in an ontology is taken into account for retrieval purpose;
(2) the query to the retrieval service is formulated by a set of concepts stemming from an ontology as opposed to a set of terms (words) as typically used
in the context of desktop search. As we are not aware of any standard test
corpora for the evaluation of an information retrieval service for the Semantic
Desktop we have created our own evaluation environment.

5.2 *The Test Corpus*

A major obstacle in the easy evaluation of Semantic Web technology based
information retrieval systems is the absence of standardized test corpora, as
they exist for text-based information retrieval.

Therefore we have built our own test corpus based on the data available in
the first release of the APOSDLE system [15]. The first version of APOSDLE
was built for the domain of Requirements Engineering. This resulted into a
domain ontology for this field and a set of documents dealing with various
topics of Requirements Engineering. The document base was provided by a
partner in the APOSDLE project, with expertise in the field of Requirements
Engineering, while the ontology was modeled by another partner. Together
these two partners sign responsible for the annotation of the document base
with concepts from the ontology. The ontology contains 70 concepts and the

document set consists of 1016 documents. 496 documents were annotated using one or more concepts. 21 concepts from the domain ontology were used to annotate documents.

In its size our test collection is comparable to test collections from early information retrieval experiments as the Cranfield or the CACM collections[3].

In addition to the absence of corpora for Semantic Web information retrieval we are unaware of any standard text-retrieval corpora for evaluating a service with characteristics similar to ours. We considered treating the ontological concepts used for querying our service equivalent to query terms of a text-retrieval system to be able to use a standard corpus. Therefore we would have needed some structure relating the terms contained in the documents, as it is the case with the ontology in our system which relates concepts. For this task we could have used a standard thesaurus. As this knowledge structure is different to the ontology originally used (and therefore different similarity measures had to be applied to it), this would have led us to evaluating a service with different properties than our original one.

We also considered the INEX[4] test collection for evaluating our service. INEX provides a document collection of XML documents which would have provided us with textual data associated with XML structure information. Unfortunately again an ontology relating the metadata used as XML markup is unavailable. This would have prevented us from employing (and evaluating) the functionality provided by the query expansion technique, which founds on the ontology.

5.3   Measures Used for Evaluation

The central problem in using classic IR measures as *recall* or *mean average precision* is that they require complete relevance judgements, which means that every document is judged against every query [4]. [10] notices that recall can not be determined precisely with reasonable effort. Finally [5] states that: *Building sets large enough for evaluation of realworld implementations is at best inefficient, at worst infeasible.*

Therefore we opted for using evaluation measures that do not require hat every document is judged against every query. We decided for using precision (P) at rank 10, 20 and 30. In addition we made use of infAP [28] which approximates the value of average precision (AP) using random sampling.

For calculating the evaluation scores we have used the `trec_eval`[5] package, which origins from the Text REtrieval Conference (TREC) and allows for calculating a large number of standard measures for information retrieval system evaluation.

---

[3] `http://www.dcs.gla.ac.uk/idom/ir_resources/test_collections/` (14.04.2008)

[4] `http://inex.is.informatik.uni-duisburg.de/` (14.04.2008)

[5] `http://trec.nist.gov/trec_eval/` (14.04.2008)

5.4   Queries Used for Evaluation

The queries that were used for the evaluation of the service are formed by sets of concepts.

The first version of the APOSDLE system presents resources to knowledge workers to allow them to acquire a certain competency. To realize search for resources that are appropriate to build up a certain competency, competencies are represented by sets of concepts from the domain ontology. These sets are used as queries for the search for resources. For the evaluation of the APOSDLE system all distinct sets of concepts representing competencies[6] were used as queries. In addition all concepts from the domain model not already present in the set of queries were used for evaluation purposes.

5.5   Collecting Relevance Judgments

8 different service configurations were tested and compared against each other based on the chosen evaluation measures. 79 distinct queries were used to query every service configuration. Queries were formed by sets of concepts stemming from the domain ontology.

For every query and service configuration the first 30 results were stored in a database table, with one row for every query-document pair. Query-document pairs returned by more than one service configuration were stored only once. The query-document pairs stored in the database-table were then judged manually by a human assessor. All query-document pairs were judged by the same person. The assessor was not involved in defining the competency to concept mappings uses as queries (c.f. section 5.4).

After relevance judgment, both, the results obtained by the different service configurations and the global relevance judgments have been stored into text files in a format appropriate for the `trec_eval` program. We then calculated the P(10), P(20), (P30) and infAP scores for the different service configurations.

5.6   The Obtained Service Configuration Ranking

Table 1 shows the calculated P(10), P(20), (P30) and infAP scores for the different service configurations. The columns *SemSim*, *TxtSim* indicate whether semantic similarity or text-based similarity was used for the search. Table 2 shows the service configuration ranking based on the obtained evaluation scores.

Configuration 1 (conf_1) is the baseline configuration of our service. The results delivered by this configuration are comparable to the use of a query language as SPARQL combined with an idf-based ranking (based on documents annotated with concepts) and no associative retrieval techniques used.

---

[6] Different competencies can be represented by the same concepts.

Table 1 Evaluation scores of service configurations calculated using P(10), P(20), P(30) and infAP

Conf.	SemSim	TxtSim	P(10)	P(20)	P(30)	infAP
conf_1	No	No	0.2418	0.2051	0.1700	0.1484
conf_2	No	Yes	0.3089	0.2778	0.2502	0.2487
conf_3	Yes (> 0.5)	No	0.3165	0.2608	0.2131	0.2114
conf_4	Yes (> 0.7)	No	0.3114	0.2582	0.2097	0.2001
conf_5	Yes (> 0.5)	Yes	0.3848	0.3405	0.3046	0.3253
conf_6	Yes (> 0.7)	Yes	0.3924	0.3494	0.3089	0.3326

Table 2 Ranking of service configurations based on P(10), P(20), P(30) and infAP

Rank	P(10)	P(20)	P(30)	infAP
1 (best)	conf_6	conf_6	conf_6	conf_6
2	conf_5	conf_5	conf_5	conf_5
3	conf_3	conf_2	conf_2	conf_2
4	conf_4	conf_3	conf_3	conf_3
5	conf_2	conf_4	conf_4	conf_4
6 (worst)	conf_1	conf_1	conf_1	conf_1

Exactly those documents are retrieved that are annotated with the concepts present in the query.

All other configurations make use of query expansion based on semantic similarity or result expansion based on text-based similarly. Configurations 3, 4, 5 and 6 perform query expansion. Configurations 2, 5 and 6 perform result expansion.

All associative search approaches employing semantic similarity (configurations 3, 4, 5 and 6), text-based similarity (configurations 2, 5 and 6) or both (configurations 2, 3, 4, 5 and 6) increase retrieval performance compared to the baseline (configuration 1). Additional relevant documents are found, which are not annotated with the concepts used to query the service.

5.7 Discussion

We now discuss the evaluation measures used and why we think that the amount of relevance judgments collected is sufficient for a proper evaluation of our service.

5.7.1 P(10), P(20) and P(30)

[3] evaluate the stability of evaluation measures. They calculate the error rate of measures based on the number of errors occurring whilst comparing two systems using a certain measure. They divide the number of errors by the total number of possible comparisons between two different systems.

Based on previous research they state that an error rate of 2.9% is minimally acceptable. They find that P(30) exactly reaches this error rate of 2.9% in their experiment with 50 queries used. Finally they suggest that the amount of queries should be increased for P(n) measures, where $n < 30$. And suggest that 100 queries would be safe if the measure P(20) is used.

We performed our experiment with 79 distinct queries and used the measures P(10), P(20) and P(30). Following the results of [3] the size of our query set should be appropriate for P(30). We are fortified in this assumption as the ranking of the 8 service configurations is identical for P(20), P(30) and infAP.

5.7.2 infAP

The Trec 8 Ad-Hoc collection consists of 528,155 documents and 50 queries which make a total amount of 26,407,750 possible relevance judgments. 86830 query-document relevance pairs are actually judged. This set of pairs is created by depth-100 pooling of 129 runs. Therefore 0.33% of the possible relevance judgments are performed.

Our collection consists of 1026 documents and 79 queries, which results in a total of 81,054 possible relevance judgments. This set of pairs is created by depth-30 pooling of 8 runs and 498 additional relevance judgments that were performed for runs that were not part of the experiment. 1938 query document pairs were actually judged. Therefore 2.39% of all possible relevance judgments were performed.

The depth-100 pool for the 8 evaluated runs would consist of 4138 query-document pairs. As we judged 1938 query-document pairs, we judged 46.83% of our potential depth-100 pool. [28] report a Kendall's tau based rank correlation of above 0.9 between infAP and AP with as little as 25% of the maximum possible relevance judgments of the depth-100 pool of the Trec 8 Ad-Hoc collection. They consider two rankings with a rank correlation of above 0.9 as equivalent.

With 46.83% of our potential depth-100 pool judged, we are confident that the infAP measure produces an estimation sufficiently accurate. Again our confidence in the results of infAP is assured by the equivalence of the ranking of the 8 service configurations for P(20), P(30) and infAP.

6 Related Work

Beagle++ [7] is a search engine for the Semantic Desktop and indexes RDF-metadata together with document content. Both [6] and [14] present an extension of the vector space model. Together with document content they index semantic annotations of documents and use this information for search. All three are very promising approaches that extend the vector space model using semantic information. None of them employs measures of semantic association.

[20] present a hybrid approach for searching the (semantic) web, they combine keyword based search and spreading activation search in an ontology for search on websites. Ontocopi [1] identifies communities of practice in an ontology using spreading activation based clustering. Both are prospective approaches employing ontology-based measures of association and evaluating them using spreading activation. They do not integrate text-based measures of association into their systems.

7   Conclusions and Future Work

We have presented an information retrieval service for the Semantic Desktop, which is based on techniques from associative information retrieval. We have evaluated the presented service using standard measures for information retrieval system evaluation. As classic measures for evaluation as recall and average precision require that every document is judged for every query we have chosen precision at ranks 10, 20 and 30 as evaluation measures. In addition we made use of the random sampling approach performed by the infAP measure. Following recent works [4] [28] in information retrieval system evaluation we are confident that our chosen approach reflects the actual relation between the service configurations as the ranking of the service configurations remains identical for the measures P(20), P(30) and infAP.

Our experiments encourage us, that the application of associative retrieval techniques to information retrieval on the Semantic Desktop is an adequate strategy. We tend to conclude that text-based methods for associative retrieval result in a higher increase in retrieval performance, therefore we want to explore the approach of attaching a set of terms to every concept in our domain ontology during modeling time to provide search results even for concepts that are not used for annotation. In addition we want to extend our research towards the application of different semantic similarity measures within our service.

Acknowledgments

While a shorter version of this paper focusing on the description of the retrieval approach was presented at I-SEMANTICS 2007 an in depth description of the approach to the evaluation of the developed service was presented at FGIR 2007 / LWA 2007. We thank the anonymous reviewers of our submissions at I-SEMANTICS 2007 and LWA 2007 for their constructive feedback.

This work has been partially funded under grant 027023 in the IST work programme of the European Community. The Know-Center is funded by the Austrian Competence Center program Kplus under the auspices of the Austrian Ministry of Transport, Innovation and Technology (www.ffg.at/index.php?cid=95) and by the State of Styria.

References

1. Alani, H., Dasmahapatra, S., O'Hara, K., Shadbolt, N.: Identifying communities of practice through ontology network analysis. IEEE Intell. Syst. 18(2), 18–25 (2003)
2. Bechhofer, S., Goble, C.: Towards annotation using DAML+OIL. In: K-CAP 2001 workshop on Knowledge Markup and Semantic Annotationq (2001)
3. Buckley, C., Voorhees, E.M.: Evaluating evaluation measure stability. In: SIGIR 2000: Proceedings of the 23rd annual international ACM SIGIR conference on Research and development in information retrieval, pp. 33–40. ACM Press, New York (2000)
4. Buckley, C., Voorhees, E.M.: Retrieval evaluation with incomplete information. In: SIGIR 2004: Proceedings of the 27th annual international ACM SIGIR conference on Research and development in information retrieval, pp. 25–32. ACM Press, New York (2004)
5. Carterette, B., Allan, J., Sitaraman, R.: Minimal test collections for retrieval evaluation. In: SIGIR 2006: Proceedings of the 29th annual international ACM SIGIR conference on Research and development in information retrieval, pp. 268–275. ACM Press, New York (2006)
6. Castells, P., Fernández, M., Vallet, D.: An adaptation of the vector-space model for ontology-based information retrieval. IEEE Trans. Knowl. Data Eng. 19, 261–272 (2007)
7. Chirita, P.-A., Costache, S., Nejdl, W., Paiu, R.: Beagle++: Semantically enhanced searching and ranking on the desktop. The Semantic Web: Research and Applications (2006)
8. Crestani, F.: Application of spreading activation techniques in information retrieval. Artif. Intell. Rev. 11, 453–482 (1997)
9. Decker, S., Frank, M.R.: The networked semantic desktop. In: WWW Workshop on Application Design, Development and Implementation Issues in the Semantic Web (2004)
10. Fuhr, N.: Information Retrieval: Skriptum zur Vorlesung im SS 06 (December 19, 2006)
11. Handschuh, S.: Semantic Web Services: Concepts, Technologies, and Applications. In: Semantic Annotation of Resources in the Semantic Web, pp. 135–155. Springer New York, Inc., Secaucus (2007)
12. Heflin, J., Hendler, J.: Searching the web with Shoe. In: Artificial Intelligence for Web Search. Papers from the AAAI Workshop. WS-00-01 (2000)
13. Hendler, J.: The dark side of the semantic web. IEEE Intell. Syst. 22, 2–4 (2007)
14. Kiryakov, A., Popov, B., Terziev, I., Manov, D., Ognyanoff, D.: Semantic annotation, indexing, and retrieval. Journal of Web Semantics 2, 49–79 (2004)
15. Lindstaedt, S.N., Mayer, H.: A storyboard of the aposdle vision. In: Nejdl, W., Tochtermann, K. (eds.) EC-TEL 2006. LNCS, vol. 4227, pp. 628–633. Springer, Heidelberg (2006)
16. Mandl, T.: Tolerantes Information Retrieval. Neuronale Netze zur Erhöhung der Adaptivität und Flexibilität bei der Informationssuche. PhD thesis, University of Hildesheim (2001)
17. McCool, R.: Rethinking the semantic web, part 1. IEEE Internet Comput. 9(6), 88–97 (2005)

18. Pammer, V., Scheir, P., Lindstaedt, S.N.: Two protégé plug-ins for supporting document-based ontology engineering and ontological annotation at document level. In: 10th International Protégé Conference, Budapest, Hungary (July 15-18, 2007)

19. Robertson, S.E., Spärck Jones, K.: Simple, proven approaches to text retrieval. Technical report, University of Cambridge, Computer Laboratory (1994)

20. Rocha, C., Schwabe, D., de Aragão, M.P.: A hybrid approach for searching in the semantic web. In: Proceedings of the 13th international conference on World Wide Web, WWW 2004 (2004)

21. Sabou, M., d'Aquin, M., Motta, E.: Using the semantic web as background knowledge for ontology mapping. In: International Workshop on Ontology Matching, OM-2006 (2006)

22. Salton, G.: Associative document retrieval techniques using bibliographic information. JACM 10, 440–457 (1963)

23. Salton, G.: Automatic Information Organization and Retrieval. McGraw Hill, New York (1968)

24. Sauermann, L., Bernardi, A., Dengel, A.: Overview and outlook on the semantic desktop. In: Proceedings of the 1st Workshop on The Semantic Desktop at the ISWC 2005 Conference (2005)

25. Scheir, P., Pammer, V., Lindstaedt, S.N.: Information retrieval on the semantic web - does it exist? In: LWA 2007, Lernen - Wissensentdeckung - Adaptivität, September 24–26, 2007. Halle/Saale (2007)

26. Spärck Jones, K.: What's new about the semantic web?: some questions. SIGIR Forum 38, 18–23 (2004)

27. Wu, Z., Palmer, M.S.: Verb semantics and lexical selection. In: Meeting of the Association for Computational Linguistics (ACL), pp. 133–138 (1994)

28. Yilmaz, E., Aslam, J.A.: Estimating average precision with incomplete and imperfect judgments. In: CIKM 2006: Proceedings of the 15th ACM international conference on Information and knowledge management, pp. 102–111. ACM Press, New York (2006)

GRISINO – A Semantic Web Services, Grid Computing and Intelligent Objects Integrated Infrastructure

Tobias Bürger, Ioan Toma, Omair Shafiq, Daniel Dögl, and Andreas Gruber

Abstract. Existing information, knowledge and content infrastructures are currently facing challenging problems in terms of scalability, management and integration of various content and services. The latest technology trends, including Semantic Web Services, Grid computing and Intelligent Content Objects provide the technological means to address parts of the previously mentioned problems. A combination of the three technologies could provide a sound technological foundation to build scalable infrastructures that provide highly automated support in fulfilling user's goals.

This paper introduces GRISINO, an integrated infrastructure for Semantic Web Services, Intelligent Content Objects and Grid computing, which may serve as a foundation for next generation distributed applications.

1 Introduction

The GRISINO[1] project investigates the use of semantic content models in service oriented architectures based on Semantic Web Services- and Grid-Technology [13] by combining three technology strands: Semantic Web Services [6], Knowledge Content Objects [2] and Grid Computing [7]. By that, GRISINO aims at defining and realizing intelligent and dynamic business processes based on dynamic service discovery and the internal state of complex objects. Advantages of this approach

Tobias Bürger, Ioan Toma, Omair Shafiq
Semantic Technology Institute - STI Innsbruck, University of Innsbruck, Austria
e-mail: {tobias.buerger,ioan.toma,omair.shafiq}@sti2.at

Daniel Dögl
Uma Information Technology GmbH, Vienna, Austria
e-mail: daniel.doegl@uma.at

Andreas Gruber
Salzburg Research Forschungsgesellschaft mbH, Salzburg, Austria
e-mail: andreas.gruber@salzburgresearch.at

[1] http://www.grisino.at

S. Schaffert et al. (Eds.): Networked Knowledge - Networked Media, SCI 221, pp. 113–128.
springerlink.com   © Springer-Verlag Berlin Heidelberg 2009

include the possibility to establish service based processes ad-hoc based on the user requirements or their alteration during run-time based on the state of the intelligent content objects. The main output of the project is a test bed for experimentation with complex processes and complex objects that takes user requirements into account and fulfils them by dynamically integrating the three underlying technologies. For this testbed, advanced prototypes of each of the technology strands are combined:

- The **Web Service Modelling Ontology (WSMO)** [11], the Web Service Modelling Language (WSML)[2] and the Web Service Modelling Execution Environment (WSMX)[3] as a framework for the description and execution of Semantic Web Services,
- **Knowledge Content Objects (KCOs)** as a model for the unit of value for content to be exchanged between services, together with its management framework, the Knowledge Content Carrier Architecture (KCCA) [2].
- The **Globus toolkit**[4] as an existing Grid infrastructure.

In this chapter we will detail the main results of the GRISINO project: its architecture (section 2) and the core parts of the architecture which realize the integration of the three technologies, i.e. a set of transformers between the protocol and description standards used (section 3 and 4). Furthermore, we provide details about the proof of concept implementation which serves to demonstrate the functionality and interoperability within the GRISINO testbed in section 5.

2   GRISINO Architecture

One of the major driving forces for the Web and its future derivatives is content which can range from multimedia data (with some metadata) to "intelligent objects", i.e. content that itself, can either exhibit behavior or at least, carry semantic information that can (and must) be interpreted by the services on the Semantic Grid. The GRISINO system architecture as shown in Figure 1 provides a set of APIs and an implementation of these APIs to ease the handling and development of applications which intend to use the three technologies together:

- the GRISINO API which gives application developers easy access to the combined functionality of the three technologies.
- the Transformer API including protocol transformations between the technologies,
- the Selector API issuing calls to transformers or the foundational API, and
- the foundational API, which is an abstracted view onto the APIs of the core technologies.

Most notably the GRISINO system architecture includes extensions to the core components that enable communication between the technologies. This includes:

---

[2] http://www.wsmo.org/wsm
[3] http://www.wsmx.org
[4] http://www.globus.org/toolkit/

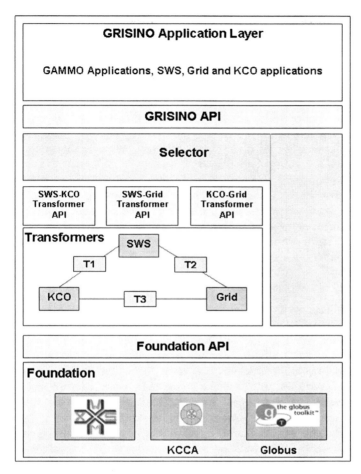

Fig. 1 GRISINO System Architecture

- an extension of WSMX for the interpretation of KCOs,
- a semantic layer for services offered by KCCA to enable their discovery and
- an extension of the Globus toolkit which extends Globus with a semantic layer in order to handle Grid services like other SWS.

The GRISINO system architecture integrates specific SWS and Grid solutions because of the existence of a wide variety of different and diverse approaches: We based our efforts on WSMO and WSMX as execution platforms because they are being well supported by an active research community to handle SWS. Furthermore we are using the Globus Toolkit as being the most widely used Grid computing toolkit which is fully compatible with the OGSA[5] - and Web Service Resource

---

[5] http://www.globus.org/ogsa/

Framework (WSRF) specifications[6]. The integration of Semantic Web Services and Grid computing includes the extension of the Semantic Web Services infrastructure to model Grid Services and resources on the Grid in order to realize the vision of the Semantic Grid. Benefits of this integration include:

- Resources on the Grid may profit from machine reasoning services in order to increase the degree of accuracy of finding the right resources.
- The background knowledge and vocabulary of a Grid middleware component may be captured using ontologies. Metadata can be used to label Grid resources and entities with concepts, e.g. for describing a data file in terms of the application domain in which it is used.
- Rules and classification-based reasoning mechanisms could be used to generate new metadata from existing metadata, for example describing the rules for membership of a virtual organization and reasoning that a potential member's credentials are satisfactory for using the VO resources.
- Activities like Grid Service discovery or negotiation of service level agreements can be potentially enhanced using the functionalities provided by Semantic Web Service technologies.
- Searches / discovery of SWS can be seamlessly extended to Grid Services.

The integration of SWS and KCO technologies will benefit from each other in several different aspects. In a KCO various kinds of information are modeled in so called semantic facets that allow to deal with KCOs in different situations. A more standardized exposition of KCO facet information would allow to base the actions that take place in a (goal-based) Web Service execution on the facet information of KCOs: for example to search for a KCO that contains certain content or to match a certain licensing schema. Also choreography could be based on facet information, e.g. to fulfill a special licensing schema where you first have to pay before you consume the content. Another benefit would be that these services could also be automatically discovered, which represents a key requirement for ad-hoc instantiation. Further benefits of the integration of SWS and KCO/KCCA include:

- Goal-based Web service execution can be based on the various kinds of information which is modeled in so called semantic facets inside KCOs; e.g. to search for a KCO that contains certain content or to match a certain licensing scheme.
- Choreography of Web services can be based on facet information, e.g. to fulfil a special licensing scheme in which you first have to pay before you consume the content.
- Plans that describe how to handle content and which are modeled inside a KCO can be automatically executed by using SWS or Grid services.

The following section will provide further details about two of the three major aspects of the integration, i.e. the integration of SWS and Grid, as well as the integration of SWS and KCOs.

---

[6] http://www.globus.org/wsrf/

3   SWS-Grid Transformer

The main task of this transformer (mentioned as T2 in the Figure 1) is the realization of the link between SWS based systems and Grid Computing systems. Our approach was to extend and refactor an existing SWS solution, namely the Web Service Modeling Ontology, Language and Execution Environment with Grid concepts in order to address Grid related requirements. The resulting modeling framework for Semantic Grid enriches the OGSA with semantics by providing a Grid Service Modeling Ontology (GSMO)[7] as an extended version of WSMO.

Based on the proposed conceptual model for Semantic Grid services, a new language called GSML (Grid Service Modelling Language) was developed that inherits the syntax and semantics of the WSML language and adds a set of additional constructs reflecting the GSMO model. Last but not least an extension of the Web Service Modeling Execution Environment (WSMX), called Grid Service Modeling Execution Environment has been proposed. More details about the conceptual model, the language and the new execution environment are available in [12].

3.1   Extensions to the WSMO Conceptual Model

The conceptual model of Semantic Web Services provides a set of guidelines or recommendations on how Semantic Web Service descriptions should look like. The Web Service Modeling Ontology (WSMO) [11] refers to the concepts it defines as its top level elements. WSMO has four top-level elements, i.e. Ontologies, Web Services, Mediators and Goals. We have extended the WSMO conceptual model to model Semantic Grid Services based on an analysis of the GLUE schema [1] for which we provided semantic annotations. The proposed extended version, called GSMO has 6 major top level entities which were either newly added to the WSMO conceptual model, are refinements of original entities or are entities which are inherited from the WSMO model. The GSMO elements are graphically represented in Figure 2: The elements *GSMO*, *Job*, *VO*, *Resources*, *Computational Resource*, *Data Resource* were newly added, the element *Grid Service* is the redefined element and finally the elements *Ontology* and *Mediator* have been inherited or adopted:

- *Job* represents the functionality requested, specified in terms of what has to be done, what are the resources needed, etc. A Job is fulfilled by executing one or more Grid Services. Ontologies can be used as domain terminology to describe the relevant aspects. Job as one of the top level entities of GSMO is adapted from WSMO Goals and is taken in GSMO as its extended version.
- *Ontologies* provide the terminology used by other GSMO elements to describe the relevant aspects of a domain. This element has been inherited from the WSMO top level entity as Ontologies.

---

[7] http://www.gsmo.org/

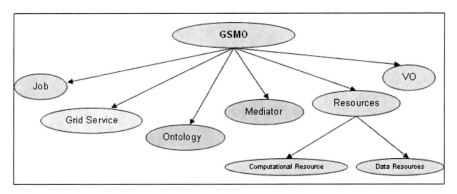

Fig. 2 Grid Service Modeling Ontology (GSMO)

- *Grid Service* describes the computational entity providing access to physical resources that actually perform the core Grid tasks. These descriptions comprise
  the capabilities, interfaces and internal working of the Grid Service. All these
  aspects of a Web Service are described using the terminology defined by the ontologies. The Grid Service top level entity has been adopted from WSMO's Web
  Services as its top level entity.
- *Mediators* describe elements that overcome interoperability problems between
  different WSMO elements. Due to the fact that GSMO is based on WSMO, it will
  be used to overcome any heterogeneity issues between different GSMO elements.
  Mediators resolve mismatches between different used terminologies (data level);
  communicate mismatches between Grid services (protocol level) and on the level
  of combining Grid Services and Jobs (process level).
- *Resources* describe the physical resources on the Grid which can be further classified into computing resources and storage resources. These computation- and
  storage-resources are key elements of the underlying Grid.
- The *Virtual Organization* element describes any combination of different physical resources and Grid Services formed as virtual organizations on the Grid. This
  element will help in automated virtual organization formation and management.

3.2 Extensions to the WSML Formal Language

Based on the conceptual model for Semantic Grid services presented in the previous
section, this section introduces the basic constructs towards a semantic language
for describing entities in the realm of the Semantic Grid. We propose a new language which is based on an existing language for Semantic Web services, namely
the WSML language. The new language called GSML (Grid Service Modelling
Language) inherits the syntax and semantic of the WSML language. Additional constructs not defined in WSML such as *VO* and *resource* can be used to semantically

describe Virtual Organizations and resources on the Semantic Grid. The constructs *webService* and *goal* from WSML are replaced by *gridService* and *job*.

As mentioned above, GSML follows the conceptual model of GSMO defining a clear syntax for each of the elements described in the previous section. The top level constructs introduced by GSML will be further described below:

A Grid service (gridService) in GSML has the following structure:

```
gridService = 'gridService' id? header* capability? interface*
    usesResources* belongsToVOs*
```

The **id, header, capability** and **interface** constructs from a Grid service definition are defined in the same way as described in WSML. Additionally the **usesResources** and **belongsToVOs** constructs with n-ary cardinality could be used to specify the resources used by the service in order to provide its functionality, respectively the VOs the service belongs to. A simplified example of a Grid service which provides movie rendering functionality is given below:

```
namespace { _"http://www.gsmo.org/movieRenderGS#",
dc_"http://purl.org/dc/elements/1.1#",
rO_"http://www.gsmo.org/renderOntology#"}
gridService_"http://www.gsmo.org/movieRenderGS.wsml"
    nonFunctionalProperties
        dc#title hasValue "Movie Render Grid service"
        dc#publisher hasValue "GSMO"
    endNonFunctionalProperties
    capability
        sharedVariables {?model}
        precondition
            definedBy ?model memberOf rO#Model.
        postcondition
            definedBy ?movie memberOf rO#1Movie and rO#
                hasModel(?movie,?model).
    interface MovieRenderServiceInterface
    choreography MovieRenderServiceChoreography
    orchestration MovieRenderServiceOrchestration
    usesResources { _"http://www.gsmo.org/resources#proc1,
        _"http://www.gsmo.org/resources#mem1}
    belongsToVOs { _"http://www.gsmo.org/resources#VO1,
        _"http://www.gsmo.org/resources#VO2}
```

The **usesResources** and **belongsToVOs** are defined as follows:

```
usesResources = 'usesResources' idlist

belongsToVOs = 'belongsToVOs' idlist
```

The **idlist** construct from the above definitions are the IDs of resources, respectively VOs. According to the principle inherited from WSML, the elements in GSML are identified mainly by IRIs, and thus idlist is a list of IRIs.

A user job (**job**) in GSML has similar structure than a **gridService** construct. Additionally an application element can be defined to explicitly specify the application to be run:

```
job = 'job' id?
     header* capability? interface* usesResources*
     belongsToVOs* application?
application = 'application' id? header* name? version?
     executable? argument* environment* input? output?
     error? working_directory?
```

Equally as in the **gridService** construct definition, the **id, header, capability** and **interface** constructs are inherited from WSML. The **usesResources** and **belongsToVOs** constructs are to be used in the same way as described above for the **gridService** construct. A simplified example of a job specification, a request for a movie rendering is given below:

```
namespace {_"http://www.gsmo.org/movieRenderGS#", dc
_"http://purl.org/dc/elements/1.1#", rO
_"http://www.gsmo.org/renderOntology#"}

job _"http://www.gsmo.org/MovieRender.wsml"
     nonFunctionalProperties
         dc#title hasValue "MovieRender Grid service"
         dc#publisher hasValue "GSMO"
     endNonFunctionalProperties
     capability
         sharedVariables {?model}
         precondition
             definedBy ?model memberOf rO#Model.
         postcondition
             definedBy ?movie memberOf rO#lMovie and rO#hasModel
             (?movie,?model).
     interface MovieRenderServiceInterface
         choreography MovieRenderServiceChoreography
         orchestration MovieRenderServiceOrchestration
     usesResources { _"http://www.gsmo.org/resources#proc1}
     belongsToVOs { _"http://www.gsmo.org/resources#VOl}
```

Equally, a job can be specified using the application element instead of the capability element. In this case the functionality is explicitly specified by naming the application that needs to be executed to fulfill the job. The elements of an application description include: the name of the application (**name**), the version of the application (**version**), the main executable file of the application (**executable**), the arguments (**argument***), any additional libraries needed to run the application (**environment**), the input data for the application specified in the input file (**input**), the output file (**output**), the error file (**error**) and finally the **working_directory**.

```
namespace {_"http://www.gsmo.org/movieRenderGS#", dc
_"http://purl.org/dc/elements/1.1#", rO
_"http://www.gsmo.org/renderOntology#"}

job _"http://www.gsmo.org/MovieRender.wsml"
     nonFunctionalProperties
          dc#title    hasValue    "MovieRender  Grid  service"
          dc#publisher hasValue "GSMO"
     endNonFunctionalProperties
          usesResources { _"http://www.gsmo.org/resources#proc1}
          belongsToVOs { _"http://www.gsmo.org/resources#VO1}
          application
     nonFunctionalProperties
          dc#title    hasValue    "MovieRender  application  "
          dc#publisher hasValue "GSMO"
     endNonFunctionalProperties
     name  movieRender
     version  0.1
     executable  /bin/usr/movieRender
     input  /home/grisino/model.mod
     output  /home/grisino/movie.avi
     error  /home/grisino/error
     working_directory  /home/grisino
```

By describing the Semantic Grid services and jobs in a symmetric manner, using terminology provided by ontologies, a semantic matchmaker will be able to determine if jobs and services hosted on the Semantic Grid match in terms of functionality, behavior, resources and VOs requested, respectively provided.

A Grid resource (**resource**) in GSML has the following structure:

```
resource = 'resource' id?
     header* hasDefinition? hasPolicy* belongsToVOs*
```

The **id**, and **header** constructs from a resource specification are defined in the same way as described in WSML. The **hasDefinition** contains a logical definition in terms of concepts and relations from ontologies describing the resource. The **hasPolicy** construct specifies the policy and access rules associated with the resource. The **belongsToVOs** construct is used to specify the VOs the resource belongs to. A **resource** could be further refined as described in the previous section in **computationalResource** and **dataResource**.

A **VO** construct in GSML has the following structure:

```
vo = 'vo' id?
     header*  hasMembers* hasDescription?
```

The **id** and **header** constructs in **VO** specification are defined in the same way as described in WSML [5].

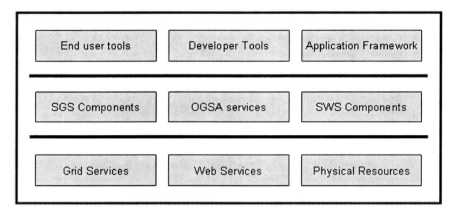

Fig. 3 Towards a Grid Services Execution Environment

3.3   Extensions to the WSMX Execution Environment

This section presents the initial architecture of the Grid Service Execution Environment which will be layered on top of OGSA based Grid toolkits (e.g. Globus Toolkit[8]). The objective of the Grid Service Execution Environment is to process the semantically enabled descriptions of Grid Services and to process the semantic descriptions of Jobs submitted on the Grid. The Grid Service Execution Environment will take care of the service execution management of user-defined applications defined at the semantics layer, which may require specific resource requirements, and imply complex interactions between services. The execution management will extend the conventional execution management of jobs on the Grid, including the execution of Web and Grid Services, with semantically enhanced descriptions of required resources.

The architecture of the proposed framework is based on the Web Services Execution Environment (WSMX) which itself is compliant to the Service Oriented Architecture (SOA) paradigm and consists of a set of loosely coupled collaborating software components. The architecture of Semantic Grid Services Execution Environment is shown in Figure 3. It will act as reference architecture for all the components (existing WSMX and Globus components and the new ones proposed based on GSMO) and integrate them inside one infrastructure. The following newly added components in the Semantic Grid Services Execution Environment are based on GSMO:

- Resource Management which deals with semantic-based resource discovery, advanced reservation, negotiation, deployment and provisioning of computational and storage resources on the Grid

---

[8] http://www.globus.org/toolkit/

- Virtual Organization Manager which deals with creation and management issues of dynamic business oriented Virtual Organizations of services, resources and users in the Grid
- Extended WSML Reasoner for GSML which addresses the knowledge representation and reasoning aspects for discovery, composition and mediation of Grid resources described in GSML.
- Extended Execution Management which covers the implementation of execution semantics of internal Grid Service Execution Environment, and also for external user-defined services and jobs including scheduling, fault-management, and support of the monitoring of execution.

Figure 3 shows the initial architecture of the Grid Service Execution Environment. It shows the initial set of required components which are grouped in three different layers: The upper layer is the problem solving layer in which end user tools, development tools and application frameworks are situated. In this layer already available development tools for WSMX will be extended. Moreover, it includes the Application Framework to support developers in building applications for the Semantic Grid based on the Grid Service Execution Environment. The middle layer (the application layer) includes the newly introduced components based on the extensions of WSMO as GSMO and WSML as GSML, i.e. the VO Manager, Resource Manager, GSML reasoner, the extended execution manager, as well as existing components in the application layer of WSMX such as discovery, selection, composition, negotiation, mediation etc., and the core OGSA services. The bottom (base) layer includes the foundation of the environment such as Grid service descriptions based on Web Services infrastructure and physical resources including computing and storage resources on the Grid.

4   The KCO-SWS Transformer

The main objective of the KCO-SWS transformer (mentioned as T1 in Figure 1) is the realization of the link between knowledge content based systems (resp. the KCCA system) and its Knowledge Content Objects with Semantic Web Service based systems (resp. WSMX). Our intention was to use information stored inside Knowledge Content Objects (KCO) for service discovery and plan execution, e.g. to automatically negotiate or to automatically enrich content and knowledge about that content during the execution of web based workflows like e.g. a document-based business process or workflow to index and enrich documents with additional knowledge. In order to do so, WSMX needs to be able to interpret KCOs and the services offered by KCCA need to be able to communicate with the other services offered by the GRISINO system.

The approach to integrate existing KCO / KCCA technology with the SWS/Grid technologies in the GRISINO system was twofold:

- Metadata descriptions that are contained inside KCOs are translated into WSMO descriptions in order to be useable for service discovery and ranking.

- The KCCA system is wrapped with Web service descriptions that describe its invoke-able functionality. These descriptions are further semantically described.

We started the integration with the investigation of the ontology of plans [8] as well as the "Description and Situation" modules embedded in the OWL DL 397[9] version of foundational ontology DOLCE. Similar work has been reported in [3] or [10]. However, no fully functional translation between DOLCE (resp. its plans extension DDPO[10]) and WSMO has been developed so far for obvious reasons: WSMO in general is a richer knowledge representation language than OWL-DL. The same holds - in principle - for DOLCE, but in order to comply with the restrictions of current semantic web machinery, DDPO has been designed for the restrictions of OWL-DL. Therefore, OWL DL has to be the lowest common denominator for WSMO and DDPO with respect to defining a mapping between the two knowledge representation schemes. This task been done partially already by the WSMO Community [9].

The remaining task was to map the concepts of the DOLCE Design and Plan Ontology (DDPO) and the regarding constructs for the KCO community facet (which are 'static' descriptions of conceptualizations over situations or states) onto WSMO descriptions. In GRISINO we used a subset of concepts defined in DDPO. The services developed within the project are focussed on fairly small parts of document processing, in which goals and plans usually described in KCOs are more generic and most likely closer to a business goal description. Furthermore, they likely include (human) agents in their description, while GRISINO is focussed on automatic manipulation of processes. The concepts 'description', 'situation', 'task', 'role', 'parameter', 'perdurant', 'endurant' and 'region' describe the community facet of the KCO and have been mapped onto WSMO elements as shown in Table 1[11]:

1. DDPO:Goal is not the same as WSMO:Goal because DDPO:Goal is a description of a very general (and semantically open) desire, whereas WSMO:Goal is a specification of a resulting situation for which several plans and executions may exist and where there is a rigid structure containing a domain ontology, a mediator, a capability and an interface. In addition, DDPO:Goal does not play a central role in the actual execution of a DDPO:plan.

   Conclusion: DDPO is epistemologically more open than WSMO. The universe of discourse for WSMO is the world of web services which is linked to the world of "goals" (i.e. desired states of the world) via mediators and whose actual queries (i.e. the goals) are formulated according to the vocabulary of arbitrary ontologies. All we want to ever express in WSMO is a desired state for which it is assumed that it can be reached by the execution of a sequence of web services. For a mapping between DDPO and WSMO it is therefore sufficient to constrain DDPO to the generative power of WSMO. Furthermore, since KCOs only require a very specific set of WSMO descriptions, we can constrain DDPO to those WSMO descriptions which cover KCCA functions for KCOs.

---

[9] http://www.loa-cnr.it/ontologies/DLP_397.owl
[10] DDPO is an acronym used for DOLCE Design and Plan Ontology
[11] The numbers in the listing below refer to the rows in the table.

Table 1 Conceptual mapping of DDPO and WSMO elements

	DDPO Concept	DDPO Features	WSMO	WSMO Features
1	DDPO	GOAL, PLAN (subclass of 'description') SITUATION, TASK, ROLE, PARAMETER, ENDURANT, PERDURANT, REGION	WSMO	GOALS, ONTOLOGIES, WEB SERVICES, MEDIATORS
2	DESCRIPTION and SITUA-TION		GOAL	constrained-by: nonFunctional-Property (=labelling) Ontology Capability Mediator Interface
3	PLAN	ROLE, TASK, PA-RAMETER	Ontology	terminology for specifying the goal
4	SITUATION	(ENDURANT, PERDURANT, REGION)	Capability	nonFunctionalProperty (=labelling) Ontology ooMediator shared-Variables axioms (Precondition) axioms(Assumption) axioms (Post condition) axioms (Effects)

2. DDPO:Plan is the conceptual and descriptive equivalent of WSMO:Goal in its ability to define and reuse concepts that can classify situations, i.e. states in the world. WSMO:Goal, via its capability section describes the specific situation that needs to be fulfilled by a web services capability. The axioms here are used to determine the pre- and/or postconditions for achieving the goal.
3. The concepts role, task and parameter are the descriptive counterparts to classify "objects", "events" and "values" of a given setting. These concepts of DDPO provide the terminology for specifying goals and to describe the domain knowledge.
4. A DDPO:Situation holds the relevant information to describe the pre- and postcondition. A particular situation can be mapped to an axiom used within a capability.

5 Use Case Example

Today information retrieval and text analytics for special interest searches are usually realized as "one of a kind" expert systems. The user input usually is a set of information sources and some definition of the "typical" users point of view, which are fed into a sequence of steps like information acquisition, processing, extraction, annotation, analysis, ... with the goal to deliver rich search and filtering capabilities based on authoritative, domain specific background knowledge. The systems tend to be mostly monolithic, with predefined processes for the specific domains in question.

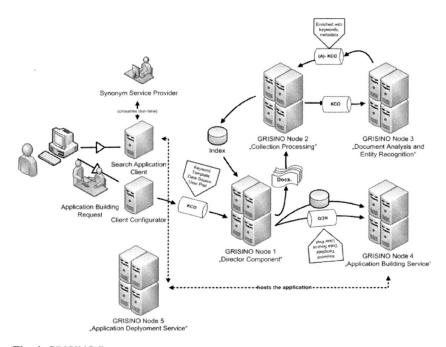

Fig. 4 GRISINO Demonstrator

While such systems deliver a good user experience, they are hard to build, because the one who builds the system has to provide in depth domain knowledge, technological knowledge, many different skills and different resources packaged as a single high quality service.

To meet the demands of the users it is considered very favorable, often even mandatory to be able to personalize the whole processing pipeline according to the needs of an user. Additionally the number of specialized providers for knowledge, content and services is growing, so that as a solution builder you have to consider integration of these providers, as it is hard if not impossible to be expert in all relevant areas, and have all needed information and specialized knowledge readily available, to answer the demands of the user. Thus we will need to build uniform but personalized solutions in the future, which transform a currently static, hardwired process into a dynamic, service oriented process. This dynamic process should be driven by the goals of the user, utilize and bundle services provided by different organizations and gather and combine information and data from different sources. The vision for such a process is to

> *"Transform the goal of the user into the appropriate process, execute it and deliver the solution automatized."*

In order to demonstrate the functionality of the integration and the interoperability between the technologies in the GRISINO test bed, a semantic search application has been designed that realizes a scalable, flexible and customizable search application generator that enables knowledge-based search in unstructured text. The search applications generated are customized and tailored to specific needs expressed by end users. The search applications include very specialized knowledge about a particular domain (e.g. football in the 19th century), collected from different knowledge bases and consolidated into one index to provide a single point of access.

To achieve this, a number of processing services deployed on the grid, are tied together to selectively collect, index and annotate content according to different knowledge bases and to generate custom search applications according to a users' input. The foundation of the users' input is his/her knowledge background or special interests. In particular the search application generator decomposes the user input (e.g. data sources of interest, specific keywords or entities considered important, etc.), into different sub goals which are used to consider different service providers for enriching the initial input. It queries these services to ask for related terms and entities, as well as authoritative information sources, such as popular websites according to the topic of interest. Using additional services, such as clustering services, the collected documents are then indexed and deployed for the use by the end user.

The goal of the scenario is amongst others to exploit as much of the GRISINO functionality as possible, e.g. to select services based on plans modeled inside KCOs or based on document types, and to parallelise indexing on the Grid. The underlying GRISINO infrastructure enables automation of the whole process of putting together the custom search application by using a number of different services from different service providers and bundling its output into a coherent application that combines knowledge and functionality from different sources. This reflects the particular and very common situation in which both knowledge found in all kinds of knowledge bases and specific skills encapsulated in special technical functionality is not found within one organization or provided by a specific technology provider, but is spread over a greater number of specialized organizations. While the benefit for the user obviously is a richer output informed by knowledge of a number of authoritative service providers, this model allows the commercial aspect of contributing specialized services as input to an open service mix by selling functionality and/or encapsulated knowledge bundled into one coherent service.

6 Conclusions

The GRISINO project brought forward the integration of three distinct technologies as detailed in this chapter. Two major sub-results of GRISINO are a new approach to realize the Semantic Grid which has been the goal of the SWS - Grid transformer and the possibility to use self-descriptions of documents for dynamic SWS discovery in order to automate and execute specific tasks. Regarding the first result, we have followed a new, and previously unexplored approach. More precisely we started from a SWS system (i.e. WSMO/L/X) and added Grid specific features and by that transformed an SWS system into a SWS-Grid system. Furthermore we support the integration of legacy systems such as Globus. The second result, might be applied in document processing, multimedia content adaptation or other similar scenarios. The semantic search application generator implemented as a proof-of-concept, shows the added value of the GRISINO system both for service providers as well as for end users.

Acknowledgements. The reported work is funded by the Austrian FIT-IT (Forschung, Innovation, Technologie - Informationstechnologie) programme under the project GRISINO - Grid semantics and intelligent objects.

References

1. Andreozzi, S., Burke, S., Field, L., Fisher, S., Konya, B., Mambelli, M., Schopf, J., Viljoen, M., Wilson, A.: GLUE schema specification (December 2005)
2. Behrendt, W., Arora, N., Bürger, T., Westenhaler, R.: A Management System for Distributed Knowledge and Content Objects. In: Proc. of AXMEDIS 2006 (2006)
3. Belecheanu, R., et al.: Business Process Ontology Framework; SUPER Deliverable 1.1 (May 2007)
4. Bürger, T.: Putting Intelligence into Documents. In: Proc. of the 1st European Workshop on Semantic Business Process Management (SBPM) held in conjunction with ESWC 2007 (2007)
5. de Bruijn, J., Lausen, H., Krummenacher, R., Polleres, A., Predoiu, L., Kifer, M., Fensel, D.: The Web Service Modeling Language WSML. Technical report, WSML. WSML Final Draft D16.1v0.21 (2005), http://www.wsmo.org/TR/d16/d16.1/v0.21/
6. Fensel, D., Bussler, C.: The Web Service Modeling Framework WSMF. Electronic Commerce Research and Applications 1(2), 127–160 (1991)
7. Foster, I., Kesselman, C.: The Grid: Blueprint for a New Computing Infrastructure. Morgan Kaufmann, San Francisco (1999)
8. Gangemi, A., Borgo, S., Catenacci, C., Lehmann, J.: Task Taxonomies for Knowledge Content. METOKIS Deliverable D07 (2004), http://metokis. salzburgresearch.at/files/deliverables/metokis_d07_task_ taxonomies_final.pdf
9. Keller, U., Feier, C., Steinmetz, N., Lausen, H.: Report on reasoning techniques and prototype implementation for the WSML-Core and WSMO-DL languages. RW2 Deliverable (July 2006)
10. Mika, P., Oberle, D., Gangemi, A., Sabou, M.: Foundations for Service Ontologies: Aligning OWL-S to DOLCE. In: Proc. of the 13th Int. World Wide Web Conf (WWW 2004). ACM Press, New York (2004)
11. Roman, D., Lausen, H. (eds.): Web service modeling ontology (WSMO). Working Draft D2v1.2, WSMO (2005), http://www.wsmo.org/TR/d2/v1.2/
12. Shafiq, O., Toma, I.: Towards semantically enabled Grid infrastructure. In: Proc. of the 2nd Austria Grid Symposium, Innsbruck, Austria, September 21-23 (2006)
13. Toma, I., Bürger, T., Shafiq, O., Dögl, D., Behrendt, W., Fensel, D.: GRISINO: Combining Semantic Web Services, Intelligent Content Objects and Grid computing. In: Proc. of EScience 2006 (2006)
14. Stärk, R., Schmid, J., Börger, E.: Java and the Java Virtual Machine. Definition, Verification, Validation.: Definition, Verification, Validation. Springer, Berlin (2001)

Collaborative Web-Publishing with a Semantic Wiki

Rico Landefeld and Harald Sack

Abstract. Semantic Wikis have been introduced for collaborative authoring of ontologies as well as for annotating textual and multimedia wiki content with semantic metadata. In this paper, we introduce a different approach for a Semantic Wiki based on an ontology metamodel that has been especially customized for the deployment within a wiki. For optimal usability client-side technologies for graphical user interface have been combined with a simple and intuitive semantic query language. Single fragments of a wiki page can be annotated in an interactive and rather intuitive way to minimize the additional effort that is necessary for adding semantic annotation. Thus, the productivity and efficiency of a Semantic Wiki system will open up for non expert users as well, which is important for fostering the popularity of Semantic Wiki systems.

1 Introduction

The very first browser to access the World Wide Web (WWW) provided an important function that soon sank into oblivion again: web pages could not only be read, but also written and thus be changed directly. Several years ago, wiki systems [14] picked up that very same idea again by providing the possibility for each visitor to change the content of wiki pages in a simple way. Wiki systems are lean content management systems that administrate HTML documents. The user of a wiki

Rico Landefeld
Jena University Language and Information Engineering (JULIE) Lab,
Friedrich-Schiller-Universität Jena, Fürstengraben 30, 07743 Jena, Germany
e-mail: Rico.Landefeld@uni-jena.de

Harald Sack
Hasso Plattner Institut for IT Systems Engineering, Universität Potsdam,
Prof.-Dr.-Helmert-Str. 2–3, 14482 Potsdam, Germany
e-mail: harald.sack@hpi.uni-potsdam.de

S. Schaffert et al. (Eds.): Networked Knowledge - Networked Media, SCI 221, pp. 129–140.
springerlink.com   © Springer-Verlag Berlin Heidelberg 2009

system is able to generate or change wiki documents only by using the facilities of a simple web browser. In this way, wiki documents are developed and maintained collaboratively by the community of all users without the need of having specialized IT expertise. Wiki systems don't give formal guidelines for generating or structuring their content. This lack of formal rules might have been responsible for their fast growth of popularity as can be seen, e.g., in the free online-encyclopedia Wikipedia[1] that is one of the most popular websites world-wide. On the other hand, if a wiki system is growing as rapidly as Wikipedia does, lack of formal rules necessitates frequent restructuring to keep the content always well arranged and usable.

Typical wiki systems only provide a limited number of functions for structuring the content. As a rule users create special pages with overviews or class systems for structuring and aggregating the wiki content. But the maintenance of this manually created categorization system becomes rather expensive. Moreover, it stimulates misusage, as e.g., you may find many categories in Wikipedia that have been created to subsume entities that share merely one special feature [23]. Similar problems have been reported for intranet wikis [4]. In general, most of the mentioned problems in wikis can be reduced to the fact that their content is encoded in HTML (Hypertext Markup Language) or some simplified version of it. HTML only formalizes formatting and (limited) structuring of documents without the possibility of formalizing any semantics that is required for automated aggregation and reuse of data.

Semantic Wikis try to combine traditional wiki systems with semantic technology as a building block of the currently emerging Semantic Web [2]. They connect textual and multimedial content with a knowledge model by formalizing the information content of a wiki page with a formal knowledge representation language. In this way, the content of wiki pages becomes machine readable and in some limited sense even machine understandable. Semantic Wikis show one possible way to overcome the aforementioned problems related to traditional wikis in general while at the same time enabling collaborative generation and maintenance of formal knowledge representations (ontologies). But, the arbitrary wiki user usually is not an expert knowledge engineer. Therefore, usability and 'ease of use' become a essential factors for the design of the user interface of a Semantic Wiki.

Current Semantic Wiki projects have chosen different ways to deploy formal knowledge representations within a wiki. From our point of view the ratio of cost and effect for the user is most important. The cost refers to the cognitive and factual work that the user has to invest to generate and maintain semantic annotations. On the other side, the effect subsumes all the advantages that the user might get from a system that deploys this semantic annotation. Cognitive and factual work is mostly determined by the design of the user interface and the underlying ontology metamodel of the Semantic Wiki. At the same time the semantic expressiveness of the annotations determines the efficiency of the achieved functionality. Therefore, the ontology metamodel of a Semantic Wiki always represents a compromise between complexity and expressiveness. In addition, the integration of semantic annotations

---

[1] http://www.wikipedia.org

into wiki systems also demands new concepts of user interaction that help to limit the necessary effort for authoring.

Existing Semantic Wiki systems have several deficiencies: either, their underlying ontology metamodel is mapping elements of the knowledge representation language directly to wiki pages, or they are using a simplified ontology metamodel that results in rather limited semantic functionality. Most projects are based on traditional wiki systems and therefore inherit also their user interaction facilities.

We propose a Semantic Wiki concept that combines the following three concepts:

1. a simplified ontology metamodel especially customized to be used within a wiki system,
2. a WYSIWYG-Editor (What You See Is What You Get) as a user interface for both text- and ontology editing, and
3. preferably a most simple semantic query language of sufficient expressiveness.

A prototype of our Semantic Wiki *Maariwa*[2] has been successfully implemented.

The paper is organized as follows: Section 2 covers related work and in particular discusses different ontology metamodels and user interaction concepts of existing Semantic Wiki systems. In Section 3 we introduce the Semantic Wiki project Maariwa, while Section 4 resumes our results and discusses future work.

2   Related Work

Traditional wiki systems administrate structured text and multimedia content connected by untyped hyperlinks. Semantic wikis complement the traditional wiki concept by providing the ability to capture additional information about the wiki pages and their relations. Besides the extension of traditional wikis with semantic annotation, we also have to consider approaches for collaborative authoring of ontologies based on wiki technology that date back to a time before the Semantic Web initiative even started (cf. [9, 21, 1]). We therefore distinguish two different Semantic Wiki approaches depending on their focus either on textual (or multimedia) wiki content or (formal) knowledge representation. The *Wikitology* paradigm [?] refers to wiki systems acting as a user interface for collaborative authoring of ontologies. There, a wiki page represents a concept and hyperlinks between wiki pages represent relationships between concepts. Thus, the wiki system acts merely as tool to author and to manipulate the ontology. In difference, so called *ontology-based wiki systems* are Semantic Wiki systems that focus on traditional wiki content, while using knowledge representations to augment navigation, searchability, and above all reusability of information.

Another differentiating factor can be determined by the adaption of the ontology metamodel for the use within the wiki and the coverage of the underlying knowledge representation languages (KRL). The *ontology metamodel* of a Semantic Wiki defines a mapping between the elements of the KRL and the application model.

---

[2] Maariwa can be accessed at http://stemnet0.coling.uni-jena.de/ Maariwa/app

Moreover, it determines the semantic expressiveness of annotation and serves as a basis for querying information. Semantic annotation can be maintained together with or separate from the textual wiki content. [17]. Next, we will introduce and discuss relevant Semantic Wiki implementations and their underlying ontology metamodel.

PlatypusWiki [5] is one of the earliest Semantic Wiki implementations. It maps a wiki system to a RDF (resource description framework) graph [12]. Wiki pages represent RDF resources and hyperlinks represent RDF properties. Semantic annotation is maintained together with the textual wiki content within a separate text field as RDF(S) (RDF Schema) [3] or OWL (Web Ontology Language) [16] in XML serialization format.

Rhizome [20] supports semantic annotations by using a special Wiki Markup Language (WikiML). The entire wiki content including text, structure, and metadata internally is encoded in RDF. Rhizome also allows direct editing of RDF data with an external RDF editor. In contrast to traditional wiki systems, Rhizome supports a fine-grained security model. Beside creation and manipulation of meta data Rhizome does not offer any functionality that utilizes this semantic annotation.

Rise [6] is customized for requirement analysis in the process of software engineering. It is based on an ontology that represents different document types and their relationships. Templates determine structure and relationships of wiki pages that represent instances of a document type. The Rise ontology can be extended by adding new templates to the existing ones. Semantic annotations can be created and edited with an extended WikiML and are used for consistency checks and navigation.

Semantic MediaWiki (SMW) [23, 13] is an extension of the popular MediaWiki[3]. The online-encyclopedia Wikipedia is the most prominent example of a MediaWiki application. SMW aims to improve the structure and searchability of the Wikipedia content by deploying semantic technologies. Therefore, SMW follows Wikipedia's user interface to attract a broad user community. SMW extends the WikiML with attributes, types, and relationships. To represent classes, SMW utilizes existing Wikipedia categories. By assigning a wiki page to a given category it becomes an instance of the class being represented by this category. Relationships between instances are implemented via typed hyperlinks. In addition, SMW provides a set of units of measurement and customizable data types. Attributes, types, and relationships are represented in seperate wiki pages. Semantic search is implemented in SMW with a proprietary query language (WikiQL) that closely reflects the annotation syntax. WikiQL allows the creation of dynamic wiki pages of automatically aggregated content. SMW has become the most popular out of all Semantic Wikis and serves as the basis for numerous domain specific Semantic Wiki applications.

MaknaWiki [7] uses RDF triples (subject, predicate, object) for annotating wiki pages. RDF triples can be added to a text page by using a customized WikiML or within a separate form. RDF triples' subjects or objects refer to textual wiki pages that represent a concept each. MaknaWiki only supports maintenance and manipulation

---

[3] http://www.mediawiki.org/

of instances, but no class definitions. It extends JSPWiki[4] and its WikiML with typed links and literals. MaknaWiki's semantic annotation is utilised for navigation based on RDF triples and for limited semantic search (e.g., for searching instances of classes with distinct properties).

IkeWiki [19] tries to bring together application experts and knowledge engineers. Therefore, the user interface offers separate views for textual content and semantic annotation. The annotation editor supports the assignment of classes to wiki pages and typed links for representing relationships. Furthermore, classes, properties, and resources can be freely created and manipulated. The ontology metamodel closely reflects the underlying KRL (RDF(S) and OWL). The user interface for editing metadata supports automatic completion of terms.

SweetWiki [4] combines social tagging [11] and semantic technologies into what they call *semantic tagging*. Wiki pages are annotated with user tags that form not only a collective index (a so called *folksonomy* [22]), but a formal ontology. This is achieved by regarding each user tag as a concept of an ontology. Relationships between concepts are not determined by users, but by designated experts. The user does not interact with the ontology directly, but is merely able to create and to assign user tags. For the user, there is no distinction between instances and classes, because tags can represent both. Sweet Wiki's user interface provides a WYSIWYG editor for manipulating the wiki content.

Besides the above mentioned projects there exist several alternative Semantic Wiki implementations that can also be arranged within the framework given by our examples ranging from wikitologies to ontology-based wikis[5]. As a rule, early implementations such as PlatypusWiki strictly separate textual content from knowledge representations, while later projects (besides IkeWiki, which separates annotations from content) integrate knowledge representations and textual content by utilizing customized WikiML. Above all, IkeWiki and MaknaWiki provide dynamic authoring support. The ontology metamodels of PlatypusWiki, MaknaWiki and IkeWiki merely provide a direct mapping of the underlying KRL without any covering. Only MaknaWiki offers (limited) semantic search facilities. If elements of RDF(S) or OWL are utilized directly, RDF query languages such as SPARQL [18] can be applied. IkeWiki enables data export via SPARQL without providing a search interface. MaknaWiki and PlatypusWiki use elements of RDF(S) and OWL without making any use of their semantic expressiveness.

Contrariwise, SMW deploys a simplified ontology metamodel based on OWL-DL with an easy to use query language (compared to SPARQL) The SMW user interface keeps the connection between classes and their attributes covered, but offers no further editing assistance to the user (beside the provision of templates). Sweet Wiki's ontology metamodel does not distinguish between classes, instances, datatype properties, or object properties – everything is mapped on tags. Therefore, information can only be provided via tag search or via direct SPARQL queries with

---

[4] http://jspwiki.org/
[5] A list of current Semantic Wiki Systems is available at
http://semanticweb.org/wiki/Semantic_Wiki_State_Of_The_Art

the consequence that information being distributed over several wiki pages can not be queried.

We propose the Semantic Wiki *Maariwa* that utilizes an ontology metamodel covering the underlying OWL-Lite language that enables information reuse as well as semantic queries over distributed information. In the following chapter we introduce Maariwa's underlying concepts and give a brief sketch on its implementation.

3   The Maariwa Concept – Architecture and Implementation

Maariwa is a Semantic Wiki project developed at the Friedrich Schiller University in Jena, Germany, with the objective to implement an augmented wiki system that enables simultaneous creation and manipulation of wiki content and ontologies. Maariwa's semantic annotation is utilized to put augmented navigation and semantic search into practice for reuse and aggregation of the wiki content. In Maariwa ontologies are used to structure the textual wiki content for providing access paths to the knowledge being represented in the wiki. Maariwa's access paths can be addressed via *MarQL*, a simple semantic query language, to enable semantic search. We therefore refer to Maariwa's underlying concept as *ontology-based web publishing*. Maariwa is geared towards user communities without expert knowledge in knowledge representations and knowledge engineering. Furthermore, Maariwa facilitates access by utilizing a WYSIWYG-editor for wiki content and ontology manipulation.

In this section, we introduce the Maariwa ontology metamodel and show how its elements are integrated into the wiki by illustrating important facets of the user interface. Finally, we comment on the implementation of ontology versioning and introduce MarQL syntax and semantics.

3.1   Maariwa's Ontology metamodel

The ontology metamodel of Maariwa is designed to enable simple and efficient searchability, as e.g., answering questions like 'What physicists were born in the 19th century' or 'Which cities in Germany have more than 100.000 inhabitants?'. To answer these questions, the ontology metamodel has to provide the semantic means to express these queries, while on the other hand it must be simple enough so that the arbitrary user without expert knowledge can comprehend and apply it. Tab. 1 shows a subset of OWL-Lite elements that are utilized for Maariwa's ontology metamodel.

We refer to datatype properties and object properties as attributes and relationships. Furthermore, both attributes and relationships are not defined as global entities but only local within their classes. This enables attributes and relationships with the same name in different classes without worrying the non-expert user with naming conflicts and disambiguation. Attributes are determined by a datatype with an optional measuring unit. For simplicity, only numbers, strings, and dates have been considered for datatypes.

Table 1 Mapping of OWL-Lite elements to Maariwa's ontology metamodel

OWL-Lite element	Maariwa element
class	class
datatype property	attribute
object property	relationship
class instance	wiki page representing a class being instantiated as an instance-page
subClassOf	superclass /subclass
individual	wiki page describing an individual, being an instance of one or more classes
datatype property instance	attribute value of an instance-page
object property instance	relationship value of an instance-page

By integrating the ontology metamodel into the wiki system, wiki pages can be annotated with concepts of ontologies to formalise their content. Classes, individuals, and sets of individuals can be described by wiki pages. A page that describes

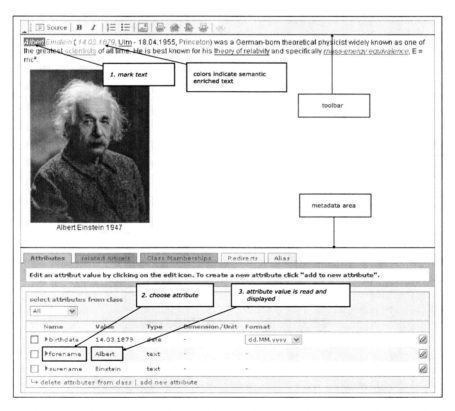

Fig. 1 Editing of a Maariwa wiki page with textual content (above) and ontologies (below)

a class is associated with attributes, relationships, and superclasses. Pages that describe individuals are associated with one class at least. To denote a set of individuals, a page has to be associated with a MarQL expression (see section 3.4). Typed links between pages may refer to relationships. Classes may use instance-pages as simple tags by associating neither a class hierarchy nor relationships, nor attributes.

3.2   Integration of Semantic Annotation and Textual Content

The graphical Maariwa editor is designed to minimize the user's additional effort for adding and maintaining semantic annotations. Repeated input of text or ontology concepts is avoided most times simply by copying existing concepts from the page context. New concepts are created in dialogue with the user. Thus, Maariwa enables the development of textual and ontological resources in parallel.

Maariwa provides two alternatives for creating semantic annotations: concepts of an ontology can be created from textual content of a wiki page, while on the other hand concepts might be defined first providing a textual description in the wiki page afterwards. Schema and instance data can be manipulated in parallel without interrupting the editing process of the wiki page (see Fig. 1). Maariwa's WYSIWIG-editor is implemented as so called *Rich Internet Application* [15] that provides desktop functionality for web applications and adopts the role of the traditional WikiML-based user interface.

In Maariwa, semantic annotations are directly displayed within the wiki page (see. Fig. 2). Different colors are used to denote the semantics of typed hyperlinks. Text that contains attribute values as well as links that represent relationships is highlighted. In addition, tooltips (i.e., pop-up information windows) display the semantic of an annotated text fragment if it is touched with the mouse pointer. Navigation within class hierarchies and class relations is enabled with a superimposed class browser. The semantic annotation of each wiki page can be separately accessed and exported in RDF/XML encoding via an own URL. Also export and import of ontologies as a whole is supported.

3.3   The MarQL Semantic Query Language

The syntax of SPARQL reflects the characteristics of the RDF data model while adopting the pattern of the simple database query language SQL. RDF statements are represented as triples and the RDF document can be interpreted as a graph. SPARQL traverses the RDF graph and as result delivers the nodes that satisfy the constraint given in the SPARQL query. Because RDFS and OWL are based on the RDF syntax SPARQL can also be used to query RDFS- and OWL-files, but without exploiting their semantic expressiveness.

MarQL is a customized semantic query language for the Maariwa ontology metamodel. MarQL syntax does not refer to RDF triples but directly addresses ontology elements such as classes, attributes and relationships. The underlying RDF encoding of the data remains hidden. In comparison to SPARQL, the syntax of MarQL is

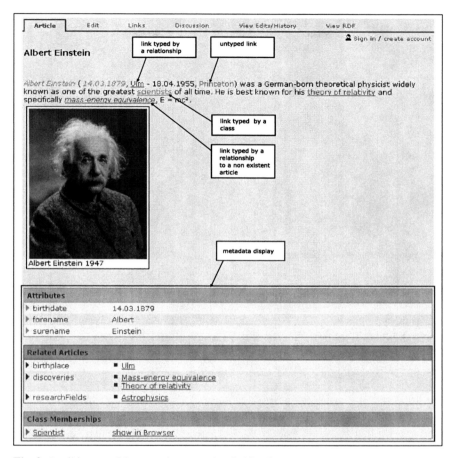

Fig. 2 A wiki page with semantic annotation in Maariwa

much more compact but less flexible. MarQL only implements a fixed set of query patterns. A MarQL query results in a set of wiki pages that refer to individuals, which satisfy the constraints of the MarQL query.

The structure of a MarQL query can be shown with an example: the expression *Scientist.institution.location.country = Germany* refers to wiki pages about scientists that work at an institute being located in Germany. *Scientist* refers to a class with a relationship *institution*. Relationships can be applied recursively and are denoted as a path expression. In this way, *institution* and *location* are connected. This means that there must exist a class, which is the target class of a relationship with *institution*, while in addition having a relationship with *location* . In the same way *location* and *country* are connected, while *country* can either be an attribute or a relationship and therefore *Germany* might denote an individual or an attribute value.

MarQL provides logical operators as well as string operators and operators for comparison. E.g., the query *City.population $\geq$ 100.000* results in a set of all wiki

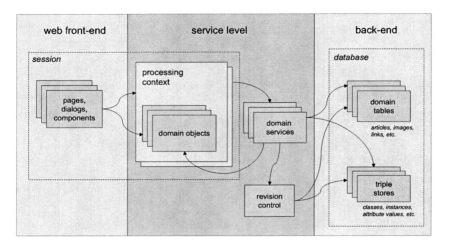

Fig. 3 Architecture of the Maariwa system

pages that describe cities with more than 100.000 inhabitants, or *Scientist.birthdate < 1.1.1900 AND Scientist.birthdate ≥ 1.1.1800* results in a set of wiki pages with scientists that are born in the 19th century. The latter example results in a list of instances of a class and can also be referred to as a simple *tag*.

3.4 Maariwa Architecture and Implementation

Maariwa does not extend one of the existing traditional wiki systems but is a proprietary development based on Java. The application core of Maariwa implements a service level that realizes a wiki system as a set of loosely coupled services (see Fig. 3). Beside data management and revision control of the wiki pages and the related ontologies, keyword search and semantic search are also implemented as services. MarQL queries are translated into SPARQL queries. Manipulation of the ontologies in the editor is organized in different dialog levels. Within a dialog updates of one or more objects can be performed. These updates can either be revoked or they will also be adopted on the subjacent levels. Therefore, the service level offers cascading manipulation levels that implement this functionality with the help of local copies and snapshots.

Back-end data processing is achieved with a relational database management system. The service level stores ontology elements in various RDF triple stores and all other objects in separate database tables. The service level is used by the components of the web front-end that constitute Maariwa's user interface. Web server and browser client communicate asynchronously via AJAX [10] to achieve better usability. The WYSIWYG-editor represents the text of the wiki page as XHTML [8] fragment and is based on Javascript.

As a rule, wiki systems deploy a *revision control system* (RCS) to prevent abuse of unprotected write access. Maariwa adapts this RCS for ontologies, too. The RCS maintains two levels: schema level and instance level. The schema level comprises all changes on classes, relationships and attributes, while the instance level covers changes of individuals including their attribute values and relationship values. Both levels are tightly coupled, because each schema update might have effects on all derived individuals. Therefore, a schema version additionally includes all updates on the instance level that occurred since the last update of the schema. Each version of a wiki page besides the update of the page content also contains updates of the associated concepts.

4 Conclusion

In this paper, we have introduced the Semantic Wiki approach Maariwa based on an ontology metamodel customized especially for the deployment within a wiki. For optimized usability recent client-side technologies have been combined with a simple semantic query language. The user can annotate text fragments of a wiki page in an interactive and rather intuitive way to minimize the additional effort that is necessary for adding semantic annotation. Thus, the productivity and efficiency of a Semantic Wiki system will open up for non-expert users as well. The ontology metamodel enables the formulation of access paths to wiki pages as well as the reuse of already implemented relationships in the underlying knowledge representation. The simple query language MarQL uses Maariwa's annotations and the contained access paths for implementing a semantic search facility. Ontology metamodel and query language are synchronized to support annotations on different levels of expressiveness. Thus, enabling simple tags as well as complex ontologies with attributes and relationships.

Currently, Maariwa is extended to include meta data also directly within the textual wiki content, as e.g., tables with self-adjusting data aggregations. Also natural language processing technology is considered to be utilized in the WYSIWYG-editor for automated suggestions as well as for translating natural language queries into MarQL. For deploying a Semantic Wiki, the user considers always the ratio of cost and effect that is caused by the additional effort of providing semantic annotation. In doing so, it is not important whether the creation of semantically annotated textual content or the creation of mere ontologies is focussed, but how both can be integrated within an system that provides a reasonable and efficient interface for user access.

References

1. Arpírez, J.C., Corcho, O., Fernández-López, M., Gómez-Pérez, A.: WebODE: a scalable workbench for ontological engineering. In: 1st Int. Conf. on Knowledge Capture (KCAP 2001), Victoria, Canada (2001)
2. Berners-Lee, T., Hendler, J., Lassila, O.: The semantic web. Scientific American 284(5), 34–43 (2001)
3. Brickley, D., Guha, R.V.: RDF Vocabulary Description Language 1.0: RDF Schema. W3C Recommendation, RDF Core Working Group, W3C (2004)

4. Buffa, M., Gandon, F.: Sweetwiki: semantic web enabled technologies in wiki. In: Proc. of the 2006 international symposium on Wikis, pp. 69–78 (2006)
5. Campanini, S.E., Castagna, P., Tazzoli, R.: Platypus wiki: a semantic wiki wiki web. In: Semantic Web Applications and Perspectives, Proc. of 1st Italian Semantic Web Workshop (2004)
6. Decker, B., Ras, E., Rech, J., Klein, B., Höcht, C.: Self-organized reuse of software engineering knowledge supported by semantic wikis. In: Workshop on Semantic Web Enabled Software Engineering (SWESE), at ISWC 2005, Galway, Ireland (2005)
7. Dello, K., Paslaru, E., Simperl, B., Tolksdorf, R.: Creating and using semantic web information with makna. In: Proc. of the 1st Workshop on Semantic Wikis – From Wiki To Semantics ESWC 2006 (2006)
8. Pemberton, S., et al: XHTML 1.0 The Extensible HyperText Markup Language. W3C Recommendation, W3C, 2nd edn. (August 2002)
9. Farquhar, A., Fikes, R., Rice, J.: The Ontolingua server: A tool for collaborative ontology construction. Int. Journal of Human-Computer Studies 46(6), 707–727 (1997)
10. Garrett, J.J.: Ajax: A new approach to web applications (2005), http://www.adaptivepath.com/publications/essays/archives/000385.php
11. Golder, S., Huberman, B.A.: The structure of collaborative tagging systems. Journal of Information Sciences 32(2), 198–208 (2006)
12. Klyne, G., Carroll, J.J.: Resource Description Framework (RDF): Concepts and abstract syntax. W3C Recommendation, W3C (February 2004)
13. Krötzsch, M., Vrandecic, D., Völkel, M.: Wikipedia and the semantic web - the missing links. In: Proc. of the 1st Int. Wikimedia Conf., Wikimania (2005)
14. Leuf, B., Cunningham, W.: The Wiki Way: Quick Collaboration on the Web. Addison-Wesley, Reading (2001)
15. Loosley, C.: Rich Internet Applications: Design, Measurement, and Management Challenges. Technical report, Keynote Systems (2006)
16. McGuinness, D.L., van Harmelen, F.: Owl Web Ontology Language: Overview. W3C Recommendation, World Wide Web Consortium (February 2004)
17. Oren, E., Delbru, R., Möller, K., Völkel, M., Handschuh, S.: Annotation and Navigation in Semantic Wikis. In: Proc. of the 1st Workshop on Semantic Wikis, Workshop on Semantic Wikis. ESWC 2006 (June 2006)
18. Prud'hommeaux, E., Seaborne, A.: SPARQL Query Language for RDF (W3C Recommendation). W3C Recommendation, W3C (January 2008)
19. Schaffert, S., Gruber, A., Westenthaler, R.: A semantic wiki for collaborative knowledge formation. In: Semantics 2005, Vienna, Austria (2005)
20. Souzis, A.: Rhizome position paper. In: Proc. of the 1st Workshop on Friend of a Friend, Social Networking and the Semantic Web (2004)
21. Swartout, B., Ramesh, P., Russ, T., Knight, K.: Toward distributed use of large-scale ontologies. In: Symp. on Ontological Engineering of AAAI, Stanford, California (March 1997)
22. Vander Wal, T.: Folksonomy Explanations (2005), http://www.vanderwal.net/random/entrysel.php?blog=1622
23. Völkel, M., Krötzsch, M., Vrandecic, D., Haller, H., Studer, R.: Semantic wikipedia. In: Proc. of WWW 2006, Edinburgh, Scottland (2006)

Collaborative Wiki Tagging

Milorad Tosic and Valentina Nejkovic

Abstract. Wikis as well as collaborative tagging have been subject of very intense research and an active discussion topic in the so-called blogsphere. In this paper, we propose Collaborative Wiki Tagging based on the idea to exploit inherent semantics of the concept of link in a wiki. The low-level integration of wiki and collaborative tagging of web resources is expected to be effective in enterprise environments particularly in the personal and group knowledge management application area. We first introduce a conceptualization of Collaborative Wiki Tagging. Then, we propose a simple scheme for using one of the existing native wiki syntax to represent tagging data. Collaborative Wiki Tagging Portal Prototype, developed as a proof of concept, is used to give illustrative practical examples of the proposed approach and illustration of the user interface.

1 Introduction

The past few years have witnessed a growing interest in the enterprise world for generating, managing and sharing knowledge. This knowledge management processes are recognized as being of crucial importance for enterprises to effectively manage innovation. The growing need for continuous innovation results fosters strategies to generate new knowledge through collaborative means at the individual, group, as well as community level. Also, traditional knowledge management technologies have not delivered as promised. In the

Milorad Tosic
University of Nis, Faculty of Electronic Engineering, Nis, Serbia
e-mail: mbtosic@yahoo.com

Valentina Nejkovic
University of Nis, Faculty of Electronic Engineering, Nis, Serbia
e-mail: valentina@elfak.ni.ac.yu

S. Schaffert et al. (Eds.): Networked Knowledge - Networked Media, SCI 221, pp. 141–153.
springerlink.com   © Springer-Verlag Berlin Heidelberg 2009

same time, new developments are becoming more attractive, such as Semantic Web, social computing, open systems, emergent semantics, etc. As a result, collaborative systems that provide technological capabilities for collaborative interaction among multiple participants with shared goals and interests across time and place, have recently gained considerable attention [1].

Collaborative tagging systems appear as a new trend in collaboration, gaining growing popularity on the Web. The purpose of those systems is to organize web pages, objects, social relations, images, people locations, etc., as a set of Web resources, representing a new paradigm of organizing information and knowledge. They are very often interchangeably referred to as Social Tagging systems due to their social nature. It was shown recently that tagging distributions tend to stabilize into power law distributions. The stable distribution is considered as an essential aspect of what might be a user consensus around the categorization of information driven by tagging behaviors [7]. Also, recent research results indicate the social aspect as one of the most influential driving motivational factors for user participation and wide acceptance of the overall concept [3]. Collaborative tagging is expected to take a leading role in knowledge work related fields such as information storage, organization and retrieval [11]. Discussions within the blogsphere on the concept of tagging, tagging applications, problems that tagging processes retrieve, social and cognitive analysis of tagging [17], tagging formats, tag-clouds, hierarchy versus tagging for information classification etc., is very active. Results of the more thorough scientific research have also been published.

In this paper, we propose low-level integration of collaborative tagging in a wiki, where tagging data is stored using the native wiki syntax. This way we are able to apply all functionalities available in the wiki on the embedded tagging data. This includes collaborative editing and full-text search of tagging data, versioning, Ajax-like interface components, etc. Resources to be tagged are not limited on the internal wiki pages only, but can be any Web resource. The idea of the deep-integration approach to collaborative wiki tagging, particularly the relation between the concepts of a link in a wiki and the semantics of a tagging act in tagging systems, is in accordance with the latest research results about the integration of semantics, wikis, and the social web. We believe that the concept of collaborative wiki tagging removes some of the inherent bottlenecks related to group and personal knowledge management.

The rest of the paper is organized in the following way. In Section 2, we introduce the concept of collaborative wiki tagging. Syntax and wiki text formatting rules for tagging data are discussed in Section 3. Section 4 gives an overview of the Collaborative Tagging Wiki Portal Prototype. Section 5 discusses some of the recent research results that relate to our work, while Section 6 concludes the paper and gives some pointers for future work.

2   Concept of Collaborative Wiki Tagging

In this paper, we introduce the concept of collaborative tagging as an interaction procedure between two (or more) resources. As is assumed by the interaction, the link of a tagged resource is memorized together with some concomitant information about the resource. The link is memorized by the tagging resource (the tagging resource is also interchangeably called agent). The link is a reification of the identity of the tagged resource in the sense of information that is sufficient to initiate and conduct an interaction protocol between participating resources[1]. The agent with memorizing capability is implemented by means of a wiki page. Note that the agent's memorizing capability does not necessarily mean an intrinsic ability of the agent to activate the memorizing process. Instead, the process may be initiated by some other agent (for example, a human agent in the case of a wiki page). A formal definition of agents, resources, interaction protocol, an agent's knowledge, and addressing has been previously given in more detail in [19].

In the case of wiki pages, information and knowledge are reified by means of wiki text stored within wiki pages. Traditionally, the content of the wiki page is interpreted as informational content primarily used by humans as a document or a Web page. We adopt an approach to reason about the system of wiki pages similar to the two-level interpretation of the Semantic Web [5]: There is a space of directed untyped links between documents, and there is a space of directed, typed relationships between the things described by the documents. In our approach, there is a space of directed untyped links between resources where each resource has an identity. The identity plays a crucial role for interaction between resources (as described above). A typed relationship between resources is considered as a composite structure consisting of an untyped directed link and an associated resource describing the link. Whether a resource represents a document, a type information about a link, or a thing described by a document is specified by associated semantics. We may distinguish two types of the associated semantics: implicit and explicit semantics. *Implicit semantics* assumes common knowledge that has a cultural nature, accumulated by a social protocol, and very loosely defined. Implicit semantics is a kind of information object called "document" since documents are assumed to be "understandable" to humans. Most of the pages on today's Web are documents, i.e. have implicit semantics and are aimed to be comprehended by humans. From the other side, *explicit semantics* does not assume any common knowledge: Explicit semantics is somehow reified and assigned to the resource representing an object (where object means a document as well as a thing described by

---

[1] Identity management of resources on the Web, as well as Semantic Web, is very important while still an open problem. Here, we adopt a definition of identity relative to the interaction protocol which is intended to be used. Informally, for an agent that has the intention to interact with some other resource using a given interaction protocol, reification of identity of the peer resource is interpreted as information sufficient for establishing this interaction protocol.

the document) with a purpose to be used in an automatic process of making conclusions by (artificial, software) agents.

Our approach to collaborative wiki tagging is based on the assignment of explicit semantics to the content of wiki pages. The design of a complete architecture of possible explicit semantics (ontology) is domain dependent and a very challenging task. Hence, we start with a minimal set of concepts that has the highest application potential, but keep the system open for future extensions with new concepts.

We identify 1) *presentation*, 2) *tag* and 3) *statement* semantic concepts that are assigned to wiki pages for the purpose of our target application. 1) *A presentation page* (a wiki page with assigned presentational semantics) is a traditional wiki page, a Web document understandable to a human, which can be created and collaboratively edited by several users. The page stores plain text using wiki syntax and text formatting rules. The presentation semantics should be considered as a placeholder for implicit semantics associated with a general Web document. 2) *A tag* is also a wiki page but with assigned tag semantics. The tag semantics is codified within the wiki page as a link to a unique meta-semantics page titled *"Tag"*. In other words, every wiki page that contains a link to a specific, predefined page *"Tag"*[2] is treated as a tag. 3) *A statement* is a wiki page that reifies semantic relations between any two wiki pages (including tag, presentation as well as content pages). The subset of wiki pages with assigned tag semantics is called *TagCloud System Repository*, or simply *TagCloud*, as shown in Fig. 1. Note that the classification, introduced in the set of all wiki pages by assignment of one of the defined explicit semantics, is not exclusive. Namely, a single wiki page may be assigned more then one explicit semantics. Which semantics is actually used for interaction with the wiki page is application dependent.

A wiki page with assigned tag semantics contains stored information about one or more links to resources that are tagged with the corresponding tag. With respect to the *TagCloud* we may identify *internal resources* (any wiki page from the set of all wiki pages containing the *TagCloud*) and *external resources* (any Web resource). The internal resources are under the same administrative control as resources representing tags. The internal resources are wiki pages too, so the system knows that it is possible to edit them, while this is not the case with the external resources. *A known resource* is a Web resource whose link is stored on a wiki page of the system. The wiki page may or may not have assigned tag semantics: If it has assigned tag semantics then we say the resource is tagged with the corresponding tag and consequently the resource belongs to the tag. A single resource may belong to zero, one or several tags. Accordingly, we may model *TagCloud* as a subset of power set of

---

[2] Note that we say nothing about the content of the *"Tag"* page, so it may be empty or may contain some additional informational content. In other words, the meta-semantics page *"Tag"* may have assigned presentational semantics itself, such that a human user may get additional explanation about the concept.

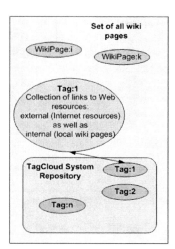

Fig. 1 TagCloud System Repository

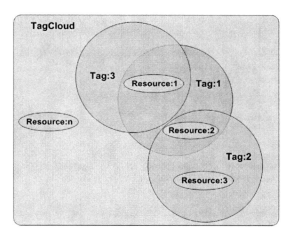

Fig. 2 *TagCloud* as a subset of power set of the set of all known resources

the set of all known resources (Fig. 2.), where each element of the *TagCloud* is a set of resources tagged with the corresponding tag.

3 Collaborative Tagging Wiki Syntax

In this paper, our aim is to store tagging data in a wiki using the native text formatting syntax of the used wiki. Consequently, we have to address two issues: 1) an ontology for tagging, and 2) a syntax for tagging data representation. We use a simple ontology for tagging based on the conceptualization developed by [19] and described in the previous paragraph. The adopted ontology is in compliance with existing ontologies for tagging [18], [14], [13], [10]. Nested bulleted lists are used as a basic element of the wiki syntax for tagging data representation. Table 1 shows some of the JspWiki text formatting rules[3] that we use as tagging format. Note that even the syntax is wiki engine dependent, the concept of nested lists is not[4]. Therefore, the proposed tagging format can be generally applied in any wiki engine.

Table 1 Some of the JspWiki text formatting rules

Rule	Description
[link]	Create hyperlink to "link", where "link" can be either an internal WikiPage or an external link (http://).
[text | link]	Create a hyperlink where the link text is different from the actual hyperlink link.
*	Make a bulleted list. For deeper indentations more (**,***) is used.

Using the JspWiki text formatting rules, a tagging data format is:

```
* [ResTitle | ResourceURL] [| Tagging]/ / Tags:[tag1][tag2][tag3]
** Clipping: content_of_clipping_if_exist
** Comment: link_to_comment_if_exist
** Posted on date&time, by [username | username_userprofile]
```

HTML preview of the tagging data is as follows:

```
<ul>
    <li>
        <a class="wikipage" href="Wiki.jsp?page= ResourceURL "> ResTitle </a>
        <a class="wikipage" href="Wiki.jsp?page= Tagging "> </a>
        <br />
        Tags:
        <a class="wikipage" href="Wiki.jsp?page=tag1">tag1</a>
        <a class="wikipage" href="Wiki.jsp?page=tag2">tag2</a>
        <a class="wikipage" href="Wiki.jsp?page=tag3">tag3</a>
        <ul>
            <li>Clipping: content_of_clipping_if_exist</li>
            <li>Comment: link_to_comment_if_exist</li>
            <li>Posted on date&time, by <a class="wikipage" href="Wiki.jsp?page=
username_userprofile"> Username </a></li>
        </ul>
    </li>
</ul>
```

Link may be a wiki-internal page link (page name reference) or an external resource link (URL address is written explicitly in the text, including the protocol prefix). In third line of the HTML code snippet, a link to an internal

---

[3] http://www.jspwiki.org

[4] WikiCreole.org is developing an universal Wiki syntax for interwiki compatibility.

wiki page is assigned the `wikipage` CSS class. In case of tagging an external resource, the link is assigned the external CSS class.

We reify the act of tagging as a segment of the wiki text. When a resource is tagged, the corresponding wiki text segment is appended to the wiki page of every used tag. For example, let us tag a presentation wiki page (or any other resource on the web) called A with a tag called B. As a result, a link to the "Tag" meta-data wiki page and a link to the resource A are appended to the content body of the tag page B. Note that there is no difference in format of the stored tagging data when we tag an internal wiki page or any other (external) web resource. The same resource can be tagged with several tags.

Following the proposed syntax of the concept of semantics reification in the wiki that we used to implement tagging, we extend the set of basic concepts in order to be able to make statements about wiki pages. For that purpose, we introduce `Statement`, `Relation`, and `Category` meta-pages. The format of the `Statement` is as follows:

```
* [WikiPage_Title | WikiPage][| Statement]
**[Relation] [Category] [Tag]
```

where, `WikiPage` is the identifier (wiki page name) of the resource that we make the statement about. The statement data is written in the wiki page called `WikiPageStatement`. The link `[| Statement]` means that the `WikiPageStatement` page is a statement saying that the `WikiPage` resource is in the `Relation` relation with the resource `Tag` from the category `Category`.

For example, when we make a statement about the wiki page `wikipageX`, a statement page `wikipageXStatement` is created (if it does not exist). Let the relation be `is_same_as`, category be `tagPage`, and tag be `tagX`. Then the format of the statement is:

```
* [WikiPage_Title | wikipageX][| Statement]
**[is_same_as] [tagPage] [tagX]
```

An unlimited number of statements can be made for a single wiki page. All statements will be written within the same statement page for that wiki page. For example, the statement says that `WikiPage` resource is also in relation `belong _ to` with the category `project`, while tag `projectName` assigns a name to the project:

```
* [WikiPage_Title | WikiPage][| Statement]
**[belong_to] [project] [projectName]
**[is_same_as] [tagPage] [tagX]
```

Making statements about resources and storing them in the wiki system is a general approach, but quite useful for folksonomy engineering. For example, let us consider a simple illustration of the more complex problem of "organizing the tags" [4]: Let us consider a collaborative wiki tagging system used in the food retail industry where each uploaded picture is tagged by users. In such a system we may expect tags such as `Fruit`, `Apples`, and `Apple`. We may want to "organize"[5] the tags so that `Apples` and `Apple` tags are considered synonyms, and `Apple` is a subclass of `Fruit`. The following text segment will be stored in the wiki page named `AppleStatement`:

```
* [Apple | Apple] [| Statement]
**[is_synonyme_with] [Tag] [Apples]
**[is_subcategory_of] [Tag] [Fruit]
```

4 Collaborative Wiki Tagging Portal Prototype

We have developed a testing prototype of the described collaborative wiki tagging system, *Collaborative Wiki Tagging Portal Prototype (CWTP)*[6]. The CWTP is aiming to support personal knowledge management, inter- and intra-community collaboration, workflow and process management, interaction, knowledge sharing and dissemination, and heterogeneous information integration (Fig. 3.). The prototype supports collaborative tagging, but it is a wiki site at the same time. This means that every page can be edited, including pages that contain tagging data as well as meta-pages. Edit rights are not publicly available but are instead regulated by an authentication and authorization mechanism at the page level.

Interaction over structure is represented by means of an automatic set of page neighborhood links (links pointing to the page and links pointing from the page) and useful drop-down menus, as well as page-specific menus. Page neighborhood links are useful for content but even more for semantic navigation. Interaction over structure is augmented with a primitive version of a *TagCloud* that is useful for navigation over presentation content. Tagging presentation wiki pages and other internal resources allows systematic (re)arrangement of the internal structure. Fig. 4. shows a tagging window for internal resource tagging, while Fig. 5. shows external resource tagging.

The differences between the two tagging windows are in statement and link. Since statements can be made for internal resources only, we provide that functionality within the tagging frame when an internal resource is being tagged. For usability purposes, this option does not exist within the tagging frame when an external resource is being tagged. Also a wiki page name is captured as the internal resource identifier instead of it's full URL as is the case for external resources. In the Tag text edit field, a user enters a set of

---

[5] The word organize is quoted due to an open nature of the tags organization problem. Namely, it depends on answers to questions such as what is the information capacity of individual tags, the information capacity of whole TagClouds, what is the relationship between folksonomy and ontology, what are dynamics of the TagCloud, etc. We use this example as an illustration only, and do not suggest any solution to the tags organization problem.

[6] http://infosys-work.elfak.ni.ac.yu/InfosysWiki-v2-1/Wiki.jsp?page=TagClouds. A proof of the concept beta site may also be found at http://www.tagleen.com.

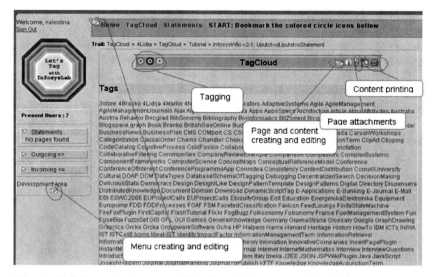

Fig. 3 Collaborative Tagging Wiki Portal Prototype (CTWP)

Welcome, Guest
Sign In

| Tag | Selection | Comment |

Add Tag to TagCloud

⊟ **Statement:**

Relation: is_same_as   Category: TagX

Title: Resource Title

Tag: tag1 tag2 tag3

Add   Close

Fig. 4 Internal resource tagging

Welcome, Guest
Sign In

| Tag | Selection | Comment |

Add http://www.citeulike.org/ to Tag

Title: CiteULike: A free online service to organise you

Tag: tag1 tag2 tag3

Add   Close

Fig. 5 External resource tagging

tags for the resource. Tags are separated by blank character, while multiple words expressions are possible by quoting them together.

Also a rudimentary tag suggestion mechanism is implemented that presents to the user a list of several existing tags based on the first letter(s) that the user has just typed in. Note that ranking and suggesting tags in collaborative tagging systems is an important research topic (see for example [20]). The suggestion algorithm guides the dynamics of the *TagCloud* towards the particular tag distribution representing a fixed point for the tagging process. Convergence of the tagging process is very important for the coherent categorization schemes that can emerge from unsupervised tagging by users [7], [6].

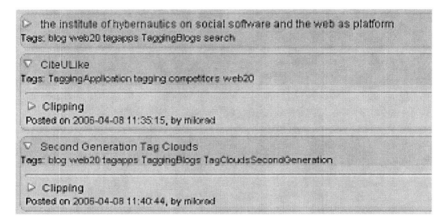

Fig. 6 Screenshot of tagging data

5 Discussion and Related Work

Probably the best known approach to organizing information within a wiki is Wikipedia[7] categories. Collaborative wiki tagging presented in this paper associates tags to wiki pages in a similar way except that the cognitive investment made by the user and his/her level of attention is much higher in the case of categorization then in the case of free tagging. As a consequence categorization requires heavy involvement of domain experts with all drawbacks that this approach brings along, such as high costs, social protocols needed to identify and prove specific expertise, etc. Instead, the collaborative tagging is a self-organizing emergent process that converges into consensus around categorization of the tagged resources.

Wikipedia categories, as well as several popular blogging and image tagging platforms that offer similar functionalities, are restricted to the classification of internal resources (wiki pages or blog posts) into internally developed

---

[7] http://www.wikipedia.org

categories. General-purpose social tagging systems (such as del.icio.us[8]) allow tagging of any web resource including web resources identifying system's internal tags. However, it is unclear how such self-tagging possibilities are further exploited. With the proposed Collaborative Wiki Tagging we are able to tag any web resource and not just an internal wiki page. We are also able to use tagging to semantically enrich the organization of local resources in an emerging manner.

Underlying models that would enable the effective integration of the powerful RDF-based Semantic Web with massively adopted user-friendly Social Web is a hot research topic at the moment [8]. A particularly interesting aspect is the relation between static and dynamic characteristics of folksonomies and ontologies: How can an ontology be automatically derived from a folksonomy? How can an ontology help driving tagging dynamics? (see for example [9]).

With respect to low-level syntax relevant research includes microformats: a simple convention for embedding semantics in HTML to enable decentralized development[9] for web resources tagging called tag-rel[10]. Both the microformats and the proposed wiki tagging syntax are based on existing, simple, and widely used mechanisms: The microformats are based on the HTML syntax while the proposed approach is based on the wiki text formatting syntax. However, the wiki tagging page can be easily edited online by an end user using a simple text area of a web browser while this is not the case with HTML based microformats.

RDFa lets XHTML authors express structured data within a document using existing XHTML attributes and a handful of new ones [2]. RDFa gets its expressive power from RDF. Like microformats, RDFa is similar to our approach in the sense that it is based on the wide adopted XHTML syntax. However, RDFa is much more along our abstract model of explicit semantics. Adoption of the RDFa into Collaborative Wiki Tagging system is one of the goals of our future research, particularly for the integration of XHTML based WYSIWYG editors.

A second relevant stream of existing research includes Semantic Web related work, particularly Semantic Wikis [15], [4]. Among the whole family of different Semantic Wikis, IkeWiki may be the closest to our approach [16]. The proposed solution is low-level and in this way a complementary approach to Semantic Wikis. Also, semantic collaborative tagging system, as proposed in [12], is based on semantic assertions that are very close to our Statements.

6 Conclusions

In this paper, we propose a concept of collaborative tagging for the organization of knowledge stored within a wiki system. The knowledge is about any

---

[8] http://del.icio.us
[9] http://microformats.org/wiki/Main_Page
[10] http://microformats.org/wiki/reltag

Web resource as well as system internal wiki pages. We first introduced the semantics of collaborative tagging that is implemented in a wiki fashion. Then, we discussed syntax and wiki text formatting rules that we use to store tagging data in a wiki system. The proposed concepts and syntax is used for the implementation of our *Collaborative Wiki Tagging Portal Prototype*. We use the prototype extensively for personal knowledge management, group knowledge interaction and project management. We have recently started testing the prototype in our undergraduate teaching practice and we experienced very promising results: improved student-teacher communication, students being more actively involved into learning processes and the management of the course being more interactive.

Acknowledgments

We would like to thank our reviewers for helpful comments. The research presented in this paper is partially funded by the Serbian Ministry of Science and Environmental Protection.

References

1. Dynamic Collaboration: Enabling Virtual Teamwork in Real Time (2003), http://www.accenture.com/xdoc/en/services/technology/research/tech_collaboration.pdf
2. Abida, B., Birbeck, M.: RDFa Primer. W3C Working Draft (October 26, 2007), http://www.w3.org/TR/xhtml-rdfa-primer/
3. Ames, M., Naaman, M.: Why We Tag: Motivations for Annotation in Mobile and Online Media. In: Proceedings of the SIGCHI conference on Human Factors in computing systems (CHI 2007), San Jose, CA, USA (2007)
4. Buffa, M., Gandon, F., Sanders, P., Faron, C., Ereteo, G.: SweetWiki: a semantic wiki. Journal of Web Semantics, Manuscript Draft (2007), http://www.websemanticsjournal.org/papers/2007119/SweetWikiBuffaV6I1.pdf
5. Berners-Lee, T., Hollenbach, J., Lu, K., Presbrey, J., Pru d'ommeaux, E., Schraefel, m.c.: Tabulator Redux: Writing Into the Semantic Web (2007) (unpublished), http://eprints.ecs.soton.ac.uk/14773/
6. Golder, S., Huberman, B.A.: Usage Patterns of Collaborative Tagging Systems. Journal of Information Science 32(2), 198–208 (2006)
7. Halpin, H., Robu, V., Shepherd, H.: The Complex Dynamics of Collaborative Tagging. In: Proceedings of the 16th International World Wide Web Conference (WWW 2007), Banff, Canada (2007)
8. Immaneni, T., Thirunarayan, K.: A Unified Approach to Retrieving Web Documents and Semantic Web Data. In: Franconi, E., Kifer, M., May, W. (eds.) ESWC 2007. LNCS, vol. 4519, pp. 579–593. Springer, Heidelberg (2007), http://www.eswc2007.org/pdf/eswc07-immaneni.pdf
9. Kim, H.L., Yang, S.K., Song, S.J., Breslin, J.G., Kim, H.G.: Tag Mediated Society with SCOT Ontology. In: The 5th Semantic Web Challenge, The 6th International Semantic Web Conference (ISWC 2007), Busan, Korea (2007)

10. Knerr, T.: Tagging Ontology - Towards a Common Ontology for Folksonomies (2006), `http://code.google.com/p/tagont/`
11. Macgregor, G., McCulloch, E.: Collaborative Tagging as a Knowledge Organisation and Resource Discovery Tool. Library Review 55(5) (2006)
12. Marchetti, A., et al.: SemKey: A Semantic Collaborative Tagging System. In: Proc. WWW 2007 Workshop on Tagging and Metadata for Social Information Organization, Banff, Canada (2007)
13. Marlow, C., Naaman, M., Boyd, D., Davis, M.: HT 2006, tagging paper, taxonomy, Flickr, academic article, to read. In: Proceedings of the Seventeenth Conference on Hypertext and Hypermedia, HYPERTEXT 2006, Odense, Denmark, August 22-25, 2006, pp. 31–40. ACM, New York (2006), `http://doi.acm.org/10.1145/1149941.1149949`
14. Newman, R.: Tag Ontology design (2007), `http://www.holygoat.co.uk/projects/tags/` (viewed December 2007)
15. Oren, E., Völkel, M., Breslin, J.G., Decker, S.: Semantic wikis for personal knowledge management. In: Bressan, S., Küng, J., Wagner, R. (eds.) DEXA 2006. LNCS, vol. 4080, pp. 509–518. Springer, Heidelberg (2006)
16. Schaffert, S.: IkeWiki: A Semantic Wiki for Collaborative Knowledge Management. In: 1st International Workshop on Semantic Technologies in Collaborative Applications STICA 2006, Manchester, UK (2006)
17. Sinha, R.: A cognitive analysis of tagging. Weblog entry, Rashmi Sinha's weblog (2005), `http://www.rashmisinha.com/archives/05_09/tagging-cognitive.html`
18. Story, H.: Search, Tagging and Wikis. Weblog entry, The Sun BabelFish Blog, (Feburary 6, 2007), `http://blogs.sun.com/bblfish/entry/search_tagging_and_wikis`
19. Tosic, M., Milicevic, V.: Semantics of The Collaborative Tagging Systems. In: 3rd European Semantic Web Conference, 2nd Workshop on Scripting for the Semantic Web, Budva, Serbia & Montenegro (2006)
20. Vojnovic, M., Cruise, J., Gunawardena, D., Merbach, P.: Ranking and Suggesting Tags in Collaborative Tagging Applications. Microsoft Research, Technical Report MSR-TR-2007-06 (2007)

O'CoP, an Ontology Dedicated to Communities of Practice

Amira Tifous, Adil El Ghali, Alain Giboin, and Rose Dieng-Kuntz

Abstract. The Palette project dedicated to learning in Communities of Practice (CoPs) aims to offer several services for CoPs, in particular Knowledge Management (KM) services based on an ontology dedicated to CoPs, the so-called O'CoP. Built from information sources about the Palette CoPs, O'CoP aims both at modelling the members of the CoP and at annotating the CoP's knowledge resources. The paper describes the structure of O'CoP, its main concepts and relations, and it reports some lessons learnt from the cooperative building of this ontology.

1 Introduction

CoPs are "groups of people who share a concern, a set of problems, or a passion about a topic, and who deepen their knowledge and expertise in this area by interacting on an ongoing basis" [1].

The objectives of the Palette IST project (`http://palette.ercim.org/`) are to develop services for CoPs: information, knowledge management (KM), and mediation services. Eleven pilot CoPs are involved in the participatory design of the Palette services. These CoPs, located in various European countries (Belgium, France, Greece, Switzerland, UK), belong to three different domains:

- Teaching: e.g. @pretic, a CoP of Belgian teachers playing the role of resources-persons to support the use of Information and Communication Technologies (ICT) in schools;
- Management: e.g. ADIRA, a French professional association gathering executives from medium to large IT companies in Rhône-Alpes region);

INRIA, Centre de Sophia Antipolis Méditerranée,
Edelweiss Project-Team
2004 route des Lucioles, BP 93,
06902 Sophia Antipolis Cedex, France
e-mail: {Amira.Tifous,Adil.El_Ghali,Alain.Giboin}@sophia.inria.fr,
         Rose.Dieng@sophia.inria.fr

S. Schaffert et al. (Eds.): Networked Knowledge - Networked Media, SCI 221, pp. 155–169.
springerlink.com   © Springer-Verlag Berlin Heidelberg 2009

- Engineering: e.g. UX-11, a CoP composed of 150 IT engineer-students practicing GNU/Linux.

The CoPs' size varies from less than ten members to more than a hundred.

KM services aim at supporting CoPs' management of their knowledge resources, so as to improve: (i) the access, sharing, and reuse of existing knowledge, and (ii) the creation of new knowledge. A knowledge resource can be either a document materialising the knowledge made explicit by CoPs' members when cooperating, or a person holding tacit knowledge. The KM services will be based on Semantic Web technologies: they will rely on an ontology (describing concepts useful about a CoP, its actors and their competences, its resources, its activities, etc.) and on the semantic annotation of the CoPs' knowledge resources w.r.t. this ontology. In [2], we proposed generic models useful for understanding a group activity, collaboration, competencies, learners' profiles, and lessons-learnt. A CoP being a specific kind of a group, the CoP-dedicated ontology, so-called O'CoP, is based on these generic models. It consists of CoP-relevant concepts and relations with which the CoPs' resources can be annotated. These CoP-relevant concepts and relations are specialisations of the high-level ontology constituted by the generic concepts used to represent the generic models. The CoP-oriented KM services will rely on the O'CoP ontology.

So what kind of CoPs' KM problems may the ontological approach help to solve, and how? Let's give two real examples from Palette CoPs. The first example concerns a knowledge capitalizing problem reported by the @pretic CoP. Members of this CoP met great difficulty in capitalizing the contents of the practice-related messages they exchange via the mailing-list of the community. These messages are under-exploited because they are poorly indexed, or because they are not expressed in a synthetic form: the yet useful information they contain is hardly retrieved when needed. The ontological approach we propose can contribute to solve this problem by enabling a semi-automatic indexation of the messages performed by a semantic annotation service using linguistic analysis techniques. An ontology-based tool annotates the messages with the concepts and relations of a reference ontology. This semantic annotation enables then a semantic navigation through the base of messages, and a semantic search providing the @pretic CoP's members with more relevant answers more easily [3]. The second example concerns a knowledge structuring problem encountered by the Did@ctic CoP. Members of this CoP need to better structure the notes they take during the meetings where they discuss their educational practices. These notes are very informal, and their underlying structure often varies from one CoP member to another. This lack of formality and of homogeneity impede the exchange of practices at a distance in time. A way of structuring the notes is to explicitly impose a predefined structure both to the note support (leading, e.g., to a template), and to the note-taking process. Another structuring way is to implicitly superimpose a more formal structure to the non-formal notes. The ontological approach allows us to implement these two ways. In the first case, an ontology may be used to elaborate a template. In the second case, the ontology may be used at the retrieval step, or between the note-taking step and the retrieval step, to annotate the non-formal notes using a commonly agreed annotation structure. These two examples illustrate the kind of KM problems encountered by Palette CoPs and possibly solved through our ontology-based approach.

After summarising our ontology development method and the ontology structure (section 2), we will describe its main concepts (section 3), the lessons learnt from its building (section 4), before concluding (section 5).

2 Ontology Development Method and O'CoP Ontology Structure

Our method for developing the O'CoP ontology includes the following steps:

1. *Proposition of generic models* enabling to define multiple semantic axes corresponding to the key notions of the O'CoP ontology [2]. Each semantic axis will be undertaken by an ontologist through a sub-ontology.
2. *Information sources collection:* selecting three main sources to be used either as corpora where picking out candidate terms, or as grids for extracting candidate terms: (i) Rough-data documents (audio records/files of CoPs' interviews, transcriptions and minutes of these interviews; the interviews were performed by Palette members that played the role of mediators between some specific CoP and the knowledge engineers); (ii) Analysed-data documents (e.g., syntheses of interviews, vignettes and scenarios structuring the CoPs' activities); (iii) Methodological and theoretical documents (e.g., our generic models and existing ontologies or thesaurus). For cooperative building of the ontology, the different ontologists analyzed the same information sources for performing steps 3) to 6), but each one focusing on his/her generic model so as to build the corresponding sub-ontology.
3. *Contextualised lexicon construction:* selecting from the corpora and w.r.t. the grids (i) the terms relevant for describing the CoPs and (ii) their respective contexts (i.e. the text surrounding the terms) to help understand the terms.
4. *Vocabulary identification*: refining the contextualised lexicon once validated by the CoPs' mediators and producing, for each term, a definition and some examples of use.
5. *Hierarchy building:* (i) identifying the terminological concepts and relations, and (ii) structuring them, and eventually adding new higher-level concepts.
6. *Formalisation of the sub-ontologies* in RDF/S, the formal language agreed in Palette.
7. *Integration of the sub-ontologies* by solving the conflicts among them and by integrating them into a single, coherent ontology.

Fig. 1 summarises the steps of the development process. A tool, called ECCO, supports these iterative steps and provides the user with mechanisms enabling to keep the traceability of the sources of the candidate terms.

The resulting O'CoP ontology is structured into three main layers (see Fig. 2):

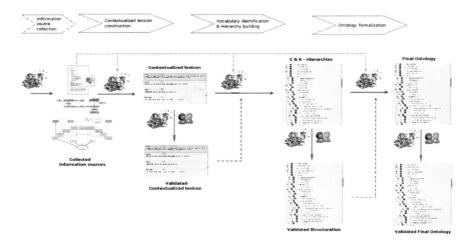

Fig. 1 Ontology development process

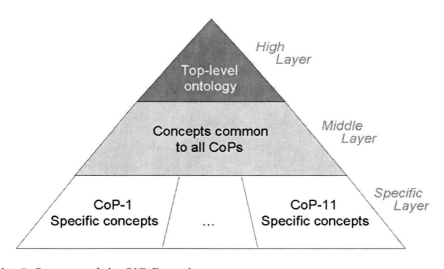

Fig. 2 Structure of the O'CoP ontology

- A *high layer* (or top-level ontology) including the concepts and relations needed to represent the generic models presented in [2]; they served as a grid for analysing the corpora and building the other layers of the ontology.
- A *middle layer* including the concepts common to all CoPs. These concepts correspond to terms confirmed by the mediators as common to all CoPs. They are specialisations of the high-level ontology concepts. Note that some concepts stemming from literature on CoPs, such as the concept of *Animator* detailed in [4], could be included in this common layer, provided that they are attested by at least the CoP corpora.
- A *specific layer* including the concepts specific to each CoP: these concepts correspond to terms confirmed by the mediators as specific to a given CoP or to very few CoPs.

3 Description of the Main Concepts of the Ontology

3.1 Community

The main concepts related to the community in the O'CoP ontology are:

- *Community*: it can be a community of interest, a community of learners, a goal-oriented community or a community of practice. In the interviews, interviewees acknowledged that the group of persons they belong to (be it so-called a community of teachers, a network of teachers, a resource-persons community, an association of companies, etc.) is a (kind of) CoP.
- *Domain* and *Field*: as defined in [5], the *Domain* is the area of knowledge that brings the community together, gives it its identity and defines the key issues that the CoP's members need to address. It is the "focus" of the CoP and evolves over its life span in response to new, emerging challenges and issues [6]. As for the *Field*, it is the "context" of the CoP; it can be referred to as the "discipline" or the "branch of knowledge" of the CoP's members (e.g. the *Domain* of ePrep[1] is the Educative use of ICT and its *Field* can be Mathematics, Physics, etc.).
- *Objective*: related to the CoP as a whole, or to a part of it (a group, a project, a team, etc. depending on the CoP's organisation and functioning modes), an objective can be Permanent or Temporary.
- *CoP's characteristics*: the CoP's identity is characterised by (i) the *Membership*: is the CoP open to any person interested in it or are there some conditions (e.g., competency, cooptation, etc.) for entering the CoP? (ii) the *Cultural Diversity* (from homogeneity to heterogeneity) of the CoP's members w.r.t. the nationality, profile, organisational culture [7].

---

[1] A CoP gathering teachers of French "Classes préparatoires aux Grandes Ecoles" interested in ICT (cf. http://www.eprep.org).

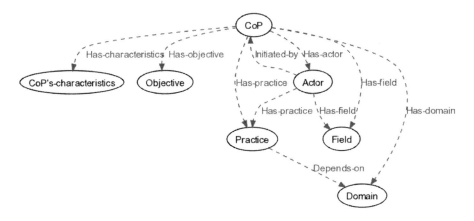

Fig. 3 Relations concerning a community

- *Organisational structure*: the organisation of CoPs varies from formal and structured (e.g. the CoP ADIRA[2] which is based on a "board of governors") to informal (e.g., the DL[3] which is based on "informal subgroups").

  Fig. 3 shows some relations concerning communities.

3.2 Actors

We define an *Actor* as "an Individual or a Legal entity intervening in the CoP". The Actors of a CoP are not only the CoP's members, but also the entities interacting with the CoP (also called the CoP environment). A *Legal entity* can be a *Professional organisation* or an *Institution* (Companies and Educational institutions). Actors can be involved in the CoP as *Members*, *Contributors* (Individuals participating in particular activities or during specific periods of the CoP's life) or *Partners* (Legal entities supporting the CoP).

Moreover, the Actors of a CoP can be defined according to their:

- *Role in the CoP*: it represents the Actor's position in the CoP, which can divided in two types:

  - *Governance role*: in order to interact, learn and share knowledge effectively, the CoP's actors (e.g. the members) need a support, which can be provided by:
    - · *Facilitator*: s/he encourages the participation of the members, facilitates the interactions among them.

---

[2] *L'Association pour la Promotion et le Développement de l'Informatique dans la région Rhône-Alpes*: http://www.adira.org/

[3] Doctoral Group Lancaster : http://domino.lancs.ac.uk/

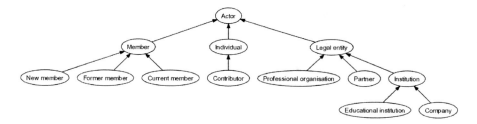

Fig. 4 Concepts describing Actors in a CoP

- · *Coordinator*: s/he organises and coordinates the activities and events of the CoP. We distinguish between Individual coordination (ensured by one main coordinator) and Collective coordination (in the case of a CoP organised per groups or teams, where the individual local coordinator belongs to a coordination group or team).
- · *Animator*: s/he guides and manages the community, ensures its development, relevance and effectiveness. An Animator thus plays both roles of Facilitator and Coordinator.
- – *Peripheral role*: represents knowledge providers and receivers. They are more or less involved or active in the CoP, their participation depends on the Actors who play these roles (personality, motivation, period, activity, etc.).

- Their *Individual profile*: identifies a CoP's Individual inside and outside the CoP. It comprises the concepts of *Competency* and *Occupation*.
- Their *Practice*: CoP's members are practitioners in an Institution, outside the CoP. They meet physically or virtually, through the CoP, which constitutes a channel for them to exchange experiences about their shared Practice (e.g. teaching practice).
- Their *Behaviour*: the *Attitude* of the member towards the CoP gives more information about his/her degree of engagement in the CoP.

3.3 Competency

A *Competency* is defined as a set of Resources to be provided or to be acquired by an *Actor* (who plays a particular *Role* in some *Environment* or *Situation*) so that the *Actor* can perform, or help some other *Actor* to perform some *Activity*. Fig. 5 gives a partial view of the *Competency-Resource* component and shows some relations concerning the *Competency* concepts.

Table 2 summarises some other relations concerning the *Competency* concepts.

Table 1 Actor-related relations

Relation	Domain	Range	Description
has-practice	Actor	Practice	An Actor of the CoP has a Practice outside the CoP.
has-field	Actor	CoP Field	A CoP, as well as an Actor has one or more Fields of knowledge.
interested-in	Actor	Domain Field Activity	An Actor can be interested in a Domain, a Field of knowledge, an Activity performed inside the CoP.
has-profile	Individual	Individual profile	An Individual has a profile, which defines him/her.
has-occupation	Individual	Occupation	An Individual has an occupation outside the CoP.
part-of-individual-profile	Occupation	Individual profile	The occupation an Individual has outside the CoP, is part of his/her profile.
employer-of	Actor	Individual	An Actor of the CoP can be the employer of another actor (an Individual) of the CoP (e.g. ADIRA).
contestant	Company	Company	A Company can be in competition with another one (both being Actors of the CoP - e.g. ADIRA)
colleague	Individual	Individual	Two Individuals of the CoP can be colleagues in their occupation outside the CoP.
has-attitude	Actor	Behaviour towards the CoP	An Actor of the CoP has a given behaviour, considering his/her motivation, satisfaction and involvement degree towards the CoP.
ordered-by	Activity	Actor	An Activity can be ordered by an Actor (a particular Role or an Institution, etc.).
assesses-activity	Actor	Activity	An Actor assesses an Activity performed in the CoP as being interesting, motivating, boring, etc.
possesses-competency	Actor	Competency	An Actor possesses a Competency linked to his/her personal characteristics and profile.

3.4 Resources

The Resources handled by a CoP are subdivided into:

- *Tools* defined according to the needs of community and their functionalities. A hierarchy describes the categorisation of these tools answering recurrent needs of a CoP including knowledge capturing (Knowledge portal), knowledge storage and sharing (Repository), collaboration (Workspace, Agenda, etc.).

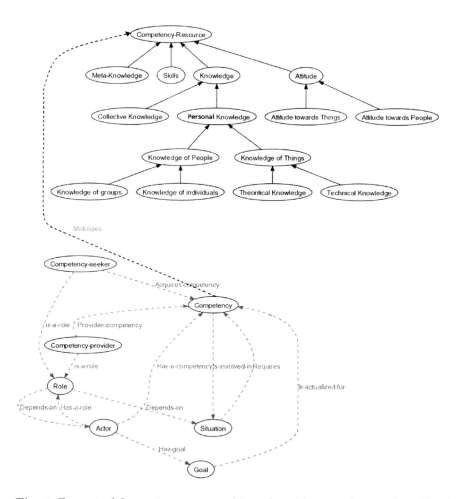

Fig. 5 Excerpt of Competency concept hierarchy and some relations describing it

- *Materialised resources* including documents or discussion. This last type of resources in the CoPs is associated to the interactions that hold within the CoP. These discussions can be synchronous (chat, audio and video conferences, etc.) or asynchronous (mail, forum, etc.). Almost all Palette CoPs are interested in easy access to these interaction traces and in archiving them.

We also characterise resources in a CoP w.r.t. the following dimensions:

- the *nature of resource*: we distinguish Documents, Tools and Interactions;
- the *access rights to a resource*;
- the *ownership of a resource*;
- the *temporal properties and versioning of a resource*.

Table 2 Some relations concerning the Competency concepts

Relation	Domain	Range	Description
is-related-to	Skill	Experience	A Skill acquired by an Actor of the CoP can be related to some Experience lived by the Actor.
is-acquired-by	Skill	Practice	An Actor acquires a Skill by Practice.
is-put-into	Knowledge	Practice	Some Knowledge acquired by an Actor can be put into practice by this Actor.
has-competency-level	Competency	Competency-level	A Competency (and consequently the Actor possessing it) has a Level (of Competency).
is-expressed-through	Practice	Practice-representation	An Actor can express a Practice through some concrete Representation of this Practice.
rises	Situation	Problem	A Problem is originated by the Situation in which an Actor is involved.
requires	Problem	Solution	A Problem occurring in some Situation requires a Solution for an Actor to achieve some Goal.
provides	Competency	Solution	A Competency is one of the resources that (can) provide a Solution to some Problem. The solution found depends on the Level of the Competency.

Fig. 6 shows some of the concepts needed to deal with the ownership of resources in a CoP.

A CoP uses and produces a number of documents of different types. Some of them are specific to CoP's life: for example, organisation policies that describe the rules organising the community life, or specific charter for the usage of the CoP's information system (e.g. in ADIRA). From a resource point of view, knowledge capitalisation takes different forms: several Palette CoPs produce (final or intermediate) reports, associated to CoP's activities. Another type of report is the logbook that can be individual or collective (e.g. in Did@ctic, where the collective logbook is called the "Meta-journal").

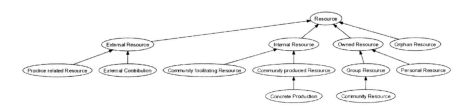

Fig. 6 Categorisation of resource ownership

Fig. 7 Excerpt of hierarchy of concepts describing documents

The CoPs' members can also produce documents related to their practices (Training reports in UX11) or scientific documents (Doctoral Lancaster). The collaboration in the production or use of documents can take the form of annotations that can be either textual or semantic depending on the tools used to produce them. Some documents are associated to a specific domain: e.g. Pedagogical documents in the education domain (Learn-Nett, Did@ctic, UX11), or Official documents useful in management domain (ADIRA).

Fig. 7 gives a global view on the hierarchy of concepts describing documents.

4   Lessons Learnt from the O'CoP Ontology Development

Concerning information sources collection, the relevance of the terms extracted from the corpora strongly depends on the relevance of the corpora. CoPs' mediators had focused their interviews on the organisation of CoPs, and had scarcely asked questions about CoPs' practices. As a consequence, the transcriptions of interviews contained very few terms related to practices. This leads to an ontology in which concepts related to practices are not very numerous.

During the terminological analysis, we found several terms common to some CoPs but used to evoke different concepts depending on the CoP: e.g. the term "platform" was used to designate a website, a workspace for the CoP, that may contain its documents and where the discussions of members are hosted, or yet a dedicated software e.g. e-learning platform. Some terms were also used ambiguously to designate concepts: e.g. CoPs use different terms to designate the persons in charge of particular tasks in the CoPs ("coordinator of the project", "local coordinator", "manager", etc.), whereas these tasks are not well described and identified. Finally, some CoPs use different terms to designate the same concepts, these synonyms must be associated to the same concept in their ontologies in order to avoid redundancy. For example, the terms Journal and Logbook are used to designate the record of activities or practices of a CoP's member. More generally, the synonym terms (either in the same CoP or in several CoPs) were recognised by the validators, during the phase of vocabulary identification and term validation. In the implementation of the ontology, the synonym terms corresponding to a given concept were indicated through the RDF/S label of this concept.

The different CoPs adopted different terminologies, sometimes quite specific to the CoP and rather different from the terminology usually found in literature on CoPs. Therefore, we did not include in the common layer of the ontology the concepts offered by literature (e.g. the taxonomy of facilitation tasks for CoPs proposed by [8]) if they were not attested by the Palette CoPs' information sources.

The O'CoP ontology building was a distributed, cooperative process between: (a) 6 ontologists focusing on different parts of the ontology since each one was guided by one generic model, (b) 11 CoPs' mediators validating from the CoPs' viewpoints. This led to the need to integrate different viewpoints. The different ontologists had various ways of modelling knowledge: e.g. the concept of *Activity* was needed for modelling *Competency* and *Resource*. Concepts related to *Activity* were thus modelled with various detail grains and various perspectives, requiring more integration work. Moreover, the integration between different concepts developed by different ontologists was often performed through the introduction of relations linking such concepts (e.g. relation between an *Actor* and an *Activity*, etc.). Notice that such kinds of relations were emphasised in the generic models that guided us. But they needed to be refined for more specialised concepts.

Our approach was both bottom-up (relying on a deep analysis of the information sources on the CoPs) and top-down (guided by our generic models).

5 Conclusions

This paper presented an original ontology composed of more than 800 concepts and 80 relations, dedicated to CoPs, and more precisely aiming at enabling the annotation of CoPs' members and the CoPs' resources. The link between CoPs and ontologies was studied in some recent related work. In [9], the authors present a method based on analysis of the relationships between instances of a given ontology in order to identify potential CoPs in an organisation. In [10], the authors develop an ontology aiming at enabling services among a civil servant CoP; [11] studies the design of situated ontologies for knowledge sharing in a CoP. [12] presents a semantic web system for open source software communities and relies on specific ontologies (Code, Bugs, Interactions, Community). In comparison to this related work, the O'CoP ontology is original through:

- the method used to build it cooperatively from the analysis of several real CoPs,
- its objective of enabling to annotate CoPs' resources in addition to modelling the notion of a CoP,
- its 3-layered structure, with a generic layer, a middle layer gathering concepts common to all CoPs and a low layer specific to a given CoP.

O'CoP was for example used by the @pretic CoP, in order to annotate the mails exchanged by the members of the CoPs about their problems in the use of ICT in schools. Our work can also be partially compared to the typology of virtual CoPs (i.e. CoPs interacting through ICT) proposed by [13] or to the typology of CoPs based on their knowledge characteristics [14] but these typologies are not materialised through ontologies.

More generally, the O'CoP ontology can be specialised for a new CoP. The high and middle layers are generic and can thus be reused for any CoP. If the new CoP is similar to one of the Palette CoPs, the low layer corresponding to this CoP can be reused. But if no Palette CoP is relevant, concepts more specific to the new CoP can be added in the low layer, possibly by relying on our method described in section 2.

As future work, after achieving the current validation of the integrated O'CoP ontology by the CoPs' mediators, we will make the ontology available to all the Palette CoPs and develop several KM services based on it: knowledge creation, annotation, retrieval, presentation, evaluation, and evolution services.

Acknowledgments

We thank very much C. Evangelou and G. Vidou that worked on subontologies not reported in this paper, the CoPs' mediators for their intensive and fruitful work of validation, P. Durville, the ECCO tool developer, F. Gandon for his methodological support, and the European Commission for funding the Palette project.

References

1. Wenger, E., McDermott, R., Snyder, W.M.: Cultivating Communities of Practice. Harvard Business School Press (2002)
2. Vidou, G., Dieng-Kuntz, R., El Ghali, A., Evangelou, C., Giboin, A., Tifous, A., Jacquemart: Towards an ontology for knowledge management in communities of practice. In: Reimer, U., Karagiannis, D. (eds.) PAKM 2006. LNCS, vol. 4333, pp. 303–314. Springer, Heidelberg (2006)
3. Makni, B., Khelif, K., Dieng-Kuntz, R., Cherfi, H.: Utilisation du web sémantique pour la gestion d'une liste de diffusion d'une CoP. In: Proc. of 8èmes Journées Francophones Extraction et Gestion des Connaissances. IN-RIA Sophia Antipolis Méditerranée, pp. 31–36 (2008)
4. Wenger, E., White, N., Smith, J.D., Rowe, K.: Guide to the implementation and leadership of intentional communities of practice. Work, learning and networked. In: Technology for Communities, CEFRIO (2005), http://www.cefrio.qc.ca/english/pdf/Guide_Final_ANGLAIS.pdf
5. Wenger, E.: Knowledge management as a doughnut: Shaping your knowledge strategy through communities of practice. Ivey Business Journal 68(3) (2004)
6. Henri, F.: CoPs: Social structures for the development of knowledge. In: PALETTE Kick-off Meeting, Lausanne, March 13-15 (2006), http://www.licef.teluq.uqam.ca
7. Langelier, L., Wenger, E. (eds.): Work, Learning and Networked. CEFRIO, Quebec (2005)
8. Tarmizi, H., de Vreede, G.-J.: A facilitation task taxonomy for communities of practice. In: Proc. of 11th Americas Conf. on Information Systems. Omaha (2005)

9. O'Hara, K., Alani, H., Shadbolt, N.: Identifying communities of practice: Analysing ontologies as networks to support community recognition. In: Proc. of IFIP 2002 (2002)

10. Bettahar, F., Moulin, C., Barthès, J.P.: Ontologies support for knowledge management in e-government environment. In: Proc. of ECAI 2006 Workshop on Knowledge Management and Organizational Memories, Riva del Garda, Italy (2006)

11. Floyd, C., Ulena, S.: On designing situated ontologies for knowledge sharing in communities of practice. In: Proc. of the 1st Workshop on Philosophical Foundations of Information Systems Engineering (PHISE 2005) in conjunction with the 17th Conference on Advanced Information System Engineering (CAiSE 2005), Porto, Portugal (2005)

12. Ankolekar, A., Sycara, K., Herbsleb, J.D., Kraut, R.E., Welty, C.A.: Supporting online problem-solving communities with the semantic web. In: Proc. of the 15th International World Wide Web Conference (WWW 2006), Edinburgh, Scotland, pp. 575–584 (2006)

13. Dubé, L., Bourhis, A., Jacob, R.: Towards a typology of virtual communities of practice. Interdisciplinary Journal of Information, Knowledge and Management 1, 69–93 (2006)

14. Klein, J.H., Connell, N.A.D., Meyer, E.: Knowledge characteristics of communities of practice. Knowledge Management Research & Practice 3(2), 106–114 (2005)

Incremental Approach to Error Explanations in Ontologies

Petr Křemen and Zdeněk Kouba

Abstract. Explanations of modeling errors in ontologies are of crucial importance both when creating and maintaining the ontology. This work presents two novel incremental methods for error explanations in semantic web ontologies and shows their advantages w.r.t. the state of the art black-box techniques. Both promising techniques together with our implementation of a tableau reasoner for an important OWL-DL subset *SHIN* are used in our semantic annotation tool prototype to explain modeling errors.

1 Introduction

The problem of error explanations turned out to be of high importance in ontology editors and semantic annotation authoring tools. Users of such tools need to be informed both about inconsistencies in the modeled ontology and about reasons for these inconsistencies to occur. Rationale for this work was formulated during the implementation and evaluation of our narrative annotation tool [10] developed within the CiPHER project [4].

This work presents two novel incremental algorithms for error explanations in ontologies. These techniques can be regarded as a compromise between glass-box error explanation methods, that are fully integrated into the reasoning algorithms and thus strongly dependent on the expressivity of the chosen semantic web language and hardly reusable for other ones, and black-box techniques, that are fully reasoner-independent, but therefore quite inefficient, especially in combination with – already EXPTIME or worse – reasoning algorithms for expressive description logics. The proposed incremental algorithms are universal enough to be reused with wide variety of reasoners, which seems to be advantageous especially in the

Petr Křemen and Zdeněk Kouba
Czech Technical University in Prague, Faculty of Electrical Engineering,
Dept. of Cybernetics, Technická 2, 16627 Prague, Czech Republic
e-mail: {kremen,kouba}@labe.felk.cvut.cz

S. Schaffert et al. (Eds.): Networked Knowledge - Networked Media, SCI 221, pp. 171–185.
springerlink.com   © Springer-Verlag Berlin Heidelberg 2009

dynamic field of semantic web languages, yet having quite tight interaction with the reasoner using the reasoner state. Proposed techniques were tested with our implementation of *SHIN* description logic tableau algorithm [1] with an incremental interface.

Section 2 surveys current black box and glass box techniques for error explanations in ontologies and shows advantages and disadvantages of these methods. Our incremental approach for error explanations is introduced in section 3 and evaluated in section 4. Section 5 briefly overviews our annotation tool prototype and this chapter is concluded by section 6.

2   Error Explanation Techniques – State of the Art

The mainstream of error explanations for description logic knowledge bases tries to pinpoint axioms in the knowledge base to localize errors. The notion of *minimal unsatisfiability preserving subterminology* (MUPS) has been introduced in [11], to describe minimal sets $\{S_i\}$ of axioms that cause a given concept to be unsatisfiable. Removing a single axiom from each of these sets turns the concept satisfiable. Similarly to defining MUPSes for concept satisfiability, in [8] the notion of *justification* for arbitrary axiom entailments has been presented. These justifications allow for explaining knowledge base inconsistencies, in our case annotation errors. Notions of MUPSes and justifications are dual, as for each concept an axiom can be found, for which the set of justifications corresponds to the set of MUPS of the concept. From now on, we use w.l.o.g. only the notion of MUPS and concept satisfiability.

At present, there are two general approaches for computing explanations of concept unsatisfiability: black-box (reasoner-independent) techniques and glass-box (reasoner-dependent) techniques. The former ones can be used directly with an existing reasoner, performing many satisfiability tests to obtain a set of MUPSes. The latter ones require a smaller number of satisfiability tests, but they heavily influence the reasoner internals, thus being hardly reusable with other reasoning algorithms.

In addition to these basic methods, [8] presents several practically interesting extensions. One of the techniques splits axioms into simpler ones (like $A \sqsubseteq B \sqcap C$ into $A \sqsubseteq B$ and $A \sqsubseteq C$) trading the error explanation granularity for the size of the knowledge base. Another technique tries to find concepts (called *root* concepts), unsatisfiability of which causes unsatisfiability of other concepts. Getting rid of unsatisfiability of these *root* concepts makes the other concepts satisfiable as well. The former technique can be used as a preprocessing and the latter as a postprocessing to all the methods described below.

2.1   Black-Box Techniques

There are plenty of black-box techniques that can be used for the purposes of error explanations. All of them have worst-case exponential time complexity in the number of axioms, as they search the power set of the axiom set – they differ in the search strategies and pruning efficiency. For each candidate set of axioms a satisfiability check is necessary to determine, whether this axiom set causes the unsatisfiability of a given concept or not.

In [5], several simple methods based on *conflict set tree* (CS-tree) notion are shown (see Fig.1). CS-trees allow for efficient and non-redundant searching in the power set of a given axiom set. Each node in a CS-tree is labeled with two sets, a set D of axioms that necessarily belong to a MUPS and a set P of axioms that might belong to a MUPS. Each node represents the set $D \cup P$ and it has $|D \cup P|$ children, each one lacking an axiom from $D \cup P$. The method (denoted as *allMUPSbb*) introduced in [5] effectively searches the CS-tree in the depth-first manner, pruning necessarily satisfiable nodes. The CS-tree structure allows for various pruning methods, like constraint set partitioning and eliminating always satisfiable constraints, see [5] for more details. However, detailed evaluation of the feasibility of these methods and their optimizations for the axiom pinpointing problem is still an open issue.

Example 1 (Basic CS-tree algorithm). Consider a knowledge base consisting of three axioms

$$1 : C \sqsubseteq B \sqcap \exists R.A,$$
$$2 : B \sqsubseteq \forall R.\neg A,$$
$$3 : C \sqsubseteq D.$$

The concept C is unsatisfiable due to the single MUPS $\{1,2\}$. The run of the basic CS-tree algorithm presented in [5] is shown in Fig.1. The algorithm starts in the root $[], [1,2,3]$ and tries to find all MUPSes in the depth-first manner. All children for $[], [1,2,3]$ are generated and the left-most node $[], [2,3]$ is used for exploration. As this node is satisfiable, all of its children are pruned, the algorithm backtracks to the node $[1], [3]$, which is also satisfiable. After pruning its child and backtracking to the $[1,2], []$ the searched MUPS is obtained. In this configuration the algorithm needs 4 satisfiability tests, while for the reversed axiom list 6 tests are needed.

An interesting black-box approach [8], [11] is based on a method for computing a single MUPS (denoted as *singleMUPSbb*) of a concept for a given axiom set. In the first phase, this algorithm starts with an empty set K and fills it with all available axioms one by one until it becomes unsatisfiable. In the second phase, each axiom is conditionally removed from K. If the new K turns satisfiable, the axiom is put back. An important observation is that *singleMUPSbb* algorithm is polynomial in the number of axioms. In the worst case we need $2n$ full consistency checks, where n is the number of axioms.

To obtain an algorithm for all MUPSes the general purpose Reiter's algorithm [13] for computing hitting sets of a given conflict set is used. This algorithm generates a tree (see Fig.2), where each node is labeled with the knowledge base and a MUPS computed for this knowledge base using *singleMUPSbb*. Starting with an arbitrary root MUPS, each of its children is generated by removing one of the MUPS axioms from the knowledge base and computing a single MUPS for the new knowledge base. The search terminates when all leaves of the tree are satisfiable. The

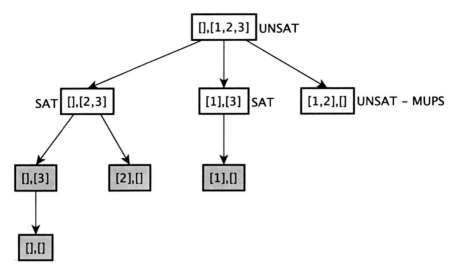

Fig. 1 An example of the basic CS-tree algorithm run searching MUPSes in a set of three axioms. Pruned nodes are darker.

advantage of this approach is that it provides also repair solutions that are repre-sented by axioms of minimal (w.r.t. set inclusion) paths starting in root. These paths correspond to hitting sets of the set of MUPSes. Due to the lack of space we refer to the works [5], [8], [11] and [12] for detailed algorithm descriptions (see Fig.2).

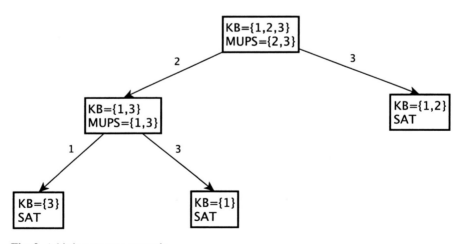

Fig. 2 A hitting set tree example

Example 2 (Single black box MUPS algorithm + Reiter's method). Let's have a knowledge base consisting of three axioms (A, B, C are concepts and R is a role.)

$$1 : B \sqsubseteq \forall R^-.\neg A,$$
$$2 : A \sqsubseteq \forall R.\neg B,$$
$$3 : C \sqsubseteq A \sqcap \exists R.B.$$

The concept C is unsatisfiable due to the MUPS set $\{\{1,3\},\{2,3\}\}$. The algorithm first uses *singleMUPSbb*, with $\{1,2,3\}$ as its input to find a MUPS, corresponding to the root of the hitting set tree in Fig.2. The *singleMUPSbb* has to perform 3 tests in the first phase $(\{1\} = SAT, \{1,2\} = SAT, \{1,2,3\} = UNSAT)$ to get an unsatisfiable set $\{1,2,3\}$ and 3 tests in the second phase $(\{2,3\} = UNSAT, \{2\}, \{3\})$.

Now, all children of the root are generated and the left-most child is being explored, calling the *singleMUPSbb* to obtain a MUPS in $\{1,2,3\} \setminus \{2\} = \{1,3\}$, which is $\{1,3\}$. As both sub-knowledge bases $\{1\},\{3\}$ of $\{1,3\}$ are satisfiable, the algorithm backtracks and tests the satisfiability of $\{1,2,3\} \setminus \{3\} = \{1,2\}$, which is satisfiable. Therefore two MUPSes $\{1,3\}$ and $\{2,3\}$ were found together with the hitting sets $\{3\}$ and $\{1,2\}$. If any of these sets is removed from the knowledge base, the concept C turns satisfiable.

As stated above all black box methods require, in general, time exponential to the number of axioms. Combining this with already (at least) exponential satisfiability checking for most description logic languages, we reach scalability problems for most real world ontologies.

2.2 Glass-Box Techniques

A fully glass-box technique for axiom pinpointing in the description logic ALC [1] is introduced in [11]. This method labels all concepts and roles in nodes of a completion tree with axioms they depend on. These labels are modified according to the applied expansion rules. Whenever no rule is applicable on any tableau, the union of labels of clashing concepts/roles builds up a superset of some MUPS. To obtain a MUPS, this set is minimized by backtracking the rule changes applied during expansions and constructing a boolean formula ϕ (so called *minimization function*) using another set of rules (see [11]). The searched MUPS is equivalent to the minimal set of axioms, conjunction of which implies ϕ. For a more formal and detailed description, see [11].

Example 3 (A glass-box technique for ALC). Let's have a knowledge base containing two axioms :

$$1 : A \sqsubseteq B \sqcup \exists R.A,$$
$$2 : A \sqsubseteq \neg A.$$

The only MUPS for the satisfiability of A is clearly $\{2\}$. The completion graphs evolve as depicted in fig.3. Inference rule applications are represented by double arrows, labeled with the type of the used rule. Right of each concept a set of axioms is shown, that is responsible for the concept appearing in the node label. The initial

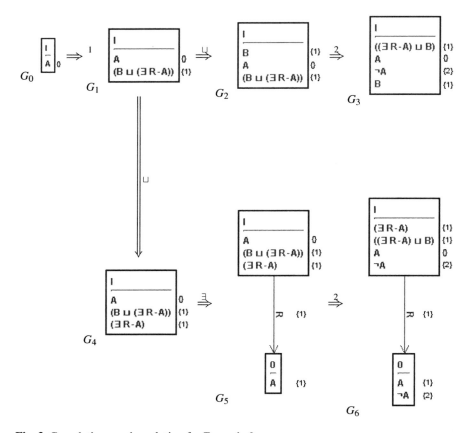

Fig. 3 Completion graph evolution for Example 3

completion graph G_0 contains a new individual I, asserted to belong to A. Rule applications result in two completion graphs G_3 and G_6, both containing a clash. The clash in G_3 is caused only by axiom 2, while the clash in G_6 is caused either by single axiom 2, or by axiom set $\{1, 2\}$. Using the minimization rules introduced in [11], the following minimization function is obtained

$$2 \wedge (2 \vee (1 \wedge 2)). \tag{1}$$

The minimal set of axioms that makes this formula valid is $\{2\}$, which corresponds to the searched MUPS.

To the best of our knowledge there is no adaptation of this approach to more expressive languages, like *SHIN*, or OWL-DL. The problem lies in possibly complex interactions between completion rules for different language constructs that have to be traced back in the minimization function. However, [8] presents a partially glass-box method for searching a single MUPS in OWL-DL. The algorithm is an extension of the first phase of the fully glass-box method described above. During completion graph expansion, concepts and roles are labeled with sets of axioms they depend on. Finding a clash in all branches, union of labels of clashing concepts/roles is a superset of a MUPS, usually much smaller than the initial axiom set. This method is then used as a preprocessing step in the first phase of the *singleMUPSbb* algorithm.

3 Incremental Approach to Error Explanations

A high number of very expensive tableau algorithm runs required for black-box methods, as well as a lack of glass-box methods, together with their poor reusability, has given rise to the idea of using incremental techniques for axiom pinpointing. These methods require the reasoner to be able both to provide its current state, and to apply a given axiom to a given state. There is, however, no other interaction with the reasoner. These features place incremental methods somewhere between black-box and glass-box approach.

3.1 *Incremental Tableau Reasoner*

Incremental tableau reasoning has already been studied in [6], where additions and deletions of ABox (concept and role) assertions are considered. While additions can be implemented in a straightforward way due to the nature of tableau algorithms, to handle deletions all completion rules applications had to be tracked. When deleting an ABox axiom a rollback of the parts of completion graph dependent on this axiom had to be performed and completion rules reapplied.

For the purpose of incremental algorithms presented in the next section we just require the reasoner to support incremental additions of all axiom types (TBox, RBox, ABox). Thus, due to the monotonicity of considered description logics (like *SHIN*) we do not need making any changes to the implementation of the tableau reasoning strategy. We only need the reasoner to provide us with its current *state*. The tableau reasoner state consists of two parts: a set of completion graphs, and an axiom set used for expanding this completion graph so far. More formally, we represent the incremental reasoner as

$$(ns, r) \leftarrow test(a, s) \qquad (2)$$

where s is the state before and ns the state after performing the incremental test, a is an axiom and r is a boolean result of the satisfiability test. Although we have tested its feasibility with a tableau algorithm for *SHIN* [7], our incremental approach can be used with a wide range of reasoning algorithms.

The following sections introduce two novel incremental methods for finding all MUPSes of a given concept. They use the above incremental reasoner interface as a black box. This makes them applicable to reasoning services (like DL tableau-algorithms) of monotonic logics without interaction with internals of these services.

3.2 Computing a Single MUPS

In this section, a novel incremental algorithm for computing a single MUPS is presented. This algorithm (see Algorithm 1) starts with an axiom list P and an empty state e of the reasoner. Axioms from P are tested one by one with the current reasoner state until the incremental test fails. The axiom $P(i)$ causing the unsatisfiability is put into the single MUPS core D, the rest of the axiom list is pruned and the direction of the search in the axiom list changes. The algorithm terminates, when all axioms are pruned.

Correctness. Correctness of the algorithm is ensured by the following invariant. Before each direction changes, D contains axioms that, together with some axioms in the non-pruned part of the axiom list, form a MUPS. Whenever an axiom i causes unsatisfiability, there must exist a MUPS that consists of all axioms in D, axiom i and some axioms in the previously searched part of the axiom list. This MUPS cannot be affected by pruning the axiom list tail that has not been explored in this iteration.

Algorithm 1. An Incremental Single MUPS Algorithm
___
```
 1: function SINGLEMUPSINC(P, e)                    ▷ P... initial axioms, e... initial state.
 2:     lower, i ← 0
 3:     upper ← length(P) − 1
 4:     D ← ∅
 5:     sD, last ← e
 6:     direction ← +1
 7:     while lower ≤ upper do
 8:         if i ≥ length(P) then
 9:             return ∅
10:         end if
11:         (incState, result) ← test(P(i), last)
12:         if result then
13:             last ← incState
14:             i = i + direction
15:         else
16:             D ← D ∪ {P(i)}
17:             (sD, result) ← test(P(i), sD)
18:             last ← sD
19:             if direction = 1 then
20:                 upper ← i − 1
21:             else
22:                 lower ← i + 1
23:             end if
24:             direction ← −direction                    ▷ +1... right, −1... left
25:         end if
26:     end while
27:     return D
28: end function
```

Example 4. Let's have an ontology containing six axioms 1 ... 6, where the unsatisfiability of some concept is caused by MUPSes $\{\{1,2,4\},\{2,4,5\},\{3,5\},6\}$. The
singleMUPSInc algorithm works as follows :

direction	input list	mups core
	[1, 2, 3, 4, 5, 6]	$D = []$
$\longrightarrow$	[1, 2, 3, 4̶,̶5̶,̶6̶]	$D = [4]$
$\longleftarrow$	[1̶, 2, 3, 4̶,̶5̶,̶6̶]	$D = [4,1]$
$\longrightarrow$	[1̶, 2̶,̶3̶,̶4̶,̶5̶,̶6̶]	$D = [4,1,2]$

Each line corresponds to a direction change. Whenever an unsatisfiability is detected, the search direction is changed, "overlapping" axioms are pruned (emphasized by strikeout) and the last axiom that caused the unsatisfiability (in bold) is
put into the MUPS core D.

Let's have n axioms. All incremental consistency checks in a single run between
two direction changes correspond approximately to one full consistency check performed for all axioms in the run. Thus, the incremental method requires in the worst
case n full consistency tests ($n(n + 1)/2$ incremental consistency tests), comparing
to worst-case $2n$ full consistency tests for the *singleMUPSb* algorithm. In the example above, 9 incremental consistency tests (effectively 3 full consistency tests) are
needed comparing to 8 full consistency tests needed by the *singleMUPSbb*.

3.3 Computing All MUPSes

An incremental algorithm that can be used to search for all MUPSes (let's denote
this algorithm as *allMUPSInc1*) is presented in [5]. This algorithm assumes that a
state of the underlying reasoner depends on the order of axiom processing. However,
tableau algorithms [1] are adopted in almost all current semantic web reasoners (for
example Pellet). In case of tableau algorithms, two different permutations of an axiom set shall result in two equivalent states. Exploiting this fact, we modified the
original algorithm to decrease the number of redundant calls to the testing procedure, resulting in Algorithm 2 (allMUPSInc2).

The original algorithm allMUPSInc1 manages three axiom lists D, T and P.
At the beginning of each recursive call, D contains axioms that must belong to all
MUPSes searched in this recursive call, P represents possible axioms that might
belong to some of these MUPSes and T represents a list of already tested axioms.
The first while cycle adds axioms from P to T while T remains satisfiable. If an
axiom that causes unsatisfiability is detected, the execution is branched. The first
recursive call tries to remove this axiom and go on adding axioms from P to T,
while the second branch tries to add the axiom to the MUPS core D found so far. If
D turns unsatisfiable, a MUPS has been found. If A does not contain a subset of this
MUPS, it is inserted into A, and A is returned.

Our modification of the original algorithm avoids executing some redundant tests
– both in the while cycle and in testing whether D turns unsatisfiable. For this purpose, we store the position of the first unsatisfiability test in P in the parameter

Algorithm 2. Modified version of *allMUPSInc1*

1: **function** ALLMUPSINC2($D, sD, P, T, sT, A,$ **cached**)   ▷ sD (sT) is the state for D (T).
2:     $result \leftarrow true$
3:     $i \leftarrow -1$
4:     **while** $result \land \exists c \in P$ **do**
5:         $i \leftarrow i + 1$
6:         **if** $c \notin T$ **then**
7:             $T \leftarrow T \cup \{c\}$
8:             $lT \leftarrow sT$
9:             **if** $i = cached$ **then**
10:                $result \leftarrow false$
11:                **break**
12:            **else**
13:                $(result, sT) \leftarrow test(c, sT)$
14:            **end if**
15:        **end if**
16:    **end while**
17:    **if** $result$ **then**
18:        **return** A
19:    **end if**
20:    $A \leftarrow allMUPSInc2(D, sD, P \setminus \{c\}, T \setminus \{c\}, lT, A, -1)$
21:    $D \leftarrow D \cup \{c\}$
22:    **if** $i = 0 \land d = t$ **then**
23:        $result \leftarrow false$
24:    **else**
25:        $(result, sD) \leftarrow test(c, sD)$
26:    **end if**
27:    **if** $\neg result$ **then**
28:        **if** $\neg \exists a \in A$ such that $a \subset D$ **then**
29:            $A \leftarrow A \cup \{D\}$
30:        **end if**
31:        **return** A
32:    **end if**
33:    **return** $allMUPSInc2(D, sD, P \setminus \{c\}, D, sD, A, i - 1)$
34: **end function**

---

cached. Let's denote $T = \{t_1, \ldots, t_a\}$ and $P = \{p_1, \ldots, p_b\}$. Then *cached* is such an index to P, that $\{t_1, \ldots, t_a, p_1, \ldots, p_{cached}\}$ is unsatisfiable and each of its subsets is satisfiable.

Correctness. Correctness of the algorithm is ensured by the same invariant as for *allMUPSInc1* presented in [5] and the fact that, the variable *cached* uses the information of the last successful test performed on T only in the second recursive call, where $D = T$. In the first recursive call, the information cannot be used, as the sets D and T differ.

Example 5. To show how *allMUPSInc2* works, assume an axiom set $\{1,2,3\}$. MUPSes for unsatisfiability of a concept are $\{\{1,2\},\{1,3\}\}$. The algorithm runs as follows :

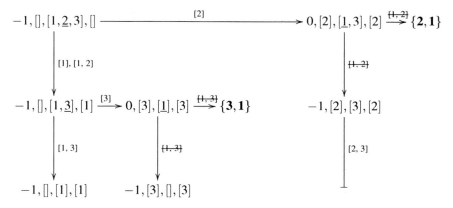

Each node in this graph represents a call to the procedure *allMUPSInc2*, with the signature $cached, D, P, T$. The search starts in the node $-1,[],[1,2,3],[]$ and is performed in the depth first manner preferring up-down direction (first recursive call) to the horizontal one (second recursive call). Axioms that cause unsatisfiability in the given recursive call are underlined. Edges are labeled with the tests that have been done before the unsatisfiability is found and struck axiom sets represent the tests that are not performed, contrary to *allMUPSInc1*. In this example *allMUPSInc2* requires 6 tests contrary to 10 tests executed by *allMUPSInc1*.

4   Experiments

First, the discussed methods have been compared with respect to the overall performance. Two ontologies have been used for the tests: the miniTambis ontology (30 unsatisfiable concepts out of 182) and the miniEconomy ontology[1](51 unsatisfiable out of 338). As shown in Tab.1, the performance of incremental methods is significantly better than the fully black box approach. Furthermore, combination of Reiter's algorithm and *singleMUPSInc1* is typically 1-2 times worse than the fully incremental approaches. However, the main advantage of the *singleMUPSInc1* in comparison to the fully incremental approaches is that it allows direct computation of hitting sets of the set of MUPSes (i.e. generating repair diagnoses), which makes them more practical.

It can be seen that our modification *allMUPSesInc2* of *allMUPSesInc1* provides just a slight increase in the performance. To evaluate the difference in more detail, see Fig.4. This figure shows the significance of caching for different MUPS configurations. The highest performance gain (over 30%) is obtained for ontologies containing a lot of MUPSes with approx. half size of the ontology size. This is caused

---

[1] To be found at http://www.mindswap.org/2005/debugging/ontologies

Table 1 Comparison of incremental and black-box algorithms

	miniTambis (time [ms])	miniEconomy (time [ms])
allMUPSbb	> 15*min.*	> 15*min.*
Reiter + singleMUPSbb	67481	> 15*min.*
Reiter + singleMUPSinc	19875	19796
allMUPSInc1	8655	14110
allMUPSInc2	7879	12970

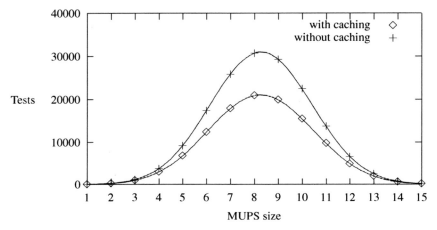

Fig. 4 Comparison of incremental algorithm with caching and without it. Different config-uration of MUPSes of 15 axioms, say $\{1, \ldots, 15\}$, were tested. For each $1 \leq k \leq 15$ (the x-axis), the set of all MUPSes is generated, so that it contains all axiom combinations of size k, thus containing $\frac{15!}{k!(15-k)!}$ MUPSes. For example, for $k = 2$, the set of MUPSes is $\{\{1,2\},\{1,3\},\ldots,\{1,15\},\{2,3\},\ldots\}$.

by the fact, that the broader the search tree of the *allMUPSesInc2* algorithm (see 5) is, the more applications of the caching occur.

Second, the performance and robustness of the incremental methods with respect to the axiom ordering were tested. As all permutations of a given axiom set are required, two small ontologies have been chosen. *TambisP* is a subset of Tambis ontology [2], restricted to the definitions of unsatisfiable concepts *metal*, *nonmetal* and *metalloid* (6 axioms). *MadCowP* is the restriction of Mad cow ontology [3] to the 7 axioms causing unsatisfiability of the concept *madCow*. The best results were obtained with *allMUPSInc2*, which is most efficient (measured by the count of IT) and robust enough to the axiom ordering (measured by test count variance). The results also show, like above, that the performance of *Reiter + singleMUPSinc* strongly depends on the axiom ordering.

[2] http://protege.stanford.edu/plugins/owl/owl-library/tambis-full
[3] http://www.mindswap.org/2005/debugging/ontologies/madcow.owl

Table 2 Comparison of discussed incremental methods. For each ontology and each unsatisfiable concept, the tests are performed for all permutations of the input axiom set. '#', 'avg' and 'var' stands for number of, average and variance of incremental tests.

tambisP	# of inc. tests	avg	var
R. + singleMUPSinc	268362	124.29	206.81
allMUPSInc1	75696	35.04	36.44
allMUPSInc2	61590	28.51	16.76

madCowP	# of inc. tests	avg	var
R. + singleMUPSinc	277200	55.00	8.00
allMUPSInc1	131040	26.00	0.00
allMUPSInc2	119520	23.04	0.50

5 Annotation Tool Prototype

Exploiting our experience with the annotation prototype based on conceptual graphs [10], we are developing an annotation tool, see Fig.5, that will integrate described reasoning services to support detecting modeling errors. The tool is aimed at authors of semantic annotations of narratives and other natural language documents.

The annotation tool prototype consists of several modules. The *ontology module* is the core of the system. Its internal model corresponds to the description

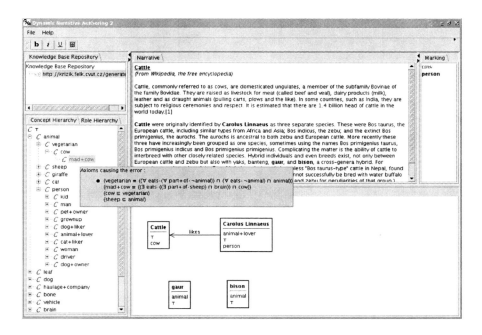

Fig. 5 Annotation Tool Prototype

logic *SHIN*. This model is connected to our implementation of a tableau reasoner for *SHIN*, allowing for concept satisfiability, subsumption, disjointness testing and knowledge base consistency checking. Furthermore, the reasoner could be run in server mode, using the DIG [3] interface for communication. The reasoner is equipped with the above introduced concept satisfiability explanation functionality.

The *annotation module* will serve to create annotations using annotation graphs. Basic form of these graphs allows for creating ABOX assertions. Further refinements are needed to provide inequality assertions, equality assertions, n-ary relations, and other.

The *document module* manages the documents to be annotated. It provides a simple text editor, that allows for visualizing the annotated parts of the document directly in the texts. Finally, the *marking module* allows for color highlighting of annotations according to the classes they belong to.

6   Conclusions and Future Work

Two novel incremental algorithms for finding minimal sets of axioms responsible for given modeling error in an ontology have been introduced. The first one is a novel incremental algorithm that searches for one such minimal axiom set (MUPS). The second one is an extension of the fully incremental algorithm presented in [5] used for searching all minimal axiom sets.

The introduced incremental methods seem promising and our experiment proved that they are also more efficient than the fully black box approaches in the context of error explanations. Although the fully incremental approaches are more efficient than the combination of single MUPS testers and Reiter's algorithm, they do not allow to compute diagnoses directly. This justifies our focus on both approaches. Efficient generation of diagnoses by the fully incremental methods is an open issue.

While it does not seem feasible to invent a sound and complete fully glass-box method that might be reused in a wide range of description logics formalisms, it seems promising to use an incomplete glass-box approach (like the one discussed in sec. 2.2) as the preprocessing step for the incremental methods discussed above. Furthermore, we would like to test several optimizations of the introduced methods, like partitioning of the axiom set.

Acknowledgements. This work has been supported by the grant No. MSM 6840770038 *Decision Making and Control for Manufacturing III* of the Ministry of Education, Youth and Sports of the Czech Republic.

References

1. Baader, F., Sattler, U.: An overview of tableau algorithms for description logics. Studia Logica 69, 5–40 (2001)
2. Baader, F., Calvanese, D., McGuiness, D., Nardi, D., Patel-Schneider, P.(eds.): The Description Logic Handbook, Theory, Implementation and Applications. Cambridge (2003)

3. Bechhofer, S., Moller, R., Crowther, P.: The DIG Description Logic Interface. In: Proc. of International Workshop on Description Logics, DL 2003 (2003)
4. CIPHER project homepage, http://cipherweb.open.ac.uk (cited December 2007)
5. De la Banda, M.G., Stuckey, P.J., Wazny, J.: Finding All Minimal Unsatisfiable Subsets. In: PPDP 2003C (2003)
6. Halaschek-Wiener, C., Parsia, B., Sirin, E.: Description logic reasoning with syntactic updates. In: Proc. of the 5th International Conference on Ontologies, Databases, and Applications of Semantics, ODBASE 2006 (2006)
7. Horrocks, I., Sattler, U., Tobies, S.: Practical Reasoning for Expressive Description Logics. In: Ganzinger, H., McAllester, D., Voronkov, A. (eds.) LPAR 1999. LNCS, vol. 1705. Springer, Heidelberg (1999)
8. Kalyanpur, A.: Debugging and Repair of OWL Ontologies. PhD thesis, University of Maryland (2006)
9. Křemen, P.: Inference Support for Creating Semantic Annotations. Technical Report GL 190/07, CTU FEE in Prague, Dept. of Cybernetics (2007)
10. Uhlíř, J., Kouba, Z., Křemen, P.: Graphical Interface to Semantic Annotations. In: Znalosti 2005, pp. 129–132. Ostrava (2005)
11. Schlobach, S., Huang, Z.: Inconsistent Ontology Diagnosis: Framework and Prototype. Technical Report, Vrije Universiteit Amsterdam (2005)
12. Schlobach, S., Huang, Z., Cornet, R., Van Harmelen, F.: Debugging Incoherent Terminologies. Technical Report, Vrije Universiteit Amsterdam (2006)
13. Reiter, R.: A Theory of Diagnosis from First Principles. Artificial Intelligence 32(1), 57–96 (1987)

Using Ontologies Providing Domain Knowledge for Data Quality Management

Stefan Brüggemann and Fabian Grüning

Abstract. Several data quality management (DQM) tasks like duplicate detection or consistency checking depend on domain specific knowledge. Many DQM approaches have potential for bringing together domain knowledge and DQM metadata. We provide an approach which uses this knowledge modeled in ontologies instead of aquiring that knowledge by cost-intensive interviews with domain-experts. These ontologies can directly be annotated with DQM specific metadata. With our approach a synergy effect can be achieved when modeling a domain ontology, e.g. for defining a shared vocabulary for improved interoperability, and performing DQM. We present five DQM applications which directly use knowledge provided by domain ontologies. These applications use the ontology structure itself to provide correction suggestions for invalid data, identify duplicates, and to store data quality annotations at schema and instance level.

1 Motivation and Goal

Data Quality Management (DQM) approaches report on the quality of data measured by defined data quality dimensions and, if desired, correct data in databases. DQM relies on domain knowledge for detecting and possibly correcting erroneous data, as data without its definition cannot be interpreted as information and is therefore meaningless. On the one hand, DQM approaches like [9] or [1] define phases where domain experts provide their knowledge for further utilization in the DQM process. On the other hand, there are domain ontologies, i.e. formal specifications of

Stefan Brüggemann
OFFIS, Germany
e-mail: b" rueggemann@offis.de

Fabian Grüning
University of Oldenburg, Germany
e-mail: fabian.gruening@informatik.uni-oldenburg.de

S. Schaffert et al. (Eds.): Networked Knowledge - Networked Media, SCI 221, pp. 187–203.
springerlink.com   © Springer-Verlag Berlin Heidelberg 2009

conceptualizations of certain domains of interest, that already provide such knowledge but remain unused in the DQM context. Our contribution is to directly use the knowledge provided by domain ontologies in the DQM context in order to improve the DQM's outcome.

This contribution is structured as follows: Firstly, we will discuss work related to this topic and secondly describe our approach which directly uses the knowledge provided by domain ontologies in the context of DQM in detail by presenting five applications, namely consistency checking, proactive management of consistency constraints, duplicate detection, metadata management, and semantic domain modeling. Finally, we will draw some conclusions and point out further work on this topic.

2   Related Work

Little work has been done on the field of using ontologies for DQM. Existing approaches can be divided into two major classes:

The first application of ontologies in the case of DQM is management of data quality problems and methods. The OntoClean Framework has been introduced in [20]. It provides a template for performing data cleaning consisting of several steps like building an ontology, translating user goals for data cleaning into the ontology query language, and selecting data cleaning algorithms.

The second application of ontologies is the use of domain ontologies. They provide domain specific knowledge needed to validate and clean data. This allows for detecting data problems, which could not be found without this knowledge. To the best of our knowledge, only [15] and [13] use domain ontologies in this way.

We extend these approaches by annotating domain ontologies with DQM-specific metadata which we show in the following section by presenting five DQM-applications that further include the usage of algorithms of the data mining domain.

3   Multiple Utilizations of Domain Ontologies for DQM

To show the advantages of using ontologies in the context of DQM and emphasize the usefulness of our approach to improve the outcome of DQM we present five applications of domain ontologies in that context: consistency checking, proactive management of consistency constraints, duplicate detection, metadata management, and semantic domain modeling.

3.1   Context-Sensitive Inconsistency-Detection with Ontologies

Data cleaning is often performed when data has to be integrated into a database. Data cleaning consists of the detection and removal of errors and inconsistencies from data [16]. We use domain specific knowledge to detect inconsistencies. Consistency is defined as the abidance of semantic rules. These rules can be described with integrity

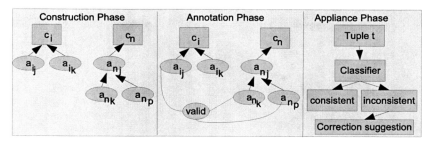

Fig. 1 Overview of an inconsistency detection algorithm using a domain ontology

constraints in relational databases for attributes on schema level. On instance level, consistency is being defined as the correct combination of attribute values. A tuple is consistent when the values from each attribute are valid in combination.

We now provide an algorithm and a data model for consistency checking.

3.1.1   Basic Idea

Figure 1 shows a graphical representation of the consistency checking algorithm. The algorithm consists of three phases. In the construction phase a domain ontology is being created. It can be learned from an existing database, created manually, or already existing ontologies can be used. The expressive power of OWL (Web Ontology Language) enables a generic semi-automatic ontology construction approach. The domain ontology can almost completely be used for DQM, only tuples have to be labeled as valid in the annotation phase. In the appliance phase tuples are being identified as being consistent or inconsistent. When an inconsistency is being detected, a correction suggestion is made. The ontology structure is used to correct invalid tuples. Other valid tuples are searched and characterized as possible corrections. The suggestions are ranked using the distance between the valid and invalid tuples. The advantage over the statistical edit/imputation-approach presented by [6] is the usage of the context of invalid attributes for correction. The statistical approach replaces invalid tuples with randomly chosen values, whilst our approach suggests context-sensitive corrections changing as few attributes as possible.

3.1.2   Data Model Used for Consistency Checking

A relation schema $R = (A_1, .., A_n)$ is defined as a list of attributes $A_1, .., A_n$. Each attribute A_i belongs to a domain $dom(A_i)$. Each domain $dom(A_i)$ defines a non-empty set of valid values. A relation r of R is a set of n-tuples $r = t_1, .., t_m$. Each tuple t_i is a set of values $t_i = (v_{i_1}, .., v_{i_n})$ with $v_{i_j} \in dom(A_j)$.

In the simplest case a tuple t_i is valid if and only if $\forall 1 \leq j \leq n : v_{i_j} \in dom(A_j)$. According to our definition, a tuple is consistent if it is valid and all v_{i_j} are combined correctly.

When validating a tuple t, using only the domain $dom(A_j)$ doesn't enable to identify inconsistencies because combinations cannot be checked. Therefore an ontology

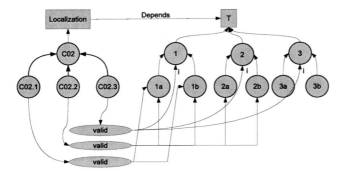

Fig. 2 Ontology containing concepts Localization and T with individuals

is being built containing all values $a_{i_k} \in dom(A_i)$ of each domain with $k = |dom(A_i)|$. Furthermore, domains often contain hierarchical, multidimensional, or other complex structures. These can be respected in an ontological structure.

An ontology consists of a concept C_i for each domain $dom(A_i)$. Attributes a_i are defined as individuals of C_i. They are arranged using "moreSpecificThan" properties to enable modeling complex structures.

Dependencies are defined between concepts. For instance, there is often no semantic dependency between the attributes "id" and "surname". Instead, in oncology, several constraints exist when combining "localization"-values and "T", "N", and "M"-values from the TNM-classification (tumour, node, metastasis) scheme [12]. Concepts have properties "valid" and "invalid" to combine attributes of different concepts and to label them as valid or invalid.

3.1.3 Example

We now provide an example from tumour classification in the cancer registry of lower saxony in Germany. Figure 2 shows an ontology containing the concept Localization, which depends on the concept T. The individuals "C02", "C02.1", "C02.2", and "C02.3" describe malignant neoplasms of the tongue, where the "C02.x" (tip, bottom, 2/3 of front) individuals are more specific than "C02". The property "moreSpecificThan" is hidden due to readability. The three "valid" nodes are introduced as blank nodes and used to describe the following three consistency rules: "C02.1" is only valid with "T"-values "1a" and "1b". "T"-values lower than 2 describe tumoursizes lower than 2cm. "C02.2" is valid with "T"-Values "1a", "1b", "2a", and "2b". "2x"-values are sizes between 2cm and 5cm. "T"-values larger than "2" describe tumour sizes larger than 5cm. "C02.3" is valid with "T"-values "1", "2", and "3". Specifying these connections with "i" defines these as inheriting connections. These connections are inherited to the children of "1", "2", and "3". Therefore the more specific values of "1", "2", and "3", namely "1a", "1b", "2a", "2b", "3a", and "3b", are also valid with "C02.3".

For instance, the tuples $< C02.3, 1 >$, $< C02.3, 3 >$, and $< C02.1, 1a >$ with the structure $< Localization, T >$ can be resolved as valid. The tuple $< C02.1, 3 >$

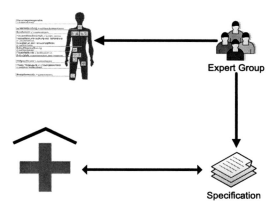

Expert Group

Specification

Fig. 3 Medical Documentation: Expert Group defines measures and publishes them in a specification. Hospitals have to send reports corresponding to this specification.

instead can be identified as invalid, but using the ontological structure, the tuples $< C02.1, 1a >$, $< C02.1, 1b >$, $< C02.3, 3 >$, $< C02.1, 3a >$, and $< C02.1, 3b >$ can be resolved as correction suggestions.

3.2 Proactive Management of Consistency Constraints

As we have seen in the previous section, a domain ontology can be either learned from an existing knowledge base or has to be created manually. We now focus on the latter case and introduce *ProCon*, which is an approach for proactive management of consistency constraints. It is described in detail in [5]. ProCon has been introduced in a scenario in the German public health sector. A German law defines that the medical quality in public health has to be measured and compared. Therefore the "BQS Bundesgeschäftsstelle Qualitätssicherung gGmbH"[1] was founded in 2000. The BQS manages and coordinates the comparison of german hospitals. Figure 3 shows the BQS-workflow: An expert group, consisting of several medics, defines quality measures for relevant medical sectors like cardiology or oncology. For each measure thresholds are being defined. For instance, when the goal is to ensure good medical quality the fatality rate for a specific operation must not exceed a predefined threshold. These measures are being delivered to hospitals and software vendors in a large document. This document describes the information needs of the BQS. The hospitals have to send quality reports with all requested data periodically.

When abnormalities in the delivered data are identified or thresholds are reached, like the mortality rate for a specific operation is beyond a specific treshold, a so called "structured dialogue" is being entered. In this dialogue the BQS analyzes whether bad data quality, insufficient medical quality, or poor medical documentation is responsible for violating the respective measure. In practice, often bad data

---

[1] http://www.bqs-online.de

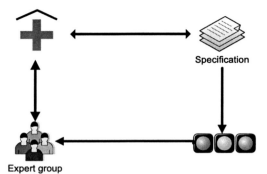

Fig. 4 BQS analyzes reported data and begins a "structured dialogue" with hospitals when measures are not matched

quality can be made responsible. This results in time-consuming communication about detecting and removing inconsistencies. The structured dialogue is shown in figure 4: Hospitals deliver the data which is requested by the specification. This data is being analyzed for inconsistencies and violations of bounds and thresholds. Based on this analysis the expert groups starts a detailed communication with the hospital, which has delivered the data. The compared data of the hospitals is then being published in quality reports. When structured dialogues with hospitals are being performed, some of their data cannot be considered for these reports. This would result in competitive disadvantages, because the data is being used for patient guides. Patients often choose hospitals with a lot of experience in a desired operation.

The described approach has some shortcomings: Although the described approach is top-down-driven, data quality is not being modeled explicitly on a high level. Information needs are being described on a high level by domain experts. They explicitly define dimensions, measures, and all required data. These domain experts are able to distinct data in valid or invalid. Therefore in ProCon we propose the role of a data quality modeler. This modeler defines attribute value combinations in the modeled information needs as valid or invalid. When hospitals know about these quality constraints, they are able to deliver data with high quality.

Another shortcoming is the technical specification itself. The BQS delivers access databases containing the defined information needs and technical rules. These databases cannot be used directly, but have to be integrated in the information system of each hospital. In our approach we propose to deliver the modeled rules using ontologies. This has the benefit that software vendors and hospitals do not need to develop rule checking software. When data has to be exported to the BQS, it can be instantiated as an instance of the ontology and reasoners are able to check the consistency of the resulting ontology automatically (compare also section 3.5).

We now present a model-driven approach of defining consistency constraints in different scenarios. The described approach fits well in the defined BQS-context, but can be applied to other domains as well. The approach is shown in figure 5. It starts with a metamodel of the domain of discourse (part A in the figure). Our

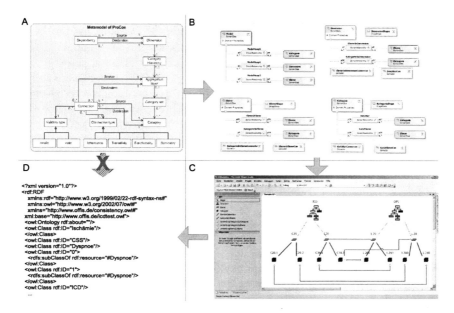

Fig. 5 Transformation steps of a modeled domain of discourse. A metamodel is being converted to a domain specific language. An editor can automatically be generated for this language. This editor can be used to create models that are then being transformed into an ontology.

scenario is based on the multidimensional data model MADEIRA [21], but other models like relational databases or XML-schemata are possible as well. Our model consists of a dimension, which contains a hierarchy of categories. These categories can be combined in aggregation layers.

This data model has been extended with the possibility of defining consistency constraints between dimensions. Therefore dependencies can be created to show where constraints are possible and where not. For instance, there is no dependency between "age" and "name", but between "localisation" and "T", as described in section 3.1. The individuals "C02", "C02.1", C02.2", and "C02.3" for localisation were defined in 3.1.3 and can be identified as categories and aggregation layer.

Connections can be defined between categories and categories, aggregation layer and aggregation layer, and categories and aggregation layer and vice versa. Each connection can be labeled as describing a valid or an invalid attribute value combination. Different connection types can be defined: An inheriting connection describes that the connection is being inherited from an aggregation layer to its children. Transitive connections can be traversed. A functional connection describes that a category or aggregation layer is only valid with the connection destination, and with no other. When such a connection is symmetric, it is modeled that both connection endings can only be connected with the respective other end.

Such a data model can simply be mapped to a domain specific language (DSL) (shown in part B). This DSL contains the same entities and relations as the meta-model. Using the Microsoft DSL tools [14], an editor can be generated automatically. This editor provides all elements defined in the metamodel (shown in part C).

This editor can be used by domain experts to model the information needs and to label combinations as valid or invalid. The graphical representation is a core benefit of this approach, because domain experts do not need to learn to use complex new tools, but can directly use the tool used for modeling the information needs.

With an appropriate transformation the modeled information needs and consistency constraints can be stored as an ontology (shown in part D). Due to the mappings the ontology is an instance of the metamodel, but must not be generated from the metamodel.

The ontology provides several benefits:

- The ontology can be used to check the consistency of the modeled constraints. It can identify contradictions in the model, which can directly be presented to the data quality modeler at design time.
- Data can be checked using this ontology. The ontology can directly be used to identify errors in data by instantiating the data which has to be checked in the ontology. A reasoner can then identify conflicts in the data.
- The ontology can be used as a technical document and can be delivered to software vendors and hospitals.

This approach can be used in situations where information needs can be previously defined on a semantically high level by domain experts. Due to the proactive creation of consistency constraints erroneous data can be avoided or repaired directly in the data sources. When this approach is being applied in the described BQS-scenario, inconsistencies can directly be avoided in hospitals. This results in a much faster publication of more and consistent data. The structured dialogue can be avoided for a couple of cases where bad data quality would initiate the structured dialogue.

3.3 Duplicate Detection

[17] presents an algorithm and its evaluation for several configurations based on [3] for detecting duplicates in databases which are multiple representations of one real world entity and therefore a major issue relevant e.g. in the scenario of integrating several databases. Figure 6 shows a graphical representation of the algorithm which uses a classification algorithm from the data mining domain and will be explained in the following.

The algorithm consists of two consecutive phases, the learning phase and the application phase. In the learning phase a classifier learns the characteristics of duplicates from labeled data, i.e. pairs of instances that are marked as duplicates or non-duplicates. The algorithm's inputs are the distances between every two of the instances' attributes and the information whether or not the instances are duplicates. The algorithm's output is a classifier that is able to distinguish between duplicates and non-duplicates by having identified the combination and grade of those

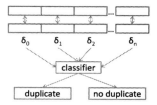

Fig. 6 A duplicate detection algorithm using a classification algorithm. In the learning phase, the algorithm's inputs are the distances between the corresponding attributes and the knowledge about whether or not the instances are duplicates, in the application phase that knowledge is deduced through the classifier that is the learning phase's output.

attributes' similarities, that are relevant for instances being duplicates. The application phase uses those classifiers for detecting duplicates in non-labeled data. The advantage over the statistical approach presented by [7] is the usage of similarity metrics, e.g. string distance metrics, to calculate the attributes' distances instead of using binary information whether two attributes have identical values or not as those metrics are more sensitive in the case of small differences.

Although the described algorithm can be used to find duplicates in any database using any data model the usage of an ontology provides a major advantage: As ontologies' concepts represent real world extracts without any normalization or considerations with respect to the performance of the database, e.g. by artificially inserting redundancy, those concepts' attributes completely describe a real world entity. Such an algorithm can therefore directly be applied to the concepts' instances as they semantically contain all information their real world counterparts are defined by. There is no "object identification problem" where real world entities are scattered around several data model elements, e.g. tables, or extended by artificial values like (primary) keys that are not relevant to the decision whether or not two instances represent the same real world entity. Therefore ontologies' conceptualizations provide an ideal basis for duplicate detection in databases. The ontology is furthermore used for the following applications: Labeling instances as correct or incorrect for using them as data that can be learned from, annotating the scales of measurement for proper preprocessing of the data, etc. Those annotations of user-defined metadata will be explained in the next section.

3.4   Metadata Annotation

Models for data quality are used to make statements about data regarding to their data quality. [2] point out that those models are a major issue for establishing a DQM approach. We show three DQM-specific metadata tasks where ontologies and especially their serializations in RDF (resource description framework) are an excellent choice for making those statements.

3.4.1 Data Provenance

Establishing a DQM approach often requires an integration of several data sources. Data provenance refers to the task of keeping track of the data's origins for correctly giving information about the data quality's state of those databases. XML Namespaces that are widely used to identify RDF's resources' origins can directly be used to point out the database the data is coming from.

To make this work data quality repositories act a little bit like data warehouses in the sense that they are used for integration of the data that's quality has to be managed. The integrated instances provide the information about their originating databases by their namespaces. These namespaces can than simply be used for compiling reports about the data quality of the respective databases by tracing back the evaluated instances to their origins.

3.4.2 Data Quality Annotations at Schema and Instance Level

Both at schema and instance level annotations are needed for DQM. At schema level DQM-algorithms might need to know the attributes' levels of measurements for proper preprocessing. At instance level several annotations like labeling for consistency checking (see section 3.1), duplicate detection (see section 3.3), or rule mining (refer to [4]) can be performed.

Again, RDF's resources provide an elegant way to make statements about data on both schema and instance level, as RDF-Schema and OWL-Ontologies are formulated as RDF-triples too. Those resources can be used as subjects in statements about data quality aspects, e.g. that a number is a nominal value (e.g. an identifier for a room) and therefore distances of two of those values cannot be calculated meaningfully. The duplicate detection algorithm must handle such information e.g. by applying "1" to the distance if those values are different and "0" otherwise.

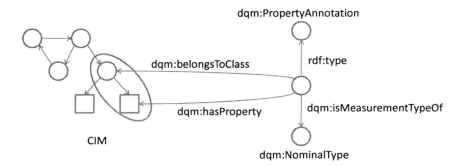

dqm: DQM-Namespace, rdf: RDF-Namespace

Fig. 7 Example of a metadata annotation at schema level

Such a metadata annotation is shown in figure 7. On the left hand side of the figure an ontology is shown which consists of concepts (circles) and datatype properties (squares). The statement we want to make about this ontology is that a certain concept's datatype property represents a nominal value (denoted by the ellipse). The right hand side of the figure shows a solution for making such a statement: An instance of a concept "PropertyAnnotation" is created that points out the concept ("belongsToClass") and the concept's datatype property ("hasProperty") in question. Finally, the statement about the level of measurements is made by the object "NominalValue" of the PropertyAnnotation's statement about its "isMeasurementTypeOf"-property.

Another example is shown in figure 8. In contradiction to the previous example, an annotation is made on instance level. The use case shown here is the marking of two instances that have been found suspicious to be duplicates by an algorithm like the one presented in section 3.1. Therefore an instance of the concept "Duplicate-Suspicion" is created and the two instances of any (domain) ontology concept under suspicion are linked to that instance by the object properties "hasSuspicion".

3.4.3 An Ontology for the DQM-Domain

The annotations introduced in the preceding section need a vocabulary. Such a vocabulary can be provided by creating a DQM-ontology. The ontology in question has to cover concepts for the following annotations: At schema level the level of measurements have to be annotated for proper preprocessing. At instance level the preprocessed values as well as time stamps for measuring the value currencies have to be annotated. Furthermore, the already mentioned labeling of consistent tuples and labels for training data mining algorithms have to be annotated. Erroneous data has to be pointed out, specifying the reason for the suspected errors like outliers and inconsistencies (also see [19]). Marking duplicates needs a special concept as several

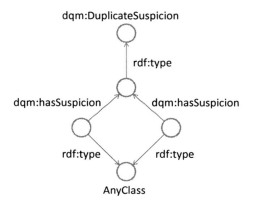

dqm: DQM-Namespace, rdf: RDF-Namespace

Fig. 8 Annotation of a duplicate suspicion

instances are involved like presented in the previous section. In this case further processing might e.g. involve the removal of one of those instances after confirming the multiple representation of a real world entity by those instances by a domain expert.

The concepts of figures 7 and 8 denoted by the namespace "dqm" belong to our proposed DQM-ontology. But not only is a vocabulary for making such statements defined by the DQM-ontology. It is further used for modeling DQM specific knowledge like presented in figure 9. There the relationships between three different DQM-aspects are defined: Data quality methods ("DataQualityMethod") describe algorithms for detecting possibly wrong data by e.g. applying statistical methods or data mining algorithms like the one presented in section 3.3 for identifying multiple representations of real world entities. Their results are interconnected with the data quality dimensions ("DQ-Dimension") which are used for measuring the quality of a database's data. Those dimensions can be defined and interconnected with the data quality algorithms outcomes by the user, so that the reports generated by the DQM fit the user's need for information regarding the data quality of his databases.

Figure 9 shows a predefined configuration where the information about the outcome of the algorithm for detecting duplicates is among others an input for the calculation of the dimension "Accuracy". In certain contexts it could be reasonable to define a dimension for duplicates itself, linking the new dimension with the outcome of the duplicate detection algorithm alone. Finally, the ontology defines the vocabulary that the algorithms use for marking suspicious data ("Suspicion"). Again, the usage of the concept used by the duplicate detection algorithm "DuplicateSuspicion" has already been introduced in the previous section and its application is shown in figure 8. With this information a report about the state of the

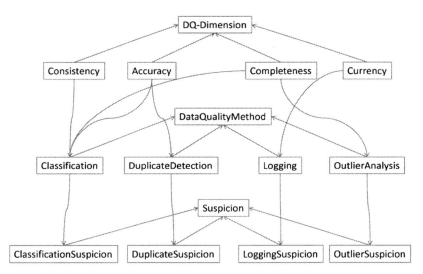

Fig. 9 Excerpt of the DQM-ontology showing interconnections between DQM-dimension, algorithms and suspicion annotations (straight lines being subclass definitions)

database's data quality can be generated as the marked suspicions can be taken into consideration for evaluating the different data quality dimensions.

3.5 Domain Ontologies as a Foundation for Correct Data

Beyond the use cases presented so far domain ontologies themselves provide semantic specifications that are useful for detecting erroneous data. We will show an example by remodeling the "Common Information Model CIM" (refer to [10]) which is an IEC standard for the utility domain by using OWL. The results presented in this section are part of a master thesis by Thomas Gebben [8] who kindly allowed us to present them here.

3.5.1 Introducing the CIM

As a domain ontology for the utility domain the CIM covers all informational aspects for that domain, like generation, protection, topology, measurements, assets, consumers, etc. The CIM itself is modeled in the Unified Modeling Language UML and is provided in several other formats like RDF by automated convert processes. As the outcome of those processes does not provide information that goes beyond that available in the source format, our approach is to build the CIM from scratch with the semantics provided by OWL.

3.5.2 Modeling the CIM as an Ontology

For the remodeling of the CIM we chose a so-called profile of the CIM which denotes a true subset of the CIM. The "Common Power System Model" CPSM [11] contains objects that are necessary to describe the topology and all assets of an electrical grid such as transformers, lines, substations, etc. A snipped of such a definition is shown in Listing 1.

```
<owl:FunctionalProperty rdf:ID="unitName">
  <rdfs:domain rdf:resource="#IdentifiedObject"/>
  <rdfs:range>
    <owl:DataRange>
      <owl:oneOf rdf:parseType="Resource">
        <rdf:first rdf:datatype="http://www.w3.org/2001/XMLSchema#string"> MW
          </rdf:first>
        <rdf:rest rdf:parseType="Resource">
          <rdf:rest rdf:parseType="Resource">
            <rdf:first rdf:datatype="http://www.w3.org/2001/XMLSchema#string"
              > MVA </rdf:first>
            <rdf:rest rdf:parseType="Resource">
              <rdf:first rdf:datatype="http://www.w3.org/2001/XMLSchema#
                string"> Count </rdf:first>
              <rdf:rest rdf:parseType="Resource">
                <rdf:rest rdf:parseType="Resource">
                  <rdf:rest rdf:parseType="Resource">
                    <rdf:first rdf:datatype="http://www.w3.org/2001/
                      XMLSchema#string"> MVAr </rdf:first>
                    <rdf:rest rdf:parseType="Resource">
                      <rdf:rest rdf:parseType="Resource">
                        <rdf:rest rdf:parseType="Resource">
```

```
                    <rdf:first rdf:datatype="http://www.w3.org/2001/
                        XMLSchema#string"> SwitchPosition </rdf:first
                        >
                    <rdf:rest rdf:resource="http://www.w3.org
                        /1999/02/22-rdf-syntax-ns#nil"/>
                </rdf:rest>
                    <rdf:first rdf:datatype="http://www.w3.org/2001/
                        XMLSchema#string"> TapPosition </rdf:first>
                </rdf:rest>
                    <rdf:first rdf:datatype="http://www.w3.org/2001/
                        XMLSchema#string"> kV </rdf:first>
                </rdf:rest>
                </rdf:rest>
                    <rdf:first rdf:datatype="http://www.w3.org/2001/XMLSchema
                        #string"> Ratio </rdf:first>
                </rdf:rest>
                    <rdf:first rdf:datatype="http://www.w3.org/2001/XMLSchema#
                        string"> PerCent </rdf:first>
                </rdf:rest>
                    <rdf:first rdf:datatype="http://www.w3.org/2001/XMLSchema#
                        string"> Amperes </rdf:first>
                </rdf:rest>
            </rdf:rest>
        </rdf:rest>
            <rdf:first rdf:datatype="http://www.w3.org/2001/XMLSchema#string">
                Degrees </rdf:first>
        </rdf:rest>
    </owl:oneOf>
    </owl:DataRange>
    </rdfs:range>
    <rdf:type rdf:resource="http://www.w3.org/2002/07/owl#DatatypeProperty"/>
</owl:FunctionalProperty>
```

Listing 1. Definition of the property "unitName" of the concept "IdentifiedObject".

We define a property "unitName" which makes a statement about ("rdfs:domain") instances of the concept "IdentifiedObject" which are e.g. measurements. The range of the property is limited to one of the denoted values ("MW", "MVA", "Count", etc.). Furthermore, the property is defined as a functional property, which means that no or exactly one value per instance is allowed. By providing this information the ontology itself will point out instances that do not satisfy those conditions and therefore prevents wrong data from the beginning. Such an ontology can be used for generating messages to exchange data between different enterprises of the utility domain. In the following examples we will show the automated detection of erroneous instances that violate the stated rules. We will use Pellet [18] as a reasoner for this task which also generates the messages that will be shown.

3.5.3 Examples for Detecting Errors with a Domain Ontology

We now give two examples of wrong instances regarding the definition stated above. Figure 10 shows an instance of "MeasurementValue" that defines two different values for the property "unitName" for the same resource.

The reasoner correctly detects a violation of the functional aspect of the property. It therefore rejects the stated definition of the shown instance. Figure 11 shows an example of the violation of the "oneOf" data range restriction of the values of the property.

```
<MeasurementValue rdf:ID="MeasurementValueTESTDATAPROPERTY">
    <unitName rdf:datatype="http://www.w3.org/2001/XMLSchema#string">MW</unitName>
</MeasurementValue>

<MeasurementValue rdf:ID="MeasurementValueTESTDATAPROPERTY">
    <unitName rdf:datatype="http://www.w3.org/2001/XMLSchema#string">MVA</unitName>
</MeasurementValue>

WARN [main] (KnowledgeBase.java:1573) - Inconsistent ontology. Reason: Individu
al http://www.owl-ontologies.com/Ontology1209114222.owl#MeasurementValueTESTDATA
PROPERTY has more than one value for the functional property http://www.owl-onto
logies.com/Ontology1209114222.owl#unitName
```

Fig. 10 Instance of "MeasurementValue" with violation of the functional property "unit-Name"

```
<MeasurementValue rdf:ID="MeasutValueTESTDATAPROPERTY">
    <unitName rdf:datatype="http://www.w3.org/2001/XMLSchema#string">MUdddA</unitName>
</MeasurementValue>

WARN [main] (KnowledgeBase.java:1573) - Inconsistent ontology. Reason: Plain literal "MUdddA"^^http://www.w
3.org/2001/XMLSchema#string does not belong to datatype not(and([not(value(literal(SwitchPosition,(),http:/
/www.w3.org/2001/XMLSchema#string))),not(value(literal(TapPosition,(),http://www.w3.org/2001/XMLSchema#stri
ng))),not(value(literal(kV,(),http://www.w3.org/2001/XMLSchema#string))),not(value(literal(MVAr,(),http://w
ww.w3.org/2001/XMLSchema#string))),not(value(literal(MVA,(),http://www.w3.org/2001/XMLSchema#string))),no
t(value(literal(MVA,(),http://www.w3.org/2001/XMLSchema#string))),not(value(literal(Count,(),http://www.w3.
org/2001/XMLSchema#string))),not(value(literal(PerCent,(),http://www.w3.org/2001/XMLSchema#string))),not(va
lue(literal(Degrees,(),http://www.w3.org/2001/XMLSchema#string))),not(value(literal(Amperes,(),http://www.w
3.org/2001/XMLSchema#string))),not(value(literal(MW,(),http://www.w3.org/2001/XMLSchema#string)))])). Liter
al value may be missing the rdf:datatype attribute.
```

Fig. 11 Instance of "MeasurementValue" with violation of the "oneOf" data range restriction of the property "unitName"

The reasoner again detects an inconsistent ontology stating that the literal "MVd-ddA" is not defined as one of the allowed values for the property in question.

Several classes of consistency checks are performed in a similar manner like checking cardinality constraints, assuring that only concepts defined by the profile are used, etc.; we use the complete expressivity of the OWL-DL dialect to create a powerful ontology that conforms to the CPSM standard and rejects as many mal-formed instantiations of its concepts as possible.

4   Conclusions and Future Work

The usage of the knowledge provided by domain ontologies can be used to improve DQM's outcomes in several ways as shown by five given examples, namely consistency checking, proactive management of consistency constraints, duplicate detection, seamless possibility of metadata annotation, and semantic domain modeling. Therefore, a synergy effect from modeling a domain ontology, e.g. for defining a shared vocabulary for improved interoperability, and DQM can be achieved. The further work will include the appliance of our described approaches on enterprise scaled databases to verify their applicability. The EWE AG (please visit www.ewe.de) partly funds the projects the presented results originate from and also provide such data for large scale tests. As described, other test scenarios are tumour classification in cancer registries and the reports for the BQS.

References

[1] Amicis, F.D., Batini, C.: A methodology for data quality assessment on financial data. Studies in Communication Sciences 4, 115–136 (2004)

[2] Batini, C., Scannapieco, M.: Data Quality. Springer, Heidelberg (2006)

[3] Bilenko, M., Mooney, J.R.: Employing trainable string metrics for information integration. In: Proceedings of the IJCAI-2003 Workshop on Information Integration on the Web, Acapulco, Mexico, pp. 67–72 (August 2003)

[4] Brüggemann, S.: Rule mining for automatic ontology based data cleaning. In: Zhang, Y., Yu, G., Bertino, E., Xu, G. (eds.) APWeb 2008. LNCS, vol. 4976, pp. 522–527. Springer, Heidelberg (2008)

[5] Brüggemann, S.: Proaktives Management von Konsistenzbedingungen im Analytischen Performance Management. In: Proceedings of DW 2008, Synergien durch Integration and Informationslogistik (2008)

[6] Fellegi, I.P., Holt, D.: A systematic approach to automatic edit and imputation. Journal of the American Statistcal Association 71, 17–35 (1976)

[7] Fellegi, I.P., Sunter, A.B.: A theory for record linkage. Journal of the American Statistical Association 64(328), 1183–1210 (1969)

[8] Gebben, T.: OWL-Reasoner basierte Gültigkeitsprüfung von CIM-Topologien gemäßCommon Power System Model (CPSM). Master thesis, Universität Oldenburg (to be published) (2009)

[9] Hinrichs, H.: Datenqualitätsmanagement in Data Warehouse-Systemen. PhD thesis, Universität Oldenburg (2002)

[10] IEC - International Electrotechnical Commission: IEC 61970:301: Energy management system application program interface (EMS-API) - Part 301: Common Information Model (CIM) Base. International Electrotechnical Commission (2003)

[11] IEC - International Electrotechnical Commission: IEC 61970: Energy Management System Application Program Interface (EMS-API) - Part 452: CIM Network Applications Model Exchange Specification. International Electrotechnical Commission (2006)

[12] International Union Against Cancer (UICC). TNM Classification of Malignant Tumours, 6th edn. John Wiley & Sons, New Jersey (2001)

[13] Kedad, Z., Métais, E.: Ontology-based data cleaning. In: Andersson, B., Bergholtz, M., Johannesson, P. (eds.) NLDB 2002. LNCS, vol. 2553, pp. 137–149. Springer, Heidelberg (2002)

[14] Microsoft Corporation: Domain Specific Language Tools, http://msdn2.microsoft.com/en-us/vstudio/aa718368.aspx/ (Feburary 12, 2009)

[15] Milano, D., Scannapieco, M., Catarci, T.: Using ontologies for xml data cleaning. In: Meersman, R., Tari, Z., Herrero, P. (eds.) OTM-WS 2005. LNCS, vol. 3762, pp. 562–571. Springer, Heidelberg (2005)

[16] Rahm, E., Do, H.H.: Data cleaning: Problems and current approaches. Bulletin of the IEEE Computer Society Technical Committee on Data Engineering 23(4), 3–13 (2000)

[17] Schünemann, M.: Duplikatenerkennung in Datensätzen mithilfe selbstlernender Algorithmen. Master thesis, Universität Oldenburg (2007)

[18] Sirin, E., Parsia, B., Grau, B.C., Kalyanpur, A., Katz, Y.: Pellet: A practical OWL-DL reasoner. Web Semantics: Science, Services and Agents on the World Wide Web 5(2), 51–53 (2007)

[19] Uslar, M., Grüning, F.: Zur semantischen Interoperabilität in der Energiebranche: CIM IEC 61970. Wirtschaftsinformatik 49(4), 295–303 (2007)

[20] Wang, X., Hamilton, H.J., Bither, Y.: An ontology-based approach to data cleaning. Technical report, Department of Computer Science, University of Regina (June 2005)

[21] Wietek, F.: Intelligente Analyse multidimensionaler Daten in einer visuellen Programmierumgebung und deren Anwendung in der Krebsepidemiologie. PhD thesis, Universität Oldenburg (2000)

Semantic Search and Visualization of Time-Series Data

Tatiana von Landesberger, Viktor Voss, and Jörn Kohlhammer

Abstract. In the economic and financial analysis domain, quick access to the right information plays a major role. Using current systems, the search for and presentation of data is very cumbersome. The data, mostly in form of time-series, is stored in various databases. In order to retrieve the searched data, the analysts need to know where to search and sometimes even the structure of the database and its coding. Then it is required to export the data, process the data and create a chart to view the data. This might take time from tens of minutes to hours. In our work we present a first prototype of an integrated search engine that takes as input a natural language query and offers graphic and text output depending on the user task. The system automatically identifies the resulting time-series and types of graphical data presentation, and shows the results in a web browser or in Excel. The knowledge-based expert system uses domain ontologies for extraction of economic terms in the search queries and specially built data type taxonomies with user task and chart type ontologies for the identification of graphical output.

1 Introduction

In the economic and financial analysis, rapid access to the right information plays a major role. Often large amounts of data have to be evaluated in short time. The data are usually stored in heterogeneous systems and in various databases. In their work, analysts combine numerical data from various sources with text (such as news, reports, etc.), and expertise that exists in the company.

Tatiana von Landesberger
Technische Universität Darmstadt, Darmstadt, Germany
e-mail: `tatiana.von.landesberger@gris.informatik.tu-darmstadt.de`

Viktor Voss and Jörn Kohlhammer
Fraunhofer IGD, Darmstadt, Germany
e-mail: `{viktor.voss,joern.kohlhammer}@igd.fraunhofer.de`

S. Schaffert et al. (Eds.): Networked Knowledge - Networked Media, SCI 221, pp. 205–216.
springerlink.com   © Springer-Verlag Berlin Heidelberg 2009

In order to collect all the required information, analysts spend a lot of time every day with information search. Analysts usually simply browse the data sources or use data search options available by the data providers. These conventional systems for searching for financial and economic information, however, do not offer naturally formulated search queries such as "GDP composition in Germany" with direct graphical output.

Finding the relevant data does not end the analysis task. In the analysis process, after the data is gathered, the data is copied by hand into another system for further processing (e.g. into Excel). Finally, appropriate graphics for data presentation are created. In addition to the domain expertise this graphics construction requires also skills in information visualization and graphics design from the analyst.

In this paper, we introduce a first prototype of a semantic-based search engine which takes as input a natural language query, searches over financial information across multiple financial data services and displays the results in a way that suits best the user tasks (e.g. as table, text or graph). A preliminary user study to examine the common analytic tasks with matching answers was conducted. In a later stage, selected end-users were asked for feedback. Results of both studies are presented.

The paper is structured as follows: Section 2 discusses relevant work about automatic graphical data representation. The third section describes current user search tasks and used systems for data search. In the Section 4, our framework is presented. Section 5 provides examples of search output visualization. After giving an overview of related visual analytics research initiatives, we conclude and discuss future work.

2   Related Work

Automatic generation of graphical data representation goes back to the work of Mackinlay [17]. The Automatic Presentation Tool (APT) defines graphical representation based on the description of visual attributes. Roth and Mattis based their SAGE system on Mackinlay's approach for designing two-dimensional static presentation of relational data [18]. Casner introduced the BOZ system taking a step further toward user-centered design with a task-analytic approach [6]. BOZ concentrated on the design of graphics that optimize human performance in information processing tasks. The idea was to replace logical inferences that are cognitively demanding with faster perceptual inferences. However, all of these approaches and their successors today are still computationally too inefficient for interactive applications. According to the visualization design methodology, we can divide current systems into two categories - constructive bottom-up and template-based top-down method [14]. The comparison of the two types of techniques can be found in [16]. In this paper, we follow a top-down approach that is strongly supported by semantics. The development of data type and visualization taxonomies was based on the work by Shneiderman [19] and by Tory et al. [22] which match different data types with visualization driven by user tasks. A general overview of analytical tasks in visualization is provided by Amar et al. [3]. Amar et al. [2] earlier described a knowledge task-based framework for design and evaluation of information visualizations. Fujishiro et Al. [13] introduce a taxonomic approach to semi-automatic design of information visualization applications using modular visualization environment. An overview of visualization techniques specialized for time-series can be found in [1]; a taxonomy for temporal data visualizations can be found in [11]. Kohlhammer [15] shows the use of domain ontologies for effective visualization. We discuss this approach in more detail in Section 4.

3   User Tasks and Current State of Data Search

In this section we describe the current working environment of financial and economic analysts and the questionnaire that was sent out to these analysts. This dialog with the application end users was very important as a guidance for our research in this area. Only those visualization solutions that are embedded in the current work flow of users will be successful solutions.

3.1   Search Tasks

In order to best meet the analytic requirements, we have asked ten financial and economic analysts from various work domains, what kind of search engines they use to find the data and what systems they use for data presentation. They were asked to provide as many data queries as possible including their desired outcomes and estimation of the time needed to accomplish the analytical task (see Table 1).

More than two thirds of the searches are aimed at getting data in tables, single data observations or data charts (see Figure 1). Chart responses included line, column, stacked column and scatter plot charts. The respondents use financial and internal databases and copy this data to create charts in Excel. The time needed to accomplish such tasks (from data search to the production of relevant data presentation) varies strongly according to the desired output and the familiarity with the searched data sources. Searching for novel information that is not accessed by the user on a regular basis is much more time-consuming than a search in well-known data sources. On average, users spend 18 minutes when creating charts, 38 minutes when creating tables and only 3 minutes when looking for a single value. However,

Table 1 Example answers to the questionnaire

Question	Expected Output	Expected Duration
Show me EUR/SKK development in the last week	Intra-day line chart	5 mins
DJI index in the past year	Tick data, line chart	5 mins
Show me GDP composition in Germany and France?	2 pie charts showing GDP components	30 mins

Fig. 1 Required search
query responses. Types of
search responses.

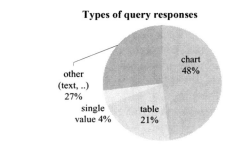

Fig. 2 Time needed for
response by type (average,
minimum and maximum)

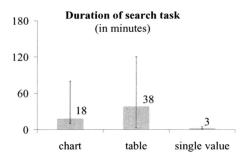

the difference between minimum and maximum time is very high (see Figure 2)
with a maximum of 4 hours for one task.

3.2 Data Sources and Currently Used Search Engines

The sources of information needed are spread across multiple databases from various data providers (e.g. Bloomberg, Reuters, DataStream, Eurostat, ECB or inter-

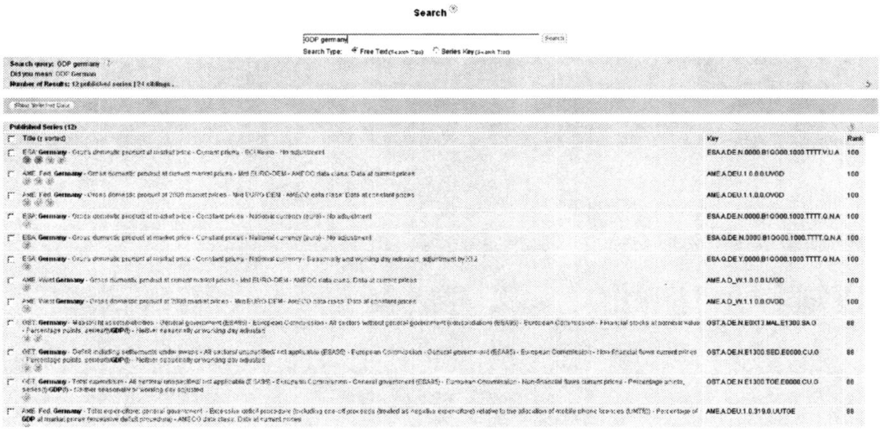

Fig. 3 Example of search results using the ECB search engine (http://sdw.ecb.int)

nal sources). Each data provider uses their own data classification system and data description system making the search for data across databases very complicated. Search in such databases is usually constrained to explorative overview of database entries by category. Some providers offer search engines which should facilitate the search for data. These search functions, in general, offer as outcome long lexicographic ranking (lists) of the names of the available time series (see Figure 3).

4   Semantics-Based Time-Series Search and Visualization

This section introduces our main concept of semantics-based search and visualization. Based on a general approach called decision-centered visualization, we designed a framework for query processing and chart generation. At the core of this approach is the semantics-based determination of a suitable chart type for the data and task at hand.

4.1   Decision-Centered Visualization

Our framework follows the decision-centered visualization (DCV) approach [15]. This approach uses knowledge representation, in particular domain ontologies and meta-databases, to filter and prioritize information and events dynamically depending on the current task type and user role. In contrast to a simple filtering algorithm, this approach takes at the same time the visualization requirements of the events

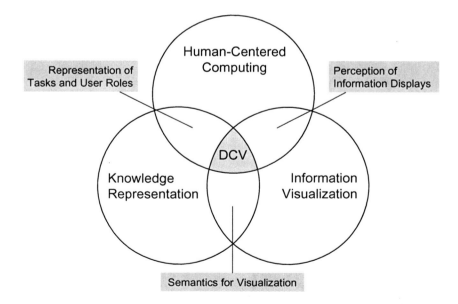

Fig. 4 Overview of the decision-centered visualization approach

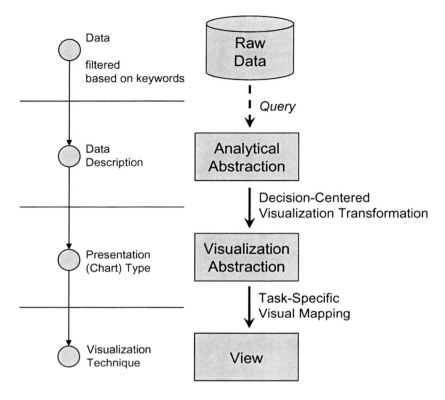

Fig. 5 DCV-specific adaptation of Chi's data state model

and information into account to be able to support effective and focused visualization (see Figure 4).

The connection between information visualization and knowledge representation is the support of determining what to visualize or how to visualize with the help of represented metadata and knowledge. The visualization of knowledge representation structures like ontologies, is not as relevant here, though these techniques are necessary on another level for creating and maintaining the ontologies. At the heart of this approach lies the mechanism to represent the presentation knowledge, i.e. the presentation requirements. These requirements are based on data types similar to those of Shneiderman, Card, and Mackinlay [19, 5] while their handling is based on an adaptation of Chi's data state model [7] (see Figure 5).

4.2   Search and Visualization Framework

Our knowledge-based system for integrated data search and visualization consists of three major parts (see Figure 6): in the first stage the input query is processed in order to extract relevant search criteria. Then a suitable set of responses is compiled and extracted from the database. Finally, the results are shown in a web browser. Our system makes use of the ConWeaver framework [9]. ConWeaver provides automated knowledge network construction, semantic integration and intelligent search in portals and intranets.

4.3   Query Processing and Response Identification

In the first stage of our framework (see Figure 5 and Figure 6), the input query is processed in order to extract economic and financial terms in the query. In our work, we have created modules that identify economic and financial terms in the data descriptions, by looking for longest terms first. For example "GDP per capita" is a different term than "GDP". Furthermore synonyms (consumer prices and inflation) and abbreviations (e.g. GDP = gross domestic product) can be extracted.

For defining the suitable data presentation type, we have analyzed visualization types used in the financial and economic domain as well as the results of our user study. A template-based system for graph generation, which can be adjusted by the users, is applied. In the system, each presentation type is described by the type of data input, visualization parameters and the analytical purpose (keywords).

For this application, a description was created for each chart type identified in the user questionnaire (line, column, pie, scatter plot, etc.). This description was saved in the chart type ontology. It describes types of data that the specific chart can use as input and the user tasks for each chart type. Each chart parameter (each chart axis) is described via the type of data that it takes as input (quantitative, nominal, ordinal, etc.) and its cardinality. For example a line chart has 3 "axis": X axis that usually takes quantitative data of cardinality ≥ 5, Y Axis takes quantitative data of cardinality ≥ 1, and Legend takes nominal data of cardinality 1-10 (more than 10 series usually make the graph overcrowded). Chart is also characterized by "keywords" indicating user task connected to the specific chart type. The keywords were

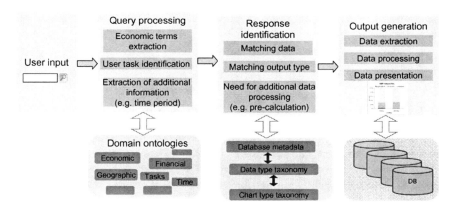

Fig. 6 Overview of the search and visualization framework

identified via matching search queries with expected responses (see in Table 1). For example, "composition" indicates pie chart, "development" indicates line chart.

4.4 Output Generation

In the output generation stage, the identified suitable response data are collected from the database. The output type is influenced by the rank of matching data, the fit of the data to specific chart types and the user task match with the chart keywords. The output data type and the identified user task are matched with the best possible chart type. The chart type match (see Figure 7) uses a weighted data type hierarchy for defining the best possible output type for a given data set to be shown. The output match is quantitatively expressed by the product of the data type hierarchy step weights. The weights are pre-set parameters, which can be adjusted. Additionally the user tasks influence the output type. Additionally, the weight of each chart type is influenced by matching the chart keywords with the search query. A ranking list of output types is thereby created. The top ones are used in the output visualization.

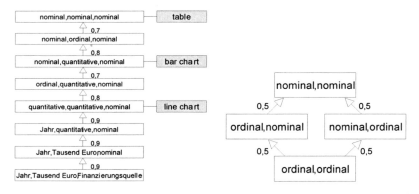

Fig. 7 Semantics-based determination of the presentation type (left) using a data set type hierarchy (right)

5   Search System Output and Evaluation

For testing the search and visualization framework, real world data sets from the Federal Reserve Bank of St. Louis (http://research.stlouisfed.org/fred2/) were used. The FRED database contains about 14,000 economic and financial time series in flat text files. It provides raw data and time series descriptions.

The engine displays the search results on a Web front-end in the form of dynamically generated HTML pages based on the search results. The data that best match the user query are displayed in a graphical form determined by the response identification phase (see Figure 8). In our prototypic application, the user may click on each chart to see the underlying data table in Excel (see Figure 9). The users can then further process the data results or design further charts in this application. Other export formats can be employed on demand. In addition to numeric data, the search engine offers a list of documents searched on the Internet using common text search engines (e.g. Google). Using the ConWeaver system [9] we can also easily include the possibility to search in internal documents, however we did not have any such data at our disposal. Thus, our approach provides a composite of relevant information in one window.

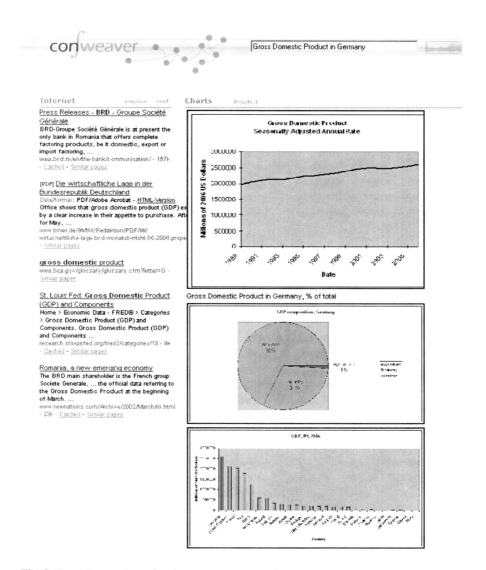

Fig. 8 Example search result using our prototype tool

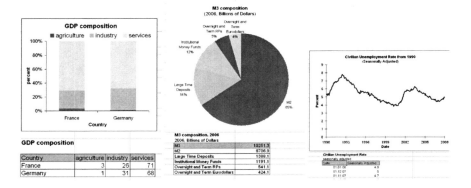

Fig. 9 Selected user responses exported to Excel

The prototype user queries were used to test the results of our search engine. The users liked very much the possibility to see an overview of the search responses in graphic/table format. It enables them to quickly identify major data developments without cumbersome data export and graph creation for each single data series. This quick and easy way of preliminary data analysis saves the analysts' time which they can use for more in-depth data analysis using e.g. data mining methods.

6 Relationship to Visual Analytics Research Initiatives

One larger extension of this work is currently implemented within the project THE-SEUS [21] – the development of a semantics-based Visual Analytics framework. This framework is intended to provide a visual connection between "high-level" (semantic) data and "low-level" data - the actual content. This connection works in both directions, each of which is necessary to support specific tasks. The primary idea of the Visual Analytics Framework is not to provide means to navigate and use semantic knowledge, but more importantly to create, develop and verify this knowledge, represented in the semantic structures. This bottom-up approach is re-alized using data analysis techniques like feature detection, similarity identification, cluster identification etc.

In a certain sense, the work on this framework extends the research by [7] to Vi-sual Analytics. Chi proposes his data state model for the analysis of visualization techniques to provide a clearer understanding of the interactions between data and operators. He describes 36 different analyzed techniques with to have a classifica-tion and a selection of how to implement different operators in a large visualization system. It can be seen that several techniques share different operating steps that could be standardized or modularized for reuse in other systems.

Visual Analytics is a field, where analysis techniques are used in conjunction with graphical displays that integrate the user and his specific abilities and interests into the process. Visual Analytics as a field requires standards, which allow a broad interchange of data of very different types between the different components of the system for the data integration, analysis and interaction. The Visual Analytics Framework will serve this purpose and will revise existing techniques in order to asses their value for a specific use-case.

7 Conclusions

A prototype system for search in economic time-series databases which takes as input a natural language query, and graphically visualizes the search results has been developed. The choice of the proper visualization type is task-dependent and is determined using a data taxonomy and a visualization type ontology. Domain ontology for economic data is used for finding suitable search responses. The system offers time savings to the potential users when looking for and presenting economic and financial data allowing them to concentrate on the data analysis part of the analytic process. Our approach overcomes the computationally time-intensive approaches for automatic graphics design by following a semantics-based top-down method, which is combined with a decision-centered visualization approach for generating visualization familiar to and expected by the user for the task at hand.

8 Future Work

In the near future we will explore the effect of different parameter setups for the hierarchy type match. It will also be necessary to expand the used financial and economic ontologies using new data sources (e.g. further classifications) and newly developed publicly available ontologies. This would allow for more accurate search responses. We would like to include automatic data processing using financial functions (e.g. calculation of indices) and widen the spectrum of used visualization techniques. It would be interesting to combine search for numeric data with search in news feeds. For example, particular patterns (peaks, strong decreases) could be matched with the relevant stock market interpretation.

Acknowledgements. This research was partly funded by the German Federal Ministry of Economics and Technology within the THESEUS project [21].

The authors thank Richard Stenzel and Thomas Kamps of ConWeaver GmbH [9]for their support and the possibility to use their engine for semantic search. We are thankful to the user study participants for their activity and helpful input. We thank Tobias Schreck for his helpful comments.

References

1. Aigner, W., Miksch, S., Muller, W., Schumann, H., Tominski, C.: Visualizing time-oriented data – A systematic view. Computers & Graphics 31(3), 401–409 (2007)
2. Amar, R., Stasko, J.: A Knowledge Task-Based Framework for Design and Evaluation of Information Visualizations. In: Proc. IEEE Symposium on Information Visualization, pp. 143–150 (2004)

3. Amar, R., Eagan, J., Stasko, J.: Low-Level Components of Analytic Activity in Information Visualization. Georgia Institute of Technology (2006)

4. Audersch, S., Flach, G.: Semantic Web Technologien fr die visuelle Exploration und Fusion multivariater Datenbestnden. In: Proceedings of Berliner XMLTage (2005)

5. Card, S.K., Mackinlay, J.D., Shneiderman, B.: Readings in Information Visualization. Using Vision to Think. Morgan Kaufmann, San Francisco (1999)

6. Casner, S.M.: A Task Analytic Approach to the Automated Design of Graphic Presentations. ACM Transactions on Graphics 10(5), 111–151 (1991)

7. Chi, E.H.: A Taxonomy of Visualization Techniques using the Data State Reference Model. In: Proceedings of the 2000 IEEE Symposium on Information Visualization, Salt Lake City, UT, USA (2000)

8. Codd, E.F.: A Relational Model of Data for Large Shared Data Banks. Communications of the ACM 13(6), 377–387 (1970)

9. ConWeaver, http://www.conweaver.de (cited December 1, 2008)

10. Cox, K., Grinter, R.E., Hibino, S.L., Jategaonkar Jagadeesan, J., Mantilla, D.: A Multi-Modal Natural Language Interface to an Information Visualization Environment. International Journal of Speech Technology (2001)

11. Daassi, C., Nigay, L., Fauvet, M.C.: A taxonomy of temporal data visualization techniques. Revue en Sciences du traitement de l'Information. Cepadues Editions 5(2), 41–63 (2005)

12. Date, C.J.: An Introduction to Database Systems, 7th edn. Addison- Wesley, Reading (2000)

13. Fujishiro, I., Ichikawa, Y., Furuhata, R., Takeshima, Y.: GADGET/IV: a taxonomic approach to semi-automatic design of information visualization applications using modular visualization environment. In: Proceedings IEEE Symposium on Information Visualization, pp. 77–83 (2000)

14. Kamps, T.M.: A Constructive Theory for Diagram Design and its Agorithmic Implementation. PhD Thesis, Technische Universitat Darmstadt (1999)

15. Kohlhammer, J.: Knowledge Representation for Decision-Centered Visualization. Dissertation, Technical University of Darmstadt. GCA-Verlag (2005)

16. Lange, S., Nocke, T., Schumann, H.: Visualisierungsdesign - ein systematischer Überblick, German, Universität Rostok (2006)

17. Mackinlay, J.: Automating the Design of Graphical Presentations of Relational Information. Transactions on Graphics 5(2) (1986)

18. Roth, S.F., Mattis, J.: Automating the Presentation of Information. In: Proceedings of the IEEE Conference on Artificial Intelligence Applications (1991)

19. Shneiderman, B.: The Eyes Have It: A Task by Data Type Taxonomy for Information Visualization. In: Proceedings IEEE Conference on Visual Languages (1996)

20. Sowa, J.F.: Principles of semantic networks. Explorations in the representation of knowledge. Morgan Kaufmann, San Francisco (1991)

21. Theseus, www.theseus-programm.de (cited on December 1, 2008)

22. Tory, M., Moller, T.: Rethinking Visualization: A High-Level Taxonomy. In: Proceedings IEEE Symposium on Information Visualization, pp. 151–158 (2004)

23. Ware, C.: Information Visualization: Perception for Design. Morgan Kaufmann, San Francisco

An Evaluation Framework and Adaptive Architecture for Automated Sentiment Detection

Stefan Gindl, Johannes Liegl, Arno Scharl, and Albert Weichselbraun

Abstract. Analysts are often interested in how sentiment towards an organization, a product or a particular technology changes over time. Popular methods that process unstructured textual material to automatically detect sentiment based on tagged dictionaries are not capable of fulfilling this task, even when coupled with part-of-speech tagging, a standard component of most text processing toolkits that distinguishes grammatical categories such as article, noun, verb, and adverb. Small corpus size, ambiguity and subtle incremental change of tonal expressions between different versions of a document complicate sentiment detection. Parsing grammatical structures, by contrast, outperforms dictionary-based approaches in terms of reliability, but usually suffers from poor scalability due to its computational complexity. This work provides an overview of different dictionary- and machine-learning-based sentiment detection methods and evaluates them on several Web corpora. After identifying the shortcomings of these methods, the paper proposes an approach based on automatically building Tagged Linguistic Unit (TLU) databases to overcome the restrictions of dictionaries with a limited set of tagged tokens.

1 Introduction

Sentiment Detection (SD) is the part of Natural Language Processing (NLP) that deals with the automated extraction of opinions (the 'sentiment') out of unstructured text. The goal is to automatically decide whether an author expresses positive or negative sentiment towards a certain topic. The appeal of this research area lies in

Stefan Gindl, Johannes Liegl, and Arno Scharl
MODUL University Vienna, Department of New Media Technology, Austria
e-mail: stefan.gindl,johannes.liegl,arno.scharl@modul.ac.at

Albert Weichselbraun
Vienna University of Economics and Business Administration,
Research Institute for Computational Methods, Austria
e-mail: albert.weichselbraun@wu-wien.ac.at

S. Schaffert et al. (Eds.): Networked Knowledge - Networked Media, SCI 221, pp. 217–234.
springerlink.com   © Springer-Verlag Berlin Heidelberg 2009

its wide range of possible applications, since reliable automated SD methods allow the analysis of large texts corpora beyond the limits of manual approaches.

The information obtained by this process may be used for several purposes. Applications include monitoring the launch and performance of commercial products, analyzing the electoral behavior of the public to guide political campaigns, or refining search engines to consider opinions. Yet, SD is a very ambitious problem to solve. NLP is one of the most challenging research areas in computer science, since natural languages are not as restrictive as formal languages. Natural language allows authors to express concepts in many different ways, which complicates automated analyses. Consider the following sentence:

The plot of the movie was banal and the actors were really clumsy.

This sentence expresses a viewer's displeasure with a particular movie. Now consider the same sentence in the following context:

The plot of the movie was banal and the actors were really clumsy. However, I enjoyed it more than any other movie I have seen in the last few months!

In this example, both sentences describe the same item (a particular movie), but differ in regard to the expressed sentiment. For an automated system, such constructs are very hard to evaluate. A human reader, by contrast, easily recognizes that the viewer liked the movie. Linguistic notions such as sarcasm or irony are even harder to spot by an algorithm.

This paper evaluates and compares several well-known SD techniques, such as the bag-of-words approach and maximum entropy modeling. Based on this analysis, we develop an alternative approach based on Tagged Linguistic Units (TLUs), annotating tokens and phrases with additional features such as part-of-speech (POS), context and topic.

The remainder of this article is structured as follows. Section 2 provides an overview of related work. Section 3 compares deep parsing strategies with approaches focusing on lexis. Section 4 describes state-of-the-art SD methods in greater detail, which are then evaluated in Section 5. After discussing the results, we identify weaknesses in current approaches and propose a novel method based on Tagged Linguistic Units in Section 6. Section 7 concludes the paper and presents an outlook on further research.

2 Related Work

The field of SD reveals emotional aspects of a written text, hinting at the opinion and intention of the author. This information can be used for several reasons: search engines can augment their results, marketing managers can find out why their product failed in a certain market, and political analysts can predict electoral behavior. The challenge of detecting sentiment in unstructured text leads to a vast amount of different approaches to tackle this task. Some of these only use binary decisions (a positive or negative sentiment), others use more sophisticated classifications.

The context of a sentiment term influences its meaning - e.g., in 'the president of the National Environment Trust', the term 'trust' refers to a large enterprise and not to 'confidence'. Wilson et al. [24] acknowledge the importance of context information by using a set of 28 features such as modifiers or adjacent terms, which are input to the AdaBoost machine learning approach.

Lexical units can also be distinguished from each other by using so called 'appraisal taxonomies' [22]. These contain information on the 'attitude' (e.g., 'appreciation' or 'affect'), the 'orientation' (positive vs. negative), the 'force' (can be increased by modifiers like 'very'), or the 'polarity' (a binary decision depending on the existence of a negation trigger) of words.

Hatzivassiloglou and McKeown [4] base predicting the sentiment of adjectives on the hypothesis that conjoined adjectives may carry the same sentiment charge. Based on this hypothesis, their proposed system assigns an adjective with unknown sentiment the same sentiment value as its conjoined adjective.

Pang et al. [12] apply machine learning methods (Naive Bayes, Maximum Entropy Model, Support Vector Machines) in combination with a bag-of-features (i.e., a collection of terms with certain characteristics such as a sentiment) framework to a data set containing reviews from the 'Internet Movie Database'. Pang and Lee present a refinement of this approach in their later work [11], where they involve a previous subjectivity classification (i.e., a method capable of discriminating sentences into subjective and objective ones). As compared to objective sentences that are only used to describe facts, subjective sentences are supposed to reflect the opinion an author intends to express. Kushal et al. [7] also apply three machine learning methods to product reviews, comparing their results to a simple baseline algorithm. Mullen and Collier [10] work with Support Vector Machines, where a list of terms and their sentiment values (i.e., a value corresponding to the general affinity of the term to express positive or negative opinion) represents the features. A generic process using Pointwise Mutual Information then determines the sentiment values of these terms.

Yu and Hatzivassiloglou [25] present an approach for subjectivity classification using a Naive Bayes classifier. Riloff and Wiebe [14] present a bootstrapping approach to automatically create large training sets in order to learn extraction patterns for subjectivity. In another work, Wiebe and Riloff [23] produce training data for the training of a Naive Bayes subjectivity classifier by employing a rule-based classifier. Subasic and Huettner [19] apply fuzzy methods to analysing affect in writings. Blitzer et al. [2] present an approach using similarities between differing domains in order to adapt a sentiment classifier to a new domain. Ding et al. [3] determine the sentiment of a sentence in regard to a specific object within this sentence (in this case, objects refer to products like cameras). Conjunction rules help accomplish this task for both the usage within a sentence as well as multiple sentences. Another feature is a distance function, which determines the correlation of sentiment terms considering their absolute distance to a specific object.

3   Lexical Approaches versus Full Parsing

Capturing the evolution of information spaces calls for a new generation of robust, language-independent and distributed natural language processing techniques optimized for throughput and scalability. From a stakeholder perspective, sentiment expressed in textual material (e.g., news media coverage) is of particular interest [17]. Automated methods to compute sentiment, however, usually belong to one of the following two categories: (i) low-overhead approaches that focus on the lexis of text, and (ii) full parsing of grammatical structures, which improves the accuracy of results but suffers from poor scalability. This paper presents a new method that falls into the first category but aims to improve the quality of results by building an adaptive databases of *tagged linguistic units*. Such a database helps ensure scalability, preserve context information and process heterogeneous data sources.

Most research projects that apply automated sentiment detection techniques such as the *US Election 2008 Web Monitor* (www.ecoresearch.net/election2008) or the *Media Watch on Climate Change* (www.ecoresearch.net/climate) typically gather a large corpus of text compiled from many sources and sampled in regular intervals. Using POS tagged and partially parsed corpora to identify relevant sketches (= co-occurrence lists for grammatical patterns provided by a grammar rule engine) improves the performance of existing SD-techniques [5, 6], but processing arbitrarily long blocks of text still requires a fundamentally new strategy. The ability to work with very short textual segments is paramount when trying to analyze the *evolution* of knowledge reflected in corpora. Longitudinal studies of specific topics or events often yield few additional occurrences of a term in a given interval, as incremental changes to existing documents are common. This complicates the analysis, because the validity of many text processing methods depends on corpus size and frequency of target terms.

Given the unresolved scalability issues of SD methods that rely on full parsing, this paper describes attempts to extend and improve lexis-based approaches with a special focus on context-aware processing. The next section will summarize standard dictionary-based SD methods and compare them to machine learning approaches.

4   Algorithm Description

This section focuses on the most common SD methods (arithmetic, machine learning based and combined), and describes a framework for evaluating them based on three different corpora compiled from Web resources available to the public:

- *Amazon* (www.amazon.com) provides customer reviews ranging from "one-star" (low recommendation) to "five-star" (high recommendation) ratings. The Amazon data set consists of 165,746 book reviews and contains 1,539,058 sentences.

- The *Internet Movie Database (IMDb)* (www.imdb.com) contains 2000 reviews comprising 69,207 sentences. The IMDb data set was also used in [12], thus the reviews already carry information on positive and negative sentiment.
- *TripAdvisor* (www.tripadvisor.com) provides reviews of holiday destinations. It contains 7554 reviews with ratings from one to five stars, where one star indicates a very low recommendation and five stars a high recommendation. This data set comprises 62,818 sentences.

Amazon and TripAdvisor rate each review on a scale from one to five stars. We generalize these ratings and consider all reviews with a rating lower than three as negative, all reviews with a rating greater than three as positive, and ratings of three as neutral. In order to avoid adulterated results, we use balanced versions of the data sets - i.e., subsets of the original data containing exactly the same number of positive and negative reviews. The Amazon data set contains 21,458 negative, 130,061 positive and 14,227 neutral reviews. The TripAdvisor data set consists of 1105 negative, 5673 positive and 776 neutral reviews. The balancing filter yields a total of 420,840 sentences from Amazon and 17,768 sentences for TripAdvisor (IMDb provides an already balanced data set).

4.1 Arithmetic Methods

The arithmetic methods are based on tagged dictionaries, which contain sentiment terms with corresponding sentiment values in a closed interval [-1,1]. For example, 'champion' is a positive word carrying the sentiment value '1', whereas 'charlatan' carries the negative value '-1'. The dictionary contains a total of 8267 sentiment terms, 5072 of them positive and 3195 negative. The tagged dictionary is not domain-specific, which helps draw conclusions on its general applicability. Subjecting the General Inquirer (www.wjh.harvard.edu/~inquirer/) dictionary to a reverse lemmatization process yielded 7302 terms, 965 additional entries were manually retrieved from a sample of online blogs. Arithmetic algorithms browse through the reviews and search for terms contained in the dictionary. The number of detected terms gives information about the overall sentiment value of a sentence. Each of the following methods calculates the overall sentiment of a review by summing up all sentiment values of the individual sentences.

- *Simple SD (SSD)* counts the values of sentiment terms in a sentence. If the sum of these values is positive (negative), the sentence is considered to have a positive (negative) sentiment. If a negation trigger such as 'not' and 'never' occurs directly before a sentiment term (e.g., 'The proposal was not approved'), the value of this term is multiplied by '-1', resulting in an inverted sentiment value. We used this method as a simple baseline approach towards SD.
- *Extended SD (ESD)*. This method incorporates other semantic components affecting the results. We extended the former detection method by so called modifier terms (e.g., 'very', 'rather'). If such a term occurs before a sentiment word, the value of the sentiment word is either increased or decreased, depending on the orientation of the modifier. The term 'very' increases a sentiment value (e.g.,

'The candidate is very charming.'); we, therefore, multiply the original value by '1.5'. In the case of the decreasing term 'little', we multiply the term's sentiment by '0.5' (e.g., 'The patient felt little pain.').

- *Adjective Detection (AD)*. Adjectives are often used to express sentiment. For that reason, we investigated the outcome of a SD method using only adjectives. In order to limit the method to adjectives, we applied the POS tagger of the OpenNLP project (opennlp.sourceforge.net).
- *Detailed Part-Of-Speech Detection (DetPOSD)*.This method applies POS tagging to determine the scope of a negation trigger. For each occurrence of a term with a semantic value, the method tries to identify a negation trigger that instructs the algorithm to multiply the sentiment value by '-1'. Certain constituents help refine this procedure and avoid negation triggers from impacting the complete sentence (although they were not meant to do so). If a noun phrase respectively a verb phrase is positioned between the sentiment word and the negation trigger, this term is regarded as negated and the original sentiment word will remain unaffected. Figure 4.1 shows an illustration of this procedure: a negation trigger occurs at the beginning of the sentence (NT) and a sentiment token at the end (SentT). Between these are placed a number of arbitrary tokens (AT; this can be determiners, adjectives etc.) that do not influence the negation. Yet, the stop token (ST; this can be either verbs or nouns) decides that the trigger does not influence the sentiment token, and thus, the sentiment token remains as being not negated.

Fig. 1 Scope determination for negation triggers (NT=Negation Trigger, AT=Arbitrary Token, ST=Stop Token, SentT=Sentiment Token)

4.2 Machine Learning Methods

In the following, we compare three different methods: a language model, a Naive Bayes classifier and a Maximum Entropy Model (Section 5). These methods do not use a tagged dictionary but build their knowledge base in a training step. The existing classification of the data sets suggests using supervised learning. Performing experiments with a generic training set and evaluating the results on a domain-specific test set sheds light on the methods' universality.

The experiments on *generic knowledge bases* trained the Language Model and Naive Bayes algorithm on the IMDb data set, using the TripAdvisor and Amazon sets for testing purposes. In a follow-up step, a model on the TripAdvisor data set was trained to be tested on the IMDb set. While this procedure allows training and testing on the complete datasets (avoiding the need to split into training and test sets), it faces domain-dependent constraints (since machine learning methods tend to strongly fit to the domain they were trained on). Training the Maximum Entropy Model with a part of each data set as a training set and the other part as a test set yields the *domain specific knowledge base*.

The LingPipe libraries[1] and OpenNLP MaxEnt Package[2] helped streamline the implementation of the different learning methods. The following itemization provides a detailed explanation of the three methods.

- *Language Model (LM)*. The evaluation uses an implementation of a LingPipe language model classifier to create a language model. A language model is a probabilistic representation of a sequence of words. We trained the language model by separately providing the classifier with positive and negative reviews (thus, the classifier *knew* the sentiment class of the presented review). In the next step, the created model had to predict the sentiment class of reviews of unknown sentiment.
- *Naive Bayes (NB)*. A Naive Bayes classifier proceeds on the conditional independence assumption, which expects the attributes allowing a classification to be independent from each other [11]. Although most real-world applications violate this assumption, the algorithm yields surprisingly good results. Zhang [26] explains this good performance by suggesting that two attributes may depend on each other in a given data set, but the dependence may be distributed evenly in each class.
- *Maximum Entropy Model (MaxEnt)*. Maximum Entropy Models can integrate features from heterogeneous information sources without posing strong independence assumptions like the Naive Bayes approach. Features correspond to constraints in the model, and the Generalized Iterative Scaling algorithm [13] outputs the model, which maximizes the entropy among the constraints. The method yields the model preserving the most uncertainty. This is desired, because every other model would add information that is not justified by empirical evidence (i.e. the training data) [1, 9]. Each data set required a unique Maximum Entropy Model, using only unigrams as features given their good performance in previous studies [12]. One-third of the reviews of the corresponding dataset were used to train the model, leaving the remaining two thirds of the dataset for evaluation purposes.

Generic approaches to sentiment detection represent a challenging problem. Domain-specific methods are generally assumed to deliver superior results. The evaluation of NB and LM across domains allowed investigating whether these methods could be used for multiple domains without having access to domain-specific training corpora. Alternatively, the Maximum Entropy Model was trained on a subset of a corpus and tested on another subset of the same corpus, yielding a model specifically fitted to the domain.

---

[1] http://alias-i.com/lingpipe/
[2] http://maxent.sourceforge.net/

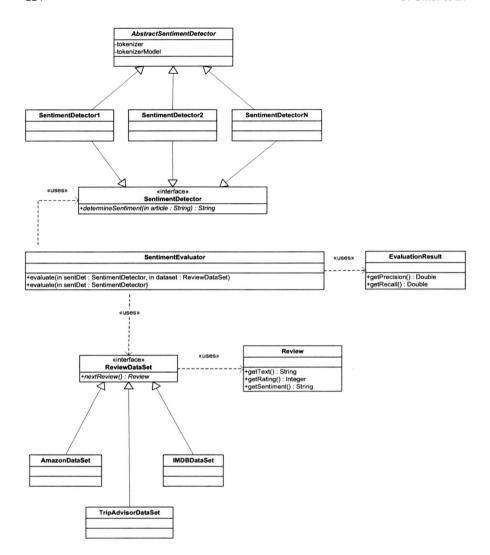

Fig. 2 UML diagram of the evaluation framework

4.3 System Architecture

The evaluation framework presented in the following allows comparing the results of the various SD methods through the SentimentDetector interface. New data is integrated by implementing the ReviewDataSet interface. To evaluate an SD method on a data set, implementations of both have to be passed to the evaluation method of the SentimentEvaluator. This component iterates the reviews in the specified data set and applies the specified SentimentDetector on each of them. The results of the SD for each review is then compared with the review's original rating and the outcome is classified as true positive, true negative, false positive or false negative for the calculation of statistical measures (see Subsection 5.2 for details on the used statistical measures). Figure 2 shows a UML diagram of the evaluation framework.

5 Evaluation

The evaluation focuses on the SD method's ability to put a review into the right polarity class (positive or negative). The IMDb data set is already divided into positive and negative classes. For the Amazon and TripAdvisor data sets, a pre-processing module maps user ratings between 'one star' and 'five stars' to the classes 'negative', 'neutral', and 'positive' as outlined in Section 4.

5.1 Statistical Properties of the Corpora and Implications

This section describes the statistical structure of the evaluation corpora. Table 1 lists the minimum, maximum, average and standard deviation of (a) the number of occurring positive and negative tokens from the tagged dictionary in the data set, (b) the absolute number of tokens, and (c) the number of sentences in a review.

The results of the descriptive statistics show that the Amazon and TripAdvisor data sets are very heterogeneous, caused by a number of extreme outliers (according to their large standard deviation, which even exceeds the average in the Amazon data set for the positive and negative tokens). The IMDb data, by contrast, presents itself as being more homogeneous. Another advantage of this set is the fact that each

Table 1 Statistical characteristics of the review data sets

Data set	Param.	Pos. Tokens	Neg. Tokens	Single Terms	Sent.
IMDb	Max	164	138	2753	124
	Min	1	1	18	1
	Avg	45.76	37.19	761.56	34.6
	StdDev	21.75	18.18	334.82	16.15
TripAdvisor	Max	70	43	1240	62
	Min	0	0	1	1
	Avg	8.66	4.81	160.11	8.04
	StdDev	7.71	5.05	130.23	5.83
Amazon	Max	260	226	4878	157
	Min	0	0	1	1
	Avg	12.32	7.86	211.91	9.81
	StdDev	12.81	9.36	209.73	8.99

sentence contains at least one sentiment token of the positive and negative class. In the case of the other two sets, this is not ensured. A number of zero sentiment tokens in a review would lead to a result of zero for the review, which is then considered as being a neutral review. We do not filter such reviews, since we assume that reviews containing no sentiment token express a neutral opinion.

5.2 Detailed Results

The evaluation considers five statistical parameters: recall, precision, accuracy, F measure, and Cohen's kappa coefficient. Recall is a measure for the completeness of a detection method - i.e., it shows how many of the requested objects could actually be found. On the other hand, precision provides a measure for the number of objects that have been identified correctly. The accuracy is the ratio of all correctly identified objects and the (either correctly or incorrectly) classified objects. The F measure combines recall and precision. Cohen's kappa coefficient is normally used to measure the inter-rater-reliability, that is, how strongly different raters classifying a number of objects agree on the classification of these objects.

We calculate recall and precision for the positive and negative class separately. Separate precision and recall results for the positive and negative class are required because the classifiers we use can also return a neutral result (namely when no sentiment token occurs in a review). Therefore, a document that is not negative is not automatically positive. This procedure also leads to separate results for Cohen's kappa value as well as the F measure. Tables 2 to 4 show the detailed evaluation results of each data set.

Using the IMDb data set resulted in fairly balanced results. The arithmetic methods achieve good results for the detection of positive as well as negative reviews

Table 2 Evaluation of the SD methods applied to the IMDb data set (FM=F measure)

| | **IMDb Data Set** | | | | | | | | | |
| | **Positive Sentiment** | | | | | **Negative Sentiment** | | | | |
Detection Method	Rec.	Prec.	Acc.	Cohen's Kappa	FM	Rec.	Prec.	Acc.	Cohen's Kappa	FM
				Generic Methods						
SSD	70.1	63.38	64.8	0.3	0.67	59.4	66.52	64.75	0.29	0.63
ESD	68.2	64.1	65	0.3	0.66	61.9	66.06	65.05	0.3	0.64
AD	67.7	60.39	61.65	0.23	0.64	55.4	63.17	61.55	0.23	0.59
DetPOSD	69.3	63.52	64.75	0.29	0.66	60.3	66.26	64.8	0.3	0.63
LM	16.1	70.61	54.7	0.09	0.26	**90.7**	53.2	55.45	0.11	0.67
NB	**96.5**	51.09	52.05	0.04	0.67	5.4	65.85	51.3	0.03	0.1
				Domain-Specific Method						
MaxEnt	80.51	**83.39**	**82.23**	**0.64**	**0.82**	83.96	**81.16**	**82.23**	**0.64**	**0.83**

Table 3 Evaluation of the SD methods applied to the TripAdvisor data set (FM=F Measure)

Detection Method	TripAdvisor Data Set									
	Positive Sentiment					Negative Sentiment				
	Rec.	Prec.	Acc.	Cohen's Kappa	FM	Rec.	Prec.	Acc.	Cohen's Kappa	FM
	Generic Methods									
SSD	93.39	65.61	72.22	0.44	0.77	30.41	86.6	62.85	0.26	0.45
ESD	92.76	66.13	72.62	0.45	0.77	32.04	85.71	63.35	0.27	0.47
AD	83.62	65.67	69.95	0.4	0.74	26.52	76.1	59.1	0.18	0.39
DetPOSD	93.21	66.15	72.76	0.46	0.77	31.67	86.63	63.39	0.27	0.46
LM	40.09	57.53	55.25	0.1	0.47	63.44	58.91	59.59	0.19	0.61
NB	49.77	**76.92**	67.42	0.35	0.6	**77.01**	67.86	70.27	0.41	**0.72**
	Domain-Specific Method									
MaxEnt	**93.89**	67.71	**74.56**	**0.49**	**0.79**	55.22	**90.04**	**74.56**	**0.49**	0.68

Table 4 Evaluation of the SD methods applied to the Amazon data set (FM=F Measure)

Detection Method	Amazon Data Set									
	Positive Sentiment					Negative Sentiment				
	Rec.	Prec.	Acc.	Cohen's Kappa	FM	Rec.	Prec.	Acc.	Cohen's Kappa	FM
	Generic Methods									
SSD	77.23	57.27	59.8	0.2	0.66	39.78	66.96	60.08	0.2	0.5
ESD	75.88	57.56	59.97	0.2	0.65	41.5	66.39	60.25	0.2	0.51
AD	59.73	55.88	56.28	0.13	0.58	39.51	63.06	58.18	0.16	0.49
DetPOSD	76.89	57.59	60.13	0.26	0.66	40.79	67.11	60.4	0.21	0.51
LM	41.67	72.9	63.09	0.26	0.53	75.17	62	64.55	0.29	0.68
NB	47.98	68.88	63.15	0.26	0.57	69.86	63.14	64.54	0.29	0.66
	Domain-Specific Method									
MaxEnt	78.73	87.02	83.49	0.67	0.83	88.26	80.58	83.49	0.67	0.84

(considering the usage of a domain-independent tagged dictionary). The TripAdvisor data set yields the best results for the detection of positive reviews. Yet, this outstanding performance is accompanied by quite poor results in the detection of negative sentences. The Amazon data set also satisfyingly identifies positive reviews at the cost of an inferior precision for negative reviews.

The better results in the positive category represents a surprising result, since the tagged dictionary contains more negative than positive sentiment tokens (5072 negative in contrast to 3195 positive ones). In spite of this, the statistical analysis in 5.1 shows that in all data sets, a larger number of positive tokens occurred. This fact leads to the assumption that customers use positive tokens more frequently than negative ones and that positive words might also be used in order to express a negative opinion towards a movie, book or holiday destination (e.g., in the case of humor or sarcasm). Additional context information would help resolve some of these cases and determine sentiment more accurately.

The Maximum Entropy Model, which entails domain-specific knowledge, outperforms the other methods. It produces results with the highest precision, recall and kappa value in the negative classification task as well as in the positive. These findings do not suggest that arithmetic SD generally provides inferior results, but that the knowledge base and the application domain play an important role in the identification of negative sentiment. Methods that consider the domain context (see Section 6) therefore have the potential to yield much better results.

5.3 Discussion

The evaluation results show that the presented SD methods have their strength in the identification of reviews with positive sentiment (high recall). Only a relatively small number was overlooked by the algorithms. On the other hand, the method's precision is less satisfactory. A rather high amount of items has been incorrectly identified as having a positive sentiment.

On the TripAdvisor data set, the methods achieve an excellent recall between 83% and 93% without any decrease in precision. We assume that the writing style of this kind of data alleviates the SD - at least for positive sentiment. The results for the detection of negative sentiment are less encouraging. It seems difficult to correctly extract reviews with negative sentiment (very low recall). Yet, precision does not decrease to the same extent. As for precision, the SD on the TripAdvisor data set again outperforms the results obtained with the other data set.

We assume that the structure of the reviews in the IMDb and Amazon data set strongly influences the outcomes. Reviews of movies and books often integrate plot summaries into the evaluation. In the case of love films, for example, a notable number of words carrying positive sentiment like 'love', or 'happy' (if the film has a happy end) will occur, even when the reviewer dislikes the product. The same consideration applies to horror films or thrillers that contain negative vocabulary in the plot summary.

The Maximum Entropy model clearly outperformed the other SD methods, particularly in detecting negative sentiment, which is not surprising given that it has been trained and tested within the same domain. This should guide future research and favors domain-specific components whenever the required context information is available. Building on this insight, the following section proposes to build databases of Tagged Linguistic Units (TLU). Such a repository contains a comprehensive list of terms of a certain language together with their significance for emotional speech (i.e., their sentiment value) and additional metadata.

6   Tagged Linguistic Units

Tagged Linguistic Units (TLUs) comprise units of linguistic content such as terms and phrases, coupled with a set of annotations (e.g., POS tags, topic or prevalent context). They combine the advantages of methods that go beyond lexis without inheriting the full complexity of grammar parsing. The following sections outline the generation of TLU databases and their application to sentiment detection.

6.1   Database Creation

As already mentioned in section 4, simple SD methods that do not use machine learning algorithms on narrowly defined domains rely on a tagged dictionary that distinguishes between positive- and negative-valued sentiment words [16]. Such dictionaries typically contain a few thousand mappings from words to their associated sentiment values - e.g., the General Inquirer [18]). They can be subjected to a reverse lemmatization procedure, adding inflections to the initial list of sentiment words. Even assuming such an extended tagged dictionary, dictionary-based approaches do not take the context of sentiment words into account, which limits their usefulness in corporate knowledge architectures.

The rest of this paper addresses this shortcoming by proposing a hybrid method based on spreading activation networks coupled with machine learning algorithms for assigning sentiment values to linguistic units. For this purpose, the following linguistic units for computing sentiment will be distinguished: *unigram* (single word), *n-gram* (multiple-word units of meaning), and *concepts* (units of meaning not tied to a particular lexical form and represented via rules or regular expressions, e.g. *climate change ⇔ global warming*).

A sentiment value and a context (e.g., part-of-speech, geographic location and named entity) are assigned to each linguistic unit. For a given amount of text, these mappings taken together are the building blocks of a *Tagged Linguistic Unit (TLU)* database. The sentiment values stored in this database are constantly being updated based on new data from the knowledge acquisition services and can be customized for specific domains, applications or users. Generating and using a TLU database instead of a tagged dictionary that only contains words and binary classifications allows a fine-grained differentiation between sentiment values associated with morphologically similar but semantically different linguistic units such as *cell, fuel cell* and *prison cell* through the consideration of contextual information like POS tags, geo tags and named entity tags.

Work by Scharl et al. [15] has demonstrated the usefulness of assigning sentiment values to geographic locations and also shows how heavily these values depend on other context dimensions. Future research will address these dependencies by combining tags with more sophisticated context information as for instance hierarchical classifications [20] or topic tags. This approach (i) is language-independent in the sense that only a small set of seed terms (e.g., 100 positive and 100 negative terms) and grammar patterns would be required to initialize the machine learning algorithm

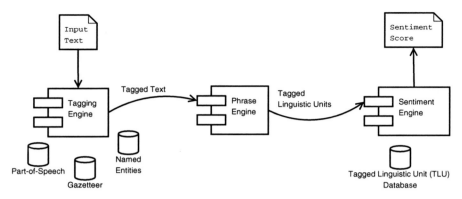

Fig. 3 Sentiment scoring based on linguistic units

and fine-tune sentiment values to any language that is decomposable into unigrams, n-grams and concepts, (ii) is not restricted to the sentiment categories 'positive' and 'negative', but supports an arbitrary number of linguistic categorizations such as weak ⟷ strong, passive ⟷ active, etc., (iii) ensures that every sentence or document can be annotated; traditional approaches often encounter sentences that do not contain any of the words listed in the tagged dictionary.

Figure 3 illustrates sentiment scoring based on linguistic units. The phrase engine identifies the linguistic units.

The tagging engine identifies part-of-speech tags, named entities, and geographic locations. The sentiment engine processes linguistic units and associated tags based on the data in the tagged linguistic units database, computing a sentiment value for the given text. Tagging provides important background information for these tasks. In the most straightforward case, the sentiment of linguistic units, as for instance the word `like`, depends on the assigned part-of-speech tag (`like/VB` versus `like/IN`). In more complex cases, named entity tags or even geo tags might be necessary to correctly identify the TLU's sentiment value (e.g., in the case of `National Environment Trust`).

6.2 *Iterative Extension and Optimization*

As outlined in the previous section, TLU databases can be easily customized to specific domains and use cases. A domain-specific corpus, language-specific grammar rules and a set of seed terms with "known" sentiment values (e.g., from conventional tagged dictionaries such as the General Inquirer repository) initialize the TLU database. The architecture identifies unknown linguistic units in the corpus and determines their sentiment value as illustrated in Figure 4.

The tagging component marks sentences with part-of-speech tags and identifies named entities such as people, organizations, and geographic locations. Combining co-occurrence analysis with a grammar rule engine yields candidate terms for

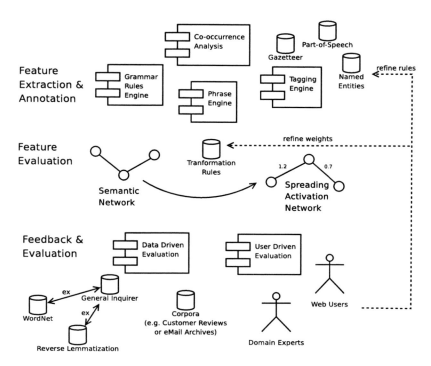

Fig. 4 Iterative fine-tuning of the Tagged Linguistic Unit (TLU) database

extending the TLU database. Annotating these terms with named entity tags and encoding characteristic grammatical patterns and known phrases creates a complex semantic network, which describes the relations between linguistic units.

Liu et al. [8] demonstrated how decomposing and translating semantic networks based on heuristic rules yield a spreading activation network for extending domain ontologies. Applying this approach to identifying and tracking tagged linguistic units builds a spreading activation network used to distribute the sentiment charges between the units based on the features and annotations generated during the annotation step. Activation of concepts with known sentiment charges in accordance to sign and strength of the charge leads to the propagation of energy pulses through the network, eventually distributing charges to all linguistic units. Analyzing the sentiment values' variance allows estimating confidence levels and identifying synonym ↔ antonym relationships.

Feedback gathered in the evaluation step adjusts and optimizes the transformation rules for a given domain and corpus, improving the quality of the TLU database with every subsequent step. Automatic data-driven evaluation on a TLU level will help assess overall performance. Using the evaluation framework outlined in Section 5 on various publicly available Web corpora will provide test cases for TLU-based sentiment detection. Automated methods will be complemented by user-driven evaluations from domain experts and Web users. The feedback gathered by the data- and user-driven evaluations will be utilized to refine the transformation rules of the feature evaluation, and to identify candidate patterns for the inclusion into the databases of the grammar rule engine and the phrase engine.

Automatically generating TLU databases faces the problem of determining the correct charge (+0.4 vs. -0.4, for example) of the sentiment value to be assigned to the linguistics unit. The problem arises from the fact that synonyms and antonyms have very similar (co-)occurrence patterns in a given corpus. Advanced relation discovery techniques developed within the AVALON project [21] will help overcome this challenge and facilitate the automation of this classification process. The machine learning algorithms will be trained and evaluated on augmented tagged dictionaries (created through reverse lemmatization and adding WordNet synonym and antonym pairs), as well as on public pre-tagged corpora.

7   Conclusion and Outlook

Simple approaches to sentiment detection based on patterns of co-occurrence with terms from tagged dictionaries scale well but provide less accurate results compared to complex methods that require a full parsing of sentence structures. The sheer volume of textual data and economic considerations, however, frequently rule out the most sophisticated approaches. Continuously updated databases of tagged linguistic units aim to balance accuracy and throughput. They add an adaptive layer to static sentiment detection approaches based on tagged dictionaries, which still tend to be compiled manually.

Preliminary results from the described approach are promising. Following a formal evaluation of different approaches to sentiment detection, recall and precision were significantly improved by adding WordNet synonyms and antonyms to the tagged dictionary (only considering synsets with high frequencies to exclude rare expressions). Currently, terms extracted from media corpora serve as candidates for assigning sentiment values via co-occurrence analysis, which will further extend the tagged dictionary.

Text mining projects have to process hundreds of thousands or millions of documents in short intervals. Thus they significantly benefit from accurate methods of determining sentiment with minimal computational requirements at run time. While the creation of tagged linguistic unit databases is computationally intense, the overhead of applying them within annotation components remains small. Improved sentiment detection algorithms will encourage their use in both academic and commercial applications. Refined versions of the sentiment detection methods presented in this paper will generate a richer set of context information (e.g., ontology concepts or explicit references to other types of structured knowledge), and consider this information in the scoring process.

Acknowledgment

This work was developed as a part of the RAVEN (Relation Analysis and Visualization for Evolving Networks; www.modul.ac.at/nmt/raven) research project funded by the Austrian Ministry of Transport, Innovation & Technology (BMVIT) and the Austrian Research Promotion Agency (FFG) within the strategic objective FIT-IT (www.fit-it.at). The authors wish to thank Arinya Eller for proof-reading the manuscript and Ivo Ponocny for his valuable feedback and suggestions.

References

1. Berger, A.L., Pietra, S.D., Pietra, V.J.D.: A maximum entropy approach to natural language processing. Computational Linguistics 22(1), 39–71 (1996)
2. Blitzer, J., Dredze, M., Pereira, F.: Biographies, bollywood, boom-boxes and blenders: Domain adaptation for sentiment classification. In: Proceedings of the 45th Annual Meeting of the Association of Computational Linguistics, Prague, Czech Republic, pp. 440–447 (June 2007)
3. Ding, X., Liu, B., Yu, P.S.: A holistic lexicon-based approach to opinion mining. In: WSDM 2008: Proceedings of the international conference on Web search and web data mining, Palo Alto, California, USA, pp. 231–240. ACM, New York (2008)
4. Hatzivassiloglou, V., McKeown, K.R.: Predicting the semantic orientation of adjectives. In: Proceedings of the eighth conference on European chapter of the Association for Computational Linguistics, Morristown, NJ, pp. 174–181. Association for Computational Linguistics (1997)
5. Kilgarriff, A., Evans, R., Koeling, R., Rundell, M., Tugwell, D.: Waspbench: A lexicographer's workbench supporting state-of-the-art word sense disambiguation. In: 10th Conference on European Chapter of the Association For Computational Linguistics, Morristown, USA, Association for Computational Linguistics (2003)
6. Kilgarriff, A., Rychl, P., Smrz, P., Tugwell, D.: The Sketch engine. In: 11th Euralex international Congress. Lorient, France (2004)
7. Kushal, D., Lawrence, S., Pennock, D.M.: Mining the peanut gallery: Opinion extraction and semantic classification of product reviews. In: WWW 2003: Proceedings of the twelfth international conference on World Wide Web, pp. 519–528. ACM Press, New York (2003)
8. Liu, W., Weichselbraun, A., Scharl, A., Chang, E.: Semi-automatic ontology extension using spreading activation. Journal of Universal Knowledge Management (1), 50–58 (2005), http://www.jukm.org/jukm_0_1/semi_automatic_ontology_extension
9. Manning, C.D., Schütze, H.: Foundations of Statistical Natural Language Processing. The MIT Press, Cambridge (1999)
10. Mullen, T., Collier, N.: Sentiment analysis using support vector machines with diverse information sources (2004)
11. Pang, B., Lee, L.: A Sentimental Education: Sentiment Analysis Using Subjectivity Summarization Based on Minimum Cuts (September 2004)
12. Pang, B., Lee, L., Vaithyanathan, S.: Thumbs up? Sentiment Classification using Machine Learning Techniques. In: Proceedings of the 2002 Conference on Empirical Methods in Natural Language Processing, EMNLP (2002)

13. Ratnaparkhi, A.: Maximum entropy models for natural language ambiguity resolution (1998)
14. Riloff, E., Wiebe, J.: Learning extraction patterns for subjective expressions. In: Proceedings of the 2003 Conference on Empirical Methods in Natural Language Processing (EMNLP 2003) (2003)
15. Scharl, A., Dickinger, A., Weichselbraun, A.: Analyzing news media coverage to acquire and structure tourism knowledge. Information Technology and Tourism 10(1), 3–17 (2008)
16. Scharl, A., Pollach, I., Bauer, C.: Determining the semantic orientation of web-based corpora. In: Liu, J., Cheung, Y.-m., Yin, H. (eds.) IDEAL 2003. LNCS, vol. 2690, pp. 840–849. Springer, Heidelberg (2003)
17. Scharl, A., Weichselbraun, A.: An automated approach to investigating the online media coverage of us presidential elections. Journal of Information Technology & Politics 5(1), 121–132 (2008)
18. Stone, P.J.: The General Inquirer: A Computer Approach to Content Analysis. The MIT Press, Cambridge (1966)
19. Subasic, P., Huettner, A.: Affect analysis of text using fuzzy semantic typing. IEEE Transaction on Fuzzy Systems 9(4), 483–496 (2001)
20. Weichselbraun, A.: Ontologiebasierende Textklassifikation mittels mathematischer Verfahren. PhD thesis, Vienna University of Economics and Business Administration (2004)
21. Weichselbraun, A., Wohlgenannt, G., Scharl, A., Granitzer, M., Neidhart, T., Juffinger, A.: Applying vector space models to ontology link type suggestion. In: 4th International Conference on Innovations in Information Technology, Dubai, United Arab Emirates, pp. 566–570. IEEE Computer Society Press, Los Alamitos (2007)
22. Whitelaw, C., Garg, N., Argamon, S.: Using Appraisal Taxonomies for Sentiment Analysis. In: Proceedings of MCLC 2005, the 2nd Midwest Computational Linguistic Colloquium, Columbus, US (2005)
23. Wiebe, J., Riloff, E.: Creating subjective and objective sentence classifiers from unannotated texts. In: Gelbukh, A. (ed.) CICLing 2005. LNCS, vol. 3406, pp. 486–497. Springer, Heidelberg (2005)
24. Wilson, T., Wiebe, J., Hoffmann, P.: Recognizing contextual polarity in phrase-level sentiment analysis. In: Proceedings of Human Language Technologies Conference/Conference on Empirical Methods in Natural Language Processing (HLT/EMNLP 2005), Vancouver, CA (2005)
25. Yu, H., Hatzivassiloglou, V.: Towards answering opinion questions: Separating facts from opinions and identifying the polarity of opinion sentences. In: Collins, M., Steedman, M. (eds.) Proceedings of EMNLP 2003, 8th Conference on Empirical Methods in Natural Language Processing, Sapporo, JP, pp. 129–136 (2003)
26. Zhang, H.: The optimality of naive bayes. In: Barr, V., Markov, Z. (eds.) FLAIRS Conference. AAAI Press, Menlo Park (2004)

Managing Ontology Lifecycles in Corporate Settings

Markus Luczak-Rösch and Ralf Heese

Abstract. Corporate Semantic Web describes the application of semantic technologies within enterprises for better knowledge management or enhanced IT service management. But, well-known cost- and process-oriented problems of ontology engineering hinder the employment of ontologies as a flexible, scalable, and cost-effective means for integrating data in small and mid-sized enterprises. We propose an innovative ontology lifecycle, examine existing tools towards the functional requirements of the lifecycle phases, and propose the vision of an architecture supporting them integratively.

1 Introduction and Related Work

Within the past years the Semantic Web community has developed a comprehensive set of standards and data formats to annotate all kinds of resources semantically. A main focus of current efforts lies on integrating publicly available data sources and publishing them as RDF on the Web. In contrast, many corporate IT areas are just starting to engage in Semantic Web technologies. Early adopters are in the areas of enterprise information integration, content management, life sciences, and government. Applying Semantic Web technologies to corporate content is known as *Corporate Semantic Web*. To employ ontologies as a flexible, scalable, and cost-effective means for integrating data in corporate contexts, *corporate ontology engineering* has to tackle cost- and process-oriented problems [10].

In Section 2 we present a set of requirements characterizing corporate settings for ontology-based information systems. We use these requirements in Section 3 to conclude the need of a new lifecycle model for continuously evolving ontologies, which we introduce afterwards. The lifecycle raises new functional requirements,

Markus Luczak-Rösch, Ralf Heese
Freie Universität Berlin, Institut für Informatik, AG Netzbasierte Informationssysteme, Takustr. 9, D-14195 Berlin, Germany
e-mail: {luczak,heese}@inf.fu-berlin.de

S. Schaffert et al. (Eds.): Networked Knowledge - Networked Media, SCI 221, pp. 235–248.
springerlink.com   © Springer-Verlag Berlin Heidelberg 2009

which we use for a comparison of some accepted tools for ontology engineering (Section 4) and a conclusion about their applicability.

Research has resulted in a wide range of ontology engineering methodologies which mainly differ in details referring to the composition of ontology engineering and application development, the range of users interacting on ontology engineering tasks, and the degree of lifecycle support. Some methodologies assume their users to be ontology experts only or at least to be knowledge workers with little technical experience while others also address users with no experience in ontology engineering at all.

METHONTOLOGY [5] transfers standards for software engineering to the task of ontology engineering and is a concept-oriented approach to build ontologies from scratch, reuse existing ontologies, or re-engineer knowledge. The lifecycle model of *METHONTOLOGY* does not respect any usage-oriented aspects.

The On-To-Knowledge methodology (*OTK*) developed by Sure and Studer [12] is less concept-oriented because it has an application-dependent focus on ontology engineering. It integrates participants which are not very familiar with ontologies in early phases of the process for identification of the use cases and competency questions. *OTK* assumes a centralized and a distributed strategy for ontology maintenance but neither presents a detailed description or evaluation of both strategies nor addresses ontology usage.

The methodologies *HCOME* [6] and *DILIGENT* [9] address the problem of ontology engineering from the viewpoint that reaching an ontology consensus is highly dependent on people with disparate skill level. Both methodologies assume a distributed setting. Every individual is free in adapting the central ontology consensus locally. The evolution of the consensual model is dependent on these local adoptions. Thus, *HCOME* and *DILIGENT* propose a human-centered approach, but they do not provide any application-dependent point of view.

Recently, the well-thought approach of agile engineering has come into focus of research in ontology engineering. In [1] the author introduces *RapidOWL* as an idea of agile ontology engineering. Auer proposes a paradigm-based approach without any phase model. *RapidOWL* is designed to enable the contribution of a knowledge base by domain experts even in absence of experienced knowledge engineers. However, the view on ontology usage is limited to the rapid feedback which is unspecific referring to the stakeholder who gives it and how it is given.

As a recent result of the NeOn project the *NeOn methodology* for ontology engineering and the NeOn architecture for lifecycle support in ontology-based systems have been developed [13]. Even though the architecture discusses usage-related ontology maintenance, both, the methodology and the toolkit, lack the agility of knowledge lifecycles and the relevance of this aspect for enterprise use.

2   Corporate Ontology Engineering

Corporate Semantic Web assumes the discrete environment of enterprises in contrast to the unlimited and heterogeneous setting of the global Web. Ontologies, in this case, are a promising artifact for improving applications on knowledge integration and management as well as efficient IT service management. When used for modeling complex domains, they can facilitate the reasoning of implicit knowledge, resulting in a better search quality, and serve as the integrating means between complementary structured but interrelated areas, e.g., the world of formal business process management and IT service management.

For this article we analyzed two specific scenarios which address the application of ontologies for enhancing knowledge management and access to knowledge. The scenarios differ with respect to their business model, thus, we call the first scenario the "ontology-based knowledge integration scenario" and the second one the "ontology-based services scenario". Ontology-based knowledge integration addresses the perspective of suppliers and adopters of such systems which are directly applied within a companies infrastructure. In contrast, ontology-based services describes the perspective of suppliers and adopters of such systems which use ontologies and semantic technologies encapsulated in the background using Web services.

2.1   Ontology-Based Knowledge Integration Scenario

Efficient access to knowledge is a bottleneck in modern enterprises because of the increasing amount of data and information spread over various heterogeneous sources of the corporate IT infrastructure. The successive application of ontology-based knowledge repositories, in which legacy data as well as new data is annotated semantically in a semi-automatic way, can help users to find the appropriate information for their specific context easier or facilitate further applications based on those knowledge representations.

Figure 1 depicts the core problem in this case: The heterogeneous systems of an existing corporate IT infrastructure are unintegratedly evolving. A semantically driven integration layer should deliver up to date knowledge. This knowledge is retrieved from the layer below represented by ontologies which have to evolve in a synchronized way.

It is obviously true that an internal application of ontologies involves a strong influence of corporate human resources, such as IT experts, knowledge managers, and possibly ontology engineers in the process of building and maintaining corporate ontologies. Especially the latter group is hard to find in small- and mid-sized enterprises due to steep learning curves of ontology development.

Another obvious aspect is that a company will never invest in ontology-based applications as single means to provide all corporate demands from its IT infrastructures. Various other – non-semantic – applications will provide an evolving amount of unstructured, semi-structured and disparate structured data in form of documents, databases etc. Thus, the cost-benefit-ratio of ontology development has to respect the possible limited benefit of the ontology-based application compared to the whole corporate IT context. Especially the effort for building an individual domain ontology and the evaluation of its benefit compared to the effort of choosing an existing ontology for the problem domain or modeling a rudimentary one is a requirement on ontology engineering methodologies to fit this setting.

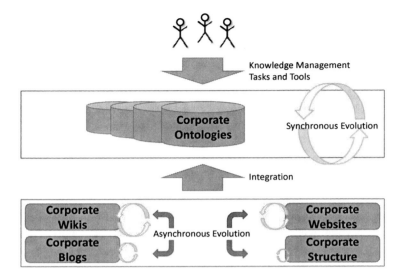

Fig. 1 Setting of a representative knowledge integration scenario

2.2 Ontology-Based Services Scenario

The missing expertise of ontology development and management is a key barrier for internal adoption of ontology-based systems in corporate infrastructures. An initial lightweight step towards broader success of semantic technologies is to offer Web services which use ontologies as the representation of background knowledge. Thus, the adopter is not overwhelmed with the efforts for developing and managing ontologies.

Figure 2 shows how clients access an ontology-based system via Web service interfaces. To facilitate an early access to the service the evolving ontology prototypes underrun an agile lifecycle at the service provider's side. These processes are hidden from the client who just benefits more and more from the improved ontology.

From the client's perspective this setting strongly needs a transparent benefit estimation model to decide for ontologies and semantic technologies compared to other approaches using databases or natural language processing etc. Moreover, with respect to the quality of the service, the question has to be answered when and why an individual or new domain ontology has to be preferred compared to existing ones.

The provider of modeling capabilities has to deal with the process of ontology engineering in an agile way because the key aspects for a purchase decision are low development costs and sufficient benefit of the ontology for its customers. Thus, it is a central goal to minimize the explicit involvement of domain experts of a company causing costs on the customer's side. In fact, the provider has to facilitate a lightweight feedback mechanism to prune and refine the ontology prototype iteratively over time.

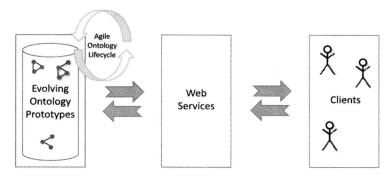

Fig. 2 Clients access to semantically-driven solutions via Web Services

2.3 Corporate Ontology Engineering Requirements

The previously described scenarios are the result of personal interviews with industrial partners of the project Corporate Semantic Web [3,4]. In addition, we collected generic crucial points of applying ontology-based information systems in corporate contexts. We raise the following main requirements:

- The influence of the ontology development and maintenance process on the workflow of domain experts have to be minimized to avoid negative influence on their productivity.
- Already existing and running systems must not be disturbed.
- The need for ontology engineers have to be minimized to reduce costs.
- The ontology has to evolve parallel to the progress of the company.

Based on this set of requirements of ontology engineering settings in a corporate environment, we derive a new viewpoint on ontology engineering processes. The widely accepted methodologies METHONTOLOGY, OTK, HCOME, DILIGENT, and NeOn regard ontology engineering loose from ontology usage. However, they agree that ontologies are undergoing lifecycles with engineering phase and usage phase, but they do not consider ontology engineering as a combination of both.

The principles of RapidOWL describe a kind of a philosophy of ontology development while explicitly described processes, methods, and tools are more or less out of scope. This is due to the nature of RapidOWL as a lightweight strategy for building application-independent and shared ontologies which underrun a continuously dynamic evolution. The latter point fits partly into our perspective, that the evolution of knowledge is the driving force behind an adequate ontology lifecycle in an enterprise and that it is strongly dependent on the usage by unexperienced people with lack of time to note, annotate, or feedback explicitly. To put emphasis to this argument, the following table shows which methodology fits which part of our requirements:

Requirements	METH	OTK	HCOME	DIL	ROWL	NeOn
Workflow integration	-	-	-	-	-	-
Independent system integration	+	+	+	+	+	+
Engineering reduction	-	-	-	-	+	-
Environment aware evolution	-	-	-	-	-	-

3   The Corporate Ontology Lifecycle Methodology – COLM

From our assumptions mentioned in Section 2, we identify a need of a new ontology lifecycle model for continuously evolving ontologies in corporate contexts: Corporate Ontology Lifecycle Methodology – COLM[1]. The model allows an intuitive understanding of raising costs per iteration and in relation to the duration and effort spent in each process phase. Due to a limited context complexity from Web-scale to corporate-scale, we assume that it is possible to converge ideas of agile software engineering and ontology engineering. But, we think it is necessary to change the scope of what is assumed as being agile.

Recent approaches such as RapidOWL focus the agile paradigms *value*, *principle*, and *practice* as a development philosophy. That accompanies agile software engineering as it is intended in the *Agile Manifesto*[2]. But, again, this focus is limited to tasks related to building ontologies, while usage is factored out. The change of requirements over the time is only one agile aspect which influences the evolution of ontology prototypes. The dynamic of knowledge in general, e.g, the evolving dimensions of legacy data and user activities, is another one.

The corporate use of ontologies requires an ontology engineering approach which respects these agile aspects and allows an comprehensible way of estimating costs for evolution steps. Both points play a key-role in our approach towards a generic corporate ontology lifecycle which is depicted in Figure 3. The two-parted cycle consists of seven phases which refer either to the outer cycle as selection/development/integration, validation, evaluation, or to the inner circle as deployment, population, feedback tracking, and reporting. The outer cycle represents an expert-oriented environmental process which consists of pure engineering tasks. The inner cycle represents a human centered concurrent process, which constitutes the ontology usage.

The process starts at the *selection/development/integration* phase. That means to start the knowledge acquisition and conceptualization, to re-use or re-engineer existing ontologies, or to commission a contractor to develop an ontology. The result of this phase is an ontology, which is *validated* against the objectives. At the intersection point between the engineering and the usage cycles, the ontology engineers and the domain experts decide whether the ontology suites the requirements or not. If this is approved the ontology is *deployed* to be used in applications. Then the ontology is *populated*, which means that a process for instance generation from structured, semi-structured and unstructured data runs up. Throughout the whole

---

[1] http://corporate-semantic-web.de/colm.html
[2] http://agilemanifesto.org/

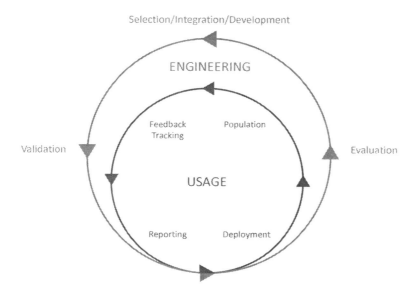

Fig. 3 The Corporate Ontology Lifecycle

feedback tracking phase, formal statements about users' feedback and behavior are recorded. A *reporting* of this feedback log is performed at the end of the usage cycle. That means that all feedback information, which was collected until a decisive point, is analyzed in respect to internal inconsistencies and their effects on the currently used ontology version. The usage cycle is left and the knowledge engineers *evaluate* the weaknesses of the current ontology with respect to the feedback log. This point may also be reached, when the validation phase results that the new ontology is weak or improper with respect to the specification. The lifecycle starts again with the implementation of the results of the evaluation.

3.1 The COLM Phases

In the last section, we described briefly how COLM is intended to run up. Now we present a detailed look at each phase. In Figure 4 we present an alternative view of COLM for better understanding of the decisive point between engineering and usage phase and the sequential dependencies of the phases.

Selection/Development/Integration

Description: The first phase of the process is characterized by a decision model grounded on available skills of the involved persons in a company. If the adopter of an ontology-based system decides to develop an ontology on her own, than she is confronted with all tasks, primitives and tools of ontology development, such as knowledge acquisition, conceptualization, ontology languages, etc. Otherwise

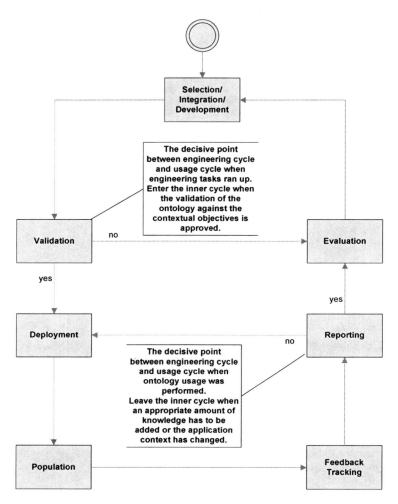

Fig. 4 Sequential representation of COLM

this tasks are provided by a service provider with explicit skills in ontology engineering, thus part of the process of purchasing and ontology from external experts. In both cases the reuse and integration of existing ontologies is an important aspect for efficient and cost-effective development. The foundation of each iteration of COLM is a requirements specification that contains competency questions and a description of the application context for the domain ontology.

Methods and tools: We propose two types of tools for this phase. First, expert-oriented tools, such as Protégé or the NeOn Toolkit. Second, intuitive non-expert tools for rapid ontology prototyping by learning fundamental concepts from folksonomies, tag clouds, or knowledge networks, e.g. by use of mind mapping tools.

Output: The result of this phase are a domain ontology and a description of the requirements on the ontology. It is intended to keep this phase as short as possible to reach high agility during the process.

Validation

Description: The explicit validation of the ontology against the objectives of the requirements specification is a decisive phase following the rapid development or selection of a prototype version. The knowledge manager needs an intuitive way to visualize and prove the correspondence of concepts and requirements. The most important aspect is the validation against application-oriented requirements to enable an appropriate usage of the ontology.

Methods and tools: An explicit model for representing the application context is needed to facilitate a validation of the ontology against it. We propose an approach adopting ideas from test-driven development. By formal unit test cases the requirements of all applications using the ontology can be validated.

Output: This phase yields a result set of passed tests and allows a decision whether the requirements are met.

Deployment

Description: We expect that corporate ontologies are in use by not a single, but various application. Thus, we distinguish between push and pull deployment meaning that the applications are configured to use the latest ontology version automatically or that the version management system notifies applications about the existence of a new ontology version.

Methods and tools: A system is needed which builds on top of well-known versioning solutions such as subversions [11], to enable an intuitive management of ontology changes and ontology versions.

Output: After deployment the newest ontology version is accessible for applications as well as any old version except refused or deleted ones.

Population

Description: If an ontology version has been deployed then the ontology has to be populated with instance data. Existing instance data relying on an old ontology version has to be upgraded to suit to a new one.

Methods and tools: This phase needs (semi-)automatic annotation tools, tools for non-experts to create annotated information intuitively, and a method to upgrade instance data to conform to a new schema with respect to the semantic difference between the two ontology versions.

Output: The result is an annotated knowledge base conforming to the newest ontology version.

Feedback Tracking

Description: The feedback tracking logs all activities with reference to the ontology. These include explicit user activities (e.g. queries and results, feedback
statements/arguments) as well as software engineering activities (e.g. updates of
the ontology by applications) and the generation of data by conventional applications (e.g. word processors not using ontologies).
Methods and tools: A management system including an API is needed to allow
easy configuration and logging of feedback messages by external applications.
Output: The output of this phase is the tracking log file.

Reporting

Description: In the reporting phase the management system analyzes the log file
to decide if a relevant amount of information has been tracked to require an evaluation of the current ontologies by ontology experts.
Methods and tools: We propose a measure for indicating the relevance of tracked
information by analyzing the whole tracking log and eliminating contradicting
or invalid information.
Output: This phase ends up with a decision whether ontology engineers have to
be invoked and a compilation of relevant information.

Evaluation

Description: The result of the reporting phase is used to evaluate the quality of the
existing ontology version. On the basis of the tracked information an ontology
engineer can detect new or missing concepts and relationships. Furthermore, the
data can be used to develop new test cases, thus, resulting in new requirements.
Methods and tools: Again, we need a test case model and an appropriate tool to run
these tests against the ontology. Furthermore, we need a tool for proposing new
ontology prototypes by learning from the old version and the tracked information.
Output: The result of this phase is a set of new requirements which have to be
respected by the next ontology version.

The innovative approach towards agile ontology engineering allows an evolution
of rapidly released ontology prototypes. We expect from our model to allow an intuitive view to ontology engineering processes and facilitate a cost estimation in the
run-up of cost-intensive evolution steps. We reach these improvements by a convergence of ontology engineering and ontology usage controlled by an innovative
versioning approach.

3.2   Functional Requirements of Corporate Ontology Engineering

From the description of the phases of COLM we derive a list of functional requirements on software tools needed to implement COLM. Because it is cumbersome to
develop new software, we use those requirements to examine existing tools.

Selection/Integration/Development: We need a dedicated decision model to decide whether to retrieve an ontology from available repositories, to outsource the work to a contractor, or to develop the ontology from scratch on its own. Global repositories of standard ontologies and ontology modeling experts could be of valuable help for this task.

Validation: Besides tools for ontology testing, we need a system to support decision making collaboratively that is suitable for experts and non-experts.

Deployment: A versioning control system should be able to provide access to different ontology versions simultaneously.

Population: Methods and tools for automatic, semi-automatic and manual annotation especially for non-expert users.

Feedback tracking: System for the integration of lightweight extended communication platforms, e.g., forums or feedback forms and automatic recovery of user behavior into a feedback log.

Reporting: The system has to analyze a snapshot of the feedback log and to extract information that helps to decide on the quality of the ontology.

Evaluation: At this point the tool should indicate advantages and disadvantages of the current ontologies.

4 Comparison of Tools

In this section we give a brief overview of some widely spread tools for ontology engineering tasks compared to the support of the different phases of our corporate ontology lifecycle. The desktop-applications Protégé and SWOOP as well as the web-applications OntoWiki and Ikewiki are in our focus. These tools represent the currently most accepted approaches to ontology engineering under the requirements of the methodologies which we introduced as state of the art in ontology engineering.

Protégé is the most common tool for ontology building. Its appearance is similar to software development environments. Protégé is rich in function and language support and very scalable in respect to its extensibility. Since Protégé contains collaborative components it is possible to develop consensual ontologies in a distributed fashion using lightweight access to the process by discussion and decision making about proposed changes. This feature does not respect any roles or permissions. Versioning control is enabled on ontology level, but not on conceptual level, enriched by annotations from a structured argumentation mechanism. Any abstraction from technical terms is missing.

To sum up, Protégé is a very useful tool for engineering ontologies in a team of experts with a lack of lifecycle support in a usage-oriented architecture.

SWOOP is a desktop environment for ontology engineering, which is a bit straightforward at the expense of functionality. The representation of the concepts allows a web-browser-like navigation. A search form supports quick searches on the recently used ontology and at least all ontologies stored. Quick reasoning support is implemented in the same fashion. However, there is no abstraction from technical primitives enabled. By definition of remote repositories, it is possible to commit versions of ontologies.

Altogether, SWOOP is a tool for ontology engineering tasks for experts and well-experienced users. It has its strengths in quick and intuitive navigation in and search within ontologies but lacks functional flexibility and lifecycle support.

OntoWiki   is a PHP-based wiki-like tool for viewing and editing ontologies. It is based on pOWL which makes use of the RAP API[3]. OntoWiki[4] allows the administration of multiple ontologies (called knowledge bases) and provides in-line editing as well as view-based editing. As an abstraction from conceptual terms OntoWiki includes an alternative visualization for geodata (Google Maps) and calendars auto-generated from the semantic statements stored. However, a general abstraction from technical primitives (e.g. class, subclass, SPARQL, etc.) in the user front-end is missing. Altogether, it allows only one single view for all users and does not respect any roles or permissions. Changes to the conceptualized knowledge have to be done manually. The ontology history is concept-oriented not ontology-oriented and implemented as known from wiki systems.

We subsume that OntoWiki is an ontology engineering tool and a knowledge base for experienced users with an academic background and that it does not support lifecycle management.

Ikewiki   implements the semantic wiki-idea and focuses annotation of wiki-pages and multimedia content. It is possible to generate an alternative graph visualization for the context of each annotated page. However, Ikewiki does not support any abstraction from technical primitives for users with less experience in the field of ontologies. Restricted views referring to roles or permissions are not provided. The ontology history is concept-oriented not ontology-oriented and implemented as known from wiki systems.

We summarize about Ikewiki, that this tool addresses familiar wiki users with technical experience which do not need any control of the conceptualization and lifecycle support.

From our experience there exists a mismatch between the described approaches and and the needs for an integrated ontology life cycle management. The tools either have an engineering-oriented perspective, which deals with the ontology application- and user-independent. Or the approaches reckoning the conceptualization on an application level for knowledge management without respecting unfamiliar users. The latter is emphasized if we note that the barriers of wiki-syntax for users without any technical background are underestimated. Thus, we subsume the support per phase of our model as follows:

Finally, we now conclude that currently no tool exists, which supports our lifecycle model completely. This is due to the fact that available tools handle engineering tasks and ontology usage separately. Some tools work with ontologies as the central artifact on an engineering level while others support the application level only. Our ongoing work is aiming at an adequate architecture for integrative lifecycle support. To reach this we start from the perspective of continuously evolving ontology

---

[3] http://www4.wiwiss.fu-berlin.de/bizer/rdfapi/
[4] http://aksw.org/Projects/OntoWiki

Table 1 Legend: + supports req., o supports req. partially or for users at different skill level, - does not support req.

Phase	Protégé	SWOOP	OntoWiki	Ikewiki
Selection/Integration/Development	+	+	+	+
Validation	+	-	-	+
Population	o	o	o	o
Deployment/Versioning	-/o	-/o	-/o	-/o
Feedback Tracking	o	-	-	o
Reporting	-	-	-	-
Evaluation	-	-	-	-

prototypes in the context of a heterogeneous corporate infrastructure. A smart version control system is needed as the central component to enable this.

5 Conclusion and Outlook

In this paper we introduced our approach towards an innovative ontology lifecycle for continuously evolving ontologies in corporate settings: Corporate Ontology Lifecycle Methodology (COLM). COLM was concluded from a gap between existing ontology engineering methodologies and the reality in corporate IT infrastructures, especially in small- and mid-sized enterprises. We held face to face interviews with cooperation partners of the project Corporate Semantic Web to found our assumptions. Furthermore, we described two core scenarios which prove the need for tool-supported agile ontology engineering – the "ontology-based knowledge integration scenario" and the "ontology-based services scenario".

From our model we derived functional requirements for an integrative tool support and compared four well-known ontology development tools with reference to these requirements. We concluded that there is yet a lack of methodological foundations as well as tool support for the agile engineering of ontologies which is strongly needed in the settings which we described.

Recently, we are working on an integrative architecture for a COLM ontology management framework. Our goal is to create a suite based on intelligent ontology versioning which allows a consistent evolution of various corporate ontologies. This will respect the whole context of a corporate IT infrastructure and provide serious support for non-experts in ontology development.

A proper evaluation of our concepts and tools is planned as a case study with an industrial cooperation partner, Ontonym GmbH, which provides ontology-based Web services as described in Section 2.

Acknowledgements. This work has been partially supported by the "InnoProfile-Corporate Semantic Web" project funded by the German Federal Ministry of Education and Research (BMBF).

References

1. Auer, S., Herre, H.: RapidOWL - An Agile Knowledge Engineering Methodology. In: Virbitskaite, I., Voronkov, A. (eds.) PSI 2006. LNCS, vol. 4378, pp. 424–430. Springer, Heidelberg (2007)
2. Abrahamsson, P., Warsta, J., Siponen, M.T., Ronkainen, J.: New Directions on Agile Methods: A Comparative Analysis. In: ICSE, pp. 244–254. IEEE Computer Society, Los Alamitos (2003)
3. Corporate Semantic Web Research Project, http://www.corporate-semantic-web.com
4. Coskun, G., Heese, R., Luczak-Rösch, M., Oldakowski, R., Schäfermeier, R., Streibel, O.: Towards Corporate Semantic Web: Requirements and Use Cases. FU Berlin Technical Report B-08-09 (2008)
5. Fernndez-Lpez, M., Gmez-Prez, A., Juristo, N.: METHONTOLOGY: From Ontological Art Towards Ontological Engineering. In: AAAI-1997 Spring Symposium on Ontological Engineering: Stanford. AAAI Press, Menlo Park (1997)
6. Kotis, K., Vouros, A.: Human-centered ontology engineering: The Hcome methodology. Knowl. Inf. Syst. 10(1), 109–131 (2006)
7. Lindvall, M., et al.: Empirical Findings in Agile Methods. In: Wells, D., Williams, L. (eds.) XP 2002. LNCS, vol. 2418, pp. 197–207. Springer, Heidelberg (2002)
8. Noy, N.F., McGuiness, D.L.: Ontology Development 101: A Guide to Creating Your First Ontology. Stanford University (2001)
9. Pinto, H.S., Tempich, C., Staab, S., Sure, Y.: Distributed Engineering of Ontologies (DILIGENT). In: Semantic Web and Peer-to-Peer. Springer, Heidelberg (2005)
10. Simperl, E., Tempich, C.: Ontology Engineering: A Reality Check. In: Meersman, R., Tari, Z. (eds.) OTM 2006. LNCS, vol. 4275, pp. 836–854. Springer, Heidelberg (2006)
11. Subversion, Open Source Version Control System, http://subversion.tigris.org/
12. Sure, Y., Studer, R.: On-To-Knowledge Methodology — Expanded Version. On-To-Knowledge deliverable, vol. 17. Institute AIFB, University of Karlsruhe (2002)
13. Tran, T., et al.: Lifecycle-Support in Architectures for Ontology-Based Information Systems. In: Aberer, K., Choi, K.-S., Noy, N., Allemang, D., Lee, K.-I., Nixon, L.J.B., Golbeck, J., Mika, P., Maynard, D., Mizoguchi, R., Schreiber, G., Cudré-Mauroux, P. (eds.) ASWC 2007 and ISWC 2007. LNCS, vol. 4825, pp. 508–522. Springer, Heidelberg (2007)

A Semantic Policy Management Environment for End-Users and Its Empirical Study

Anna V. Zhdanova, Joachim Zeiß, Antitza Dantcheva, Rene Gabner, and Sandford Bessler

Abstract. Policy rules are often written in organizations by a team of people in different roles and technical backgrounds. While user-generated content and community-driven ontologies become common practices in the semantic environments, machine-processable user-generated policies have been underexplored, and tool support for such policy acquisition is practically non-existent. We defined the concept and developed a tool for policy acquisition from the end users, grounded on Semantic Web technologies. We describe a policy management environment (PME) for the Semantic Web and show its added value compared to existing policy-related developments. In particular, we detail a part of the PME, the policy acquisition tool that enables non-expert users to create and modify semantic policy rules. An empirical study has been conducted with 10 users, who were new to the semantic policy acquisition concept and the developed tool. The main task for the users was to model policies of two different scenarios using previously unknown to them. Overall, the users successfully modeled policies employing the tool, with minor deviations between their performance and feedback. Observation-based, quantitative and qualitative feedback on the concept and the implementation of the end-user policy acquisition tool is presented.

1 Introduction

Community-driven services and portals unifying physical and virtual realities, such as 43things.com, SecondLife, YouTube, LinkedIn and Facebook, or the Web 2.0 developments, are currently at their popularity peak attracting millions of users. The

Anna V. Zhdanova, Joachim Zeiß, Rene Gabner, and Sandford Bessler
Telecommunications Research Center Vienna (ftw.), Donau-City Strasse 1,
A-1220 Vienna, Austria
e-mail: {zhdanova,zeiss,gabner,bessler}@ftw.at

Antitza Dantcheva
Eurecom, BP 193, F-06904 Sophia Antipolis, France
e-mail: antitza.dantcheva@eurecom.fr

S. Schaffert et al. (Eds.): Networked Knowledge - Networked Media, SCI 221, pp. 249–267.
springerlink.com   © Springer-Verlag Berlin Heidelberg 2009

existing portals with their community and personal data management environments, while collecting and attempting to manage large amounts of user-generated content, are still highly limited in providing the functionality assisting adequate management and sharing of the submitted data. In most cases, the users still cannot specify the provisioning conditions of the generated content or services (i.e., policies with whom and for what they want to share), as well as set up automatic execution of arbitrary actions provided that the certain conditions are met (e.g., notifications about appearance of specific information, products, services or user groups). Ability to define and employ policies would lead to efficient personal information management, decrease amounts of electronic spam, and increase revenues for targeted provisioning of content and services.

Semantic Web and social software technologies have proved to be a success in resolving the knowledge acquisition bottleneck. Approaches such as Semantic Wikis [7], [10] enable acquisition of large quantities of arbitrary ontology instance data. Community-driven ontology management [11], [12] shows feasibility of acquisition of ontology classes, properties and mappings from the end user communities. Meanwhile on the large-scale light-weight and tag-based social Web, *user-generated policies* have not yet gained a broad usage. The latter is largely due to the complexity of this problem w.r.t. the user perspective [5] and a lack of practices and tools for policy acquisition from the non-expert users.

The main contributions of the presented work are:

- Definition of a user-driven policy management environment for open, sharable infrastructures such as for Web or mobile services,
- A concept and a tool for policy acquisition from the end users, grounded on Semantic Web technologies.
- An empirical study conducted to test the approach and the tool. The study shows that end users modeled the policies employing the tool successfully and are inclined to use similar tools in the future.
- Observation-based, quantitative and qualitative feedback on the end user policy acquisition concept and the implementation is presented. Requirements towards the design of an improved policy acquisition tool are drawn.

The paper is structured as follows. In Section 2, we describe our approach of a policy management for the Social Semantic Web. In Section 3, we describe the potential research applicability and the related work. The implementation of the policy acquisition tool is presented in Section 4. The user study and tool evaluation settings are described in Section 5. In Section 6, the results of the study are presented, and lessons for the future construction of policy acquisition tools are drawn. Section 7 concludes the paper.

2   Semantic Policy Management

The following paragraphs describe the basic components of our architecture. The architecture is strongly related to conventional ontology and policy management services [2], [5], [9], but is enriched with end-user generated policy acquisition and advanced policy communication. The basic model is that of an open system in which policy rules can be shared, adapted to individual needs and enriched with facts and instance combinations.

A **Policy Storage and Query** component is provided to efficiently store and query parts of policy data and metadata by providing indexing, searching and query facilities for ontologies. In addition to conventional policy management services and practices [2], [5], [9], we propose to enrich the existing search and query components with community-generated policy information. This would improve their performance and make the search, reasoning and consistency checking features mature and more attractive to use.

As the users of the environment are generally not bound to a single community or application, they must be able to publish personal and community-related policies in a multi-accessible way. The current focus in semantic policy storage and querying is thus maintaining distributed repositories with functionalities for aggregation, decomposition and discovery of information in simple ways.

A **Policy Editing** component is introduced for creating and maintaining policies and instance data. The front-end, a user-friendly interface, helps users to easily add and modify policy-like rules on the basis of existing imported ontology classes and properties shared among several users and communities, policies and instances. The back-end consists of a storage and query system. A Policy Editor enables sharable editing for multiple users and tight integration with semantic publishing, delivery and visualization components, allowing the involved parties to observe the evolution of policy settings. These requirements are due to the elevated degree of flexibility required by community-oriented environments as the Social Semantic Web and its members to freely evolve schemata, policies and to influence community processes.

A **Policy Versioning** component is introduced to maintain different versions of policy definitions, as communities, content and relationships change over time. The user should be able to easily adapt policies to new scenarios and communities without losing previous definitions. Earlier versions can be reused for definitions of new policies. Also users could experiment with more restricting policy definitions and roll back to previous versions wherever practical. A Policy Versioning component interacts with existing versioning systems like svn [3] to provide a versioning service to the user. Semantic metadata describes the necessary versioning information inside the policy definition itself.

A **Policy User Profile and Personalization** component is responsible for the users' access to the environment and it connects the policies with the user profiles. At a more advanced level, the component helps to share and communicate policies across the users' profiles, apply policies dependent on the user profiles and recommend policies based on the user profiles. In particular, access and trust policies can be implemented taking into consideration community and social networking information provided by the users [7].

Our *overall* ontology-based **policy management** approach features: *user-driven policy construction*, meaning that the system extensively assists the users to model the policies correctly (e.g., proactive suggestion of the ontology items that can be combined in a policy, consistency checking for the modelled policy solutions);

policy semantic representation and sharing across communities, essential for the further extension for the rules layer of the Semantic Web; ontology import and ***policy creation on the basis of shared ontologies,*** the user is free to input any ontologies he/she likes and define policies on them.

Thus, ontology-based and community-oriented policy management is an advance over a conventional policy management. The advantages are gained by introducing an infrastructure that enables the communities to manage their policies.

3   Research Applicability and Related Work

In this section, we discuss the applicability of the approach and related work in the field of (semantic) policy editing.

3.1   Applicability of Policy Acquisition Tools

In a distributed environment, such as the Internet, there is a need to set policies for sharing user information and for providing access to services on the Web. However, the ways to model and operate with Semantic policies are currently very limited, and there are *little or no approaches for policy acquisition from the end users.* Meanwhile enabling the end users to define and share policies is crucial for widespread, acceptance and the growth of the rule-based Semantic Web.

The types of users of a Policy Acquisition Tool (PAT) include (but are not restricted to) the following:

- Individual users who have one or several profiles and have to manage them on several systems (related to single-sign-on systems);
- Owners of web services who want to sell or offer their functionality to others and need to specify the conditions under which the service can be used;
- Users who manage the physical reality or link physical and virtual worlds. For instance, such user activities include setting policies on forwarding phone calls from the user's phone to his/her mobile phone while he/she is on vacations, or sending an SMS to a remotely-located mother if her baby wakes up and starts to cry, employing integration with the sensor technology.

The customers of a PAT would be companies or institutions:

- providing single-sign-on applications;
- providing identity management and security systems (e.g., for the users who want to specify different user groups on an instant messenger and show their location information only to some of these groups);
- developing aggregation solutions for systems with similar functionalities for users to have one profile and a possibility to set various policies for various systems (such as Trillian for instant messaging);

- providing (semantic) web service publication space and (semantic) web service search engines for web service owners to annotate their services with specific service features (e.g., conditions for execution);
- providing community sites where people add/create content (such as Flickr for pictures), so that the users set policies on how and by whom their content can be accessed and used;
- providing systems for the management of physical environments such as semi-automatic policy-based assistance in a hospital on observing patients, notifying nurses, etc.

3.2 Related Work

In current software products, policy editing often can be performed in a simple manner, in particular, via checkboxes and scroll-down forms. Well-known examples of such policy management include a policy editing interface for handling files in Microsoft Word. For the user convenience multiple templates are offered for selection of the rule type that the user may want to edit, e.g., in Microsoft Outlook. Also, in the contemporary applications, web-portals and online shops, the users are often asked to commit to agreements or copyrights written in a natural language by clicking an "accept" button. Such agreements or copyright statements in particular may contain policies on the e-mail addresses and personal data sharing, the users' preferences, etc.

Techniques from the following *research* fields are relevant for the user-oriented policy acquisition:

- Policies on the Semantic Web in general: state of the art in this area and the new trends (e.g., automated trust negotiation) are overviewed by Bonatti et al. [2]. As a particular effort, Attempto Project[1] has developed tool support for transferring statements (possibly, user rules) specified in controlled natural language (English) to the OWL format [10];
- Policies have been applied to web services [13], and "The Web Service Policy Framework"[2] is an example of an industry-led effort in this area;
- Ontologies for defining policies: a number of works are driven in this area, for instance, an ontology for defining business rules in OWL [12];
- Knowledge acquisition methods for ontology construction, including knowledge acquisition principles in ontology editors, community portals [16];
- Editing of policies: an editor developed by Karat et al. [8] is one of the advanced works most strongly related to our work and therefore we go into detail when explaining the differences to our work. The editor has three variations of policy editing for the end user: (i) Unguided and (ii) Guided Natural Language interfaces, and (iii) Structured List method, i.e., the interface allowing composition

---

[1] Attempto Project: http://attempto.ifi.unizh.ch/site/description/index.html
[2] The Web-Service Policy Framework:
    http://www-106.ibm.com/developerworks/library/ws-polfram/

of policies out of pre-existing items, via web form-based selection. One of the test-based observations was that "Structured List methods helped the users create more complete rules for all element categories except Conditions as compared to the Unguided NL method".

In our work, the ontology-based policy engine is structurally similar to the Structured List method, i.e., major part of the policy specifications are created from pre-existing components (in our case, ontology and instance data items). Using ontology technologies brings an added value to the conventional policy editing technologies due to the following factors:

1. End users define the policies easier and faster than with other methods (including both natural language and Structured List methods) due to the fact that the users are familiar with ontologies employed for their profile or context information;
2. The composed set of rules is even more complete than the most complete set obtained by now (with the Structured List method);
3. While the Structured List method was shown to be the most preferred method by users, the ontology-based method would become the first preferred method.

There are also *tools* enabling the users to edit the policies:

- PERMIS[3] has similar policy editing functionalities as addressed here (e.g., for personal data protection), however, the tool does not include semantic policies, there the policies are specified in XML which hinders referencing or reusing items of already existing ontologies. In addition, the tool is restricted for scenarios specific for settings of certain existing platforms (namely, Apache Web server, Globus Toolkit, Shibboleth, .Net, Python interfaces);
- WebSphere Policy Editor[4] is an Eclipse plug-in tool for generating, creating, and editing cache policies (as in PERMIS, based on XML) for the dynamic cache service of WebSphere® Application Server;
- P3PWiz[5] is an online commercial tool by Net-Dynamics allowing website owners to design P3P[6] compliant policies via graphical interfaces. Mainly predefined selection forms are used, which are tightly compliant to the fixed P3P specification. Other similar online services are P3PEdit[7] and P3PWriter[8]. IBM's P3P Policy Editor[9] is also a similar tool which is downloadable as a JAR file.

The policy constructions supported in these tools are restricted to certain domains and do not allow inclusion of arbitrary ontology-based vocabularies. As new domains and ontologies always appear and evolve in community user-driven systems, inability to support construction of policies in dynamic semantic environments is a severe bottleneck of the aforementioned tools.

---

[3] PERMIS: http://sec.cs.kent.ac.uk/permis/

[4] WebSphere Policy Editor: http://www.alphaworks.ibm.com/tech/cachepolicyeditor

[5] P3PWiz: http://www.p3pwiz.com

[6] P3P, the Platform for Privacy Preferences: http://www.w3.org/P3P/

[7] P3PEdit: http://p3pedit.com

[8] P3PWriter: http://www.p3pwriter.com

[9] IBM P3P Policy Editor: http://www.alphaworks.ibm.com/tech/p3peditor

4   Policy Acquisition Tool

In this section, we describe the implemented tool for a policy acquisition from end users: its general overview, functionality and user interfaces.

4.1   Tool Overview

The implemented policy acquisition infrastructure is designed as a component for a community Semantic Web portal, providing policy management facilities to the community members and managers. The infrastructure is built as a Web-based application using JSON technology[10] [4] and exploiting Python version of Euler [6] for manipulating ontology schemata, instance data and policies in a N3 format [1]. A policy is modeled as a rule in a typical form of one or more *conditions* followed by one or more *conclusions*.

The architecture of the community-driven policy acquisition infrastructure is shown in Figure 1. The policy acquisition tool (PAT) is facilitated by another major block, the policy engine (PE). A PAT server is a component interacting with the end-user over a GUI, and the policy engine is a component responsible for the "logical" side of the system, accomplishing integration of external and internal information, reasoning and rule production. The PAT server is the active component addressing the policy engine with requests whenever the user loads a policy, selects the policy building blocks or saves a policy.

The policy engine (PE) is a stateless request-/response-based server that deals with any kind of requests expressed in N3 [1]. The policy engine has associated a *Decision Space*, a set of files containing N3 triplets as well as rule objects, i.e., parsed N3 statements, kept in memory. The files contain persistent semantic data like ontology definitions, instance data and rules. Volatile semantic data relevant for the current policy request are added to the N3 objects in memory. The *Request Processor* is the part of the PE that extracts data from the request (out of a SIP message, a http GET/POST message or a SMS) and inserts it into the decision space. The policy engine may also extract data from a user profile, user context such as location, or policy data via an additional context interface. The *Reasoner*, the heart of PE, is a N3 rule engine that is invoked with the receipt of a request and uses all semantic data made available in the decision space as reasoning input. The reasoner is based on the python implementation of Euler (backward-chaining, enhanced with Euler path detection).

The tool applications comprise usage and population of domain-dependent and domain-independent ontologies, and service support for the portals' data and metadata. The prototype is using N3 notation due to its simplicity and efficiency in representation of the rules, and a straightforward integration with other semantic languages such as RDF/S and OWL. Technically, the rules written in N3 syntax are compact enough to be effectively executed and managed even on devices with limited computational and storage capacities, such as mobile phones. The overall

---

[10]   JSON: http://www.json.org

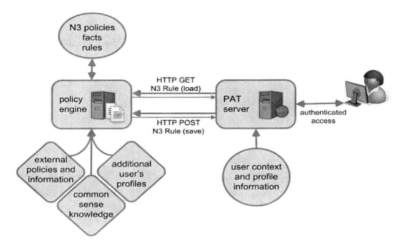

Fig. 1 Policy management infrastructure

size of the ontologies employed in the presented user studies comprised 94 ontology items (including classes, properties, instances) for the Eshop case study and 127 items for the Etiquette case study.

4.2 Tool Functionality

Below we list the functions and features of PAT. Currently most of the listed functions and features have been implemented and the others are being implemented.

PAT functions are as follows:

- *Viewing a policy/user rule:* With PAT, the user can view all the constructed policies, possibly divided into groups of rules;
- *Dynamic user interface generation:* The user interface is generated directly from the ontologies and the instance data that are imported by PAT. The ontologies and data can be provided by the end user(s) or deduced by the policy engine based on defined business logic. For example in Table 1, PAT recommends the user to choose between the objects that are compatible with the subject "Maria";
- *Modifying a policy/user rule:* By loading a specific rule from the policy engine's decision space, it is possible to modify existing rules, i.e., either rules generated by the current user or by other users;
- *Assisted fill out:* User profiles and context ontologies are employed to assist the user in filling out the policy items when modeling a rule. In the shown experiments here, PAT interface offers only combinable data according to the context ontologies;
- *Deleting a policy/user rule:* Alike to modifying a rule, it is also possible to delete the currently edited rule from the policy engine space. A deletion immediately effects queries from other clients;

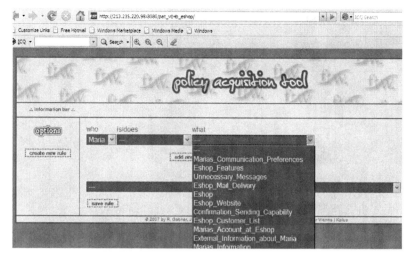

Fig. 2 Interface for policy acquisition: policy construction

- *Saving a policy/user rule:* Once the user has successfully finalized a rule, the rule can be saved directly to the policy engine's decision space (as a N3 language notation text file). Figure 3 shows an example of the finalized rule;
- *Human readable format rule tagging:* Naming a policy or user rule for further reference and search enables the users to retrieve the rules employing the rules' human-readable names or tags. All PAT functions (viewing, modifying, deleting and saving) benefit from this human readable format.

An example of a policy that is valid for an Eshop case study on online shopping (see Appendix) and is applicable to a hypothetical online customer Maria is "We might receive information about you from other sources and add it to our account information". This policy is being designed in PAT's user interface in Table 1 and 3, and its N3-based representation is as follows:

```
Maria a :Customer.
Eshop a :Eshop.
External_Info_about_Maria a :External_Customer_Info.
Marias_Account_at_Eshop a :Eshop_Customer_Account.
{
  Maria :has Marias_Account_at_Eshop.
  Eshop :receives External_Info_about_Maria
} => {
  External_Info_about_Maria
      :is_added_to :Marias_Account_at_Eshop
}
```

Fig. 3 Interface for policy acquisition: policy is finalized

As the user in question is already familiar with Eshop and its terms, the selection of the relevant concepts for the rule in the interface becomes easier for him/her.

5   User Study

In this section, we explain the goals of the user study, and its set up and procedure. **Goals of the user study** were as follows:

- To investigate the users' attitudes towards applying a semantic policy editor;
- To identify usability problems of the evaluated PAT interface and to derive suggestions for its improvement;
- To gain a first evidence about the following theses:

  - The editor will contribute to the widespread use of policies on the web and in mobile environments (e.g., the users use the tool well and think that they will use it in the future);
  - Policy acquisition tools are applicable both to private data management and business/company settings;
  - The editor reduces the costs of policy construction and makes the policy design accessible to non-professionals;
  - The editor is helpful in reducing the mistakes in the process of the rule construction;
  - The editor makes the user more aware of policies and encourages the user to observe, construct and manage them.

User study setup. The user study contained 10 test persons of different age (between 25 and 35), sex (4 female, 6 male), education and technical experience. To obtain a MOS (mean opinion score) the subjects were asked to rate the criteria on a five grade scale (1-fully disagree, 3-undecided, 5-fully agree) with half-grades possible (e.g., 2.5). The working device was a laptop placed on a table. The test took place in the HTI (human telecommunication system interface) lab, simulating a home atmosphere. A test conductor was observing and taking notes during the test.

User study procedure. The test with each user lasted for up to one hour and comprised the steps as specified in Table 1. A questionnaire, the log files created automatically during the test and the filled out observation forms were used to analyze the user study.

6 Results

In this section we present and discuss the results obtained in the user study. The first subsection discusses findings derived from the observation of the users while they were constructing policies. The second subsection summarizes and discusses the results obtained from questioning the users about the policy acquisition tool.

6.1 Observation-Based Results

In this subsection we present the user study results (obtained via observation of the users) and a discussion. In particular, we explored how people started and proceeded with the construction of a policy, to which extent the resulting policies were complete and correct, how much time the user spent on the construction of a policy, how well the user understood modeling of more complex rules comprising such constructions as multiple conditions and consequences, active vs. passive voice or negation.

All the users without any exception started to model the first policy beginning with the first graphical condition field ("subject" of the statement), see Figure 2. Later on, the users discovered that modeling a rule from any graphical slot was possible: 4 out of 10 users discovered and started using this feature themselves and the remaining 6 users did the same after an indication from a test conductor. The feature of constructing the complex rules (the "Add sentence" button) on the interface drew more attention: 7 out of 10 users noticed it and started to try it out proactively themselves, and only 3 users ignored the feature until the test conductor drew the attention to it.

Another interesting observation was that the users were generally inclined to model at first the consequence of the rule and then the condition(s), which was contradictory to the modeling suggested by the user interface. Six out of 10 users started modeling with representing the consequence prior to the condition: as a rule they modeled the consequence statement in the condition statement, realizing their misplacement afterwards. Three users demonstrated such behavior persistently, i.e., more than once during the test.

Table 1 Test roadmap

Step	Details
1. Introduction to the user study	A description of the type of assessment, the opinion scale and the presentation of the policy editor were given in oral form before the beginning of the test.
2. Explanation of the Eshop use case scenario to the user	The first scenario Eshop is detailed in Appendix, Case Study 1.
3. First scenario: Eshop policy modeling tasks done by the user	The user was told that he/she was to act as a manager of an internet shop (Eshop). He/she had to model the on-line shop's privacy policies in a policy editor for the customers. The customer was named Maria: "Maria regularly shops online, likes special offers and recommendations, but wants to keep her personal profile information under control." The user was to model 5 policies according to that case within the policy management environment.
4. Explanation of the Etiquette use case to the user	The second scenario Etiquette is detailed in Appendix, Case Study 2.
5. Second scenario: Etiquette policy modeling tasks done by the user	Here the user had to implement behaviour rules valid for different situations. The policies modelled would be incorporated in software to help children, foreigners, robots or automated personal assistants to make choices about their behaviour. Again 5 tasks for various situations (e.g., tea party, restaurant) were to be implemented by the user.
6. Overall questionnaire answered by the user	A final questionnaire provides the information about the attitude of users towards the policy editor.
7. Test completion	

In ca. 90% of the cases policies were modeled correctly by the user, i.e., the resulting rule conveyed the same meaning as the one offered to be modeled. Almost every second "correctly modeled" rule was represented using the same vocabulary and the same level of precision as implied initially with the given ontology. Assuming the community-driven rule development and sharing, incomplete rules (e.g., lacking certain condition or consequence statements) or rules using alternative vocabularies could achieve the same level as the completely and precisely modeled rules after being augmented with the context information and ontology mappings repositories [15]. The ratios for the users' policy modeling success rates are presented in Figure 4.

An *average amount of time* spent on the construction of a policy was 3.5 minutes (Table 1, 3 and 5). Major delays have been caused by a reasoner taking the time to upload the matching statements and the recurring need to re-model certain parts of the rules when they proved to be unfitting. As the users were getting familiar with the tool, the time spent on a construction of a typical policy approached one minute. Technically, the spent time can be reduced by a more scalable implementation.

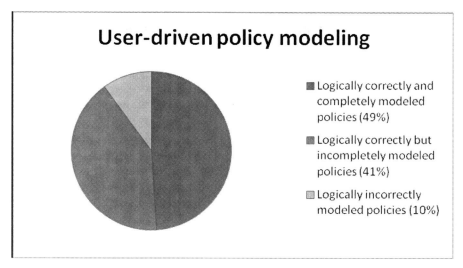

Fig. 4 Policy modeling success rates obtained in the experiment

Additionally, we observed the users' reaction on various logical constructions: (i) rules containing negation statements, (ii) rules containing statements represented both in active and in passive voices, (iii) rules containing more than one condition statement, and (iv) rules containing more than one conclusion statement.

For the rules containing *negation statements*, the test included not appearing vocabulary to express a statement with a negation and the test leader suggested the user to model the same statement using a positive construction. In this case, 6 out of 10 users modeled the statement positively indicating that the modeled statement is not the same as the provided statement, while the other 4 users modeled the statement without indicating the difference.

A mix between *active and passive voices* in the statement descriptions and models has been explicitly indicated as an obstacle by 2 users out of 10. Though the majority of the users (8) did not raise the issue of an active and passive voice mix, in many cases the users were expecting usage of a certain voice in the statement model. Occasionally incorrect expectations in the voice matter lead to longer rule construction times as the users had to re-model the rule parts that are constructed based on wrong assumptions.

Rules with *more than one condition* or *more than one consequence* were modeled more imprecisely than rules with only one condition or consequence by the user in 5 and 1.5 cases out of 10, accordingly. There, in 3 and 6 cases out of 10, the user still modeled the rule with only one condition or consequence, though giving a note on imprecise modeling. Only in 1 and 2.5 cases out of 10, the users implemented such complex rules logically correctly via finding less trivial tool usages, in particular, representing the semantic of one rule by modeling two separate rules.

6.2 Questionnaire-Based Results

The questionnaire consisted of nine questions. Users were asked to answer all the questions. Under every question they could give a grade from 1-5, where 1 meant "fully disagree" and 5 meant "fully agree". Below we present the questions, the feedback and the recommendations provided by the test persons verbally. In Figure 5, the mean scores of the grades for all the questions and all the users are summarized.

Question 1: Was it easy for you to understand the system?
Users liked the minimalistic design (that led to fewer distractions) and that learning to use the tool is easy. After about 2 accomplished modelled policies, the users were familiar with the policy editor and could fully concentrate on the more complex tasks and not the system itself. They rated the easiness of use of the system with a MOS of 3.6 (standard deviation of 1.15) that is above average. The difficulties in understanding the system mentioned by users were:

- Active / passive or negatively formulated policies: the policy linguistic description and its ontology vocabulary could mismatch the users' expectations;
- Distinguishing between "condition" and "consequence" of the policy: the users attempted to model a consequence in a place of a condition;
- Two users mentioned that editing the rules starting from the consequence is easier;
- Speed: the loading time was too long, what made users impatient to try further combinations;
- Clarity and visibility: an overview of the selectable conditions and results at each time was desired.

Recommendations: Keeping a simple and overview giving design; clear and obvious explanations of entities; clear labeling; improving the speed (e.g., via caching of

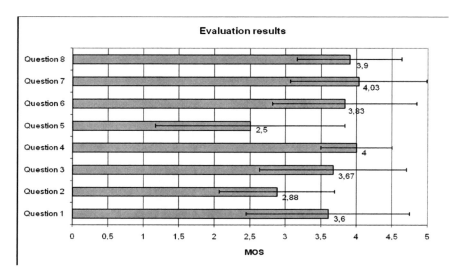

Fig. 5 Policy modeling success rates obtained in the experiment

information); specialized and customized policy editors, kept simple with well-chosen expressions.

Question 2: Was it easy to find/select the right terms to express your ideas?
The obtained MOS was 2.88 (standard deviation of 0.81) for the expressive potential of the policy editor. Here users again pointed out that they would like to be able to change all the fields anytime and that they had difficulties in choosing the direction of implementing ("subject – predicate – object" or "object – predicate – subject"). Also the negation containing policy and synonym search were confusing.
Recommendations: flexible system, easy to edit and change yet simple; well-chosen available vocabularies.

Question 3: Was it easy to combine simple sentences in more complex constructions to express your ideas?
Users found it easy to combine simple sentences in more complex constructions (MOS=3.67, standard deviation 1.03). The question here appearing is about the logical operation relating to two or more conditions. However, 5 persons left out this part, needing more explanations and time than they were willing to spend;
Recommendations: anticipate complex structures and define the connections and relations between the options to choose properly.

Question 4: Did the system work in such a way as you expected?
The novelty of policy acquisition tools as a class leads to mostly no exact expectations. One person was expecting to write the policies in words without further options, whereas another thought of a graphical display. Still in general the implemented system matched the test subjects, resulting in a MOS grade 4.00 (standard deviation 0.5) for this question.
Recommendations: considering a tree-like design: visualization of conditions and rules as a graph overview.

Question 5: Do you think that the amount of time you generally have spent on the construction of a policy is adequate?
The most improvable factor of the policy editor is time consumption. The loading time on the one hand and users' thinking time on the other hand are influencing this aspect, leading to the rating of 2.50 (standard deviation 1.33).
Recommendations: More simplicity and technical optimization are desired here.

Question 6: Can you imagine yourself using the policy editor in the future for managing your personal data?
With a MOS rating of 3.83 (standard deviation 1.01), the users were willing to use policy editors in future for managing their personal data. After experiencing the system 9 users had positive attitudes towards policy editors, one user could not think of an application. The only concerns were related to privacy, security and transparency of the tool.
Recommendations: Privacy and security procedure should be declared.

Question 7: Can you imagine using the policy editor professionally or privately for defining policies to offer services or sell products?

Here users were very convinced (highest rating of the study: 4.03; standard deviation 0.96) that they would chose to use the policy editor. The difference to the previous question is that the application did not concern employment of personal data (but offering a service / selling a product). In addition, the users would like to have more options to select, simpler (well-constrained) domains and to have a "testing function" in order to make sure that the modeled policies lead precisely to the result that the user wants.

Recommendations: testing function; available overview of all implemented rules.

Question 8: Do you think that you would use this or a similar system regularly in the future?

Again a relatively high MOS grade of 3.9 (standard deviation 0.74) was given. The users are eager to use policy acquisition tools on a regular basis in the future. They found the system "helpful".

Recommendations: topic-related customization.

Question 9: What was your overall impression of the policy editor? What are its strengths and weaknesses? Please provide us any kind of feedback in a free form.

The overall perception was very positive. Users liked the idea of editing policies, the user interface and found the usage easy after a short period of learning how the tool works. Again the problems noted above were pointed out (improveable speed, difficulties with more complex rule structures, "consequence vs. condition" modeling, more intuitive vocabularies. An improved PAT is expected to be received even better.

Recommendations: Improved modelling principles, in particular, based on recommendations from further user tests or placing the system online.

7 Conclusions and Discussion

We see the following value being added by an ontology-based policy management compared to conventional policy practices:

1. *Spreading of policies*, freedom in policy distribution and sharing, annotation of the end users' data and services, easyness in reading other people's and organizations' policies; all this would be difficult without the semantic practices.
2. *Reduction of costs for policy construction*: existing similar policies may be available and easy to reuse elsewhere. For example, most of the internet shops have very similar polices on how to deal with the customer data and they would not need to redefine all the policies from scratch. One could also advance eGovernment visions by provisioning machine readable laws, e.g., on data protection,
3. *Reduction of the mistakes in the user-generated policy modeling* as the system's storage, query and reasoning service as well as sharing of policies within communities act as controllers for policy correctness.
4. *Better awareness of the end users about policies, rules and regulation:* With the suggested system the policies are easily retrieved and presented to the users.

Policy editors emerge as new assistance tools, allowing users to define rules in various settings. The user study showed that users came along very well with the policy modeling tasks, without special preparations or much prior knowledge of the concept. Overall, the test subjects felt consistently very positive about the introduced policy acquisition tool. In particular, the standard deviation of users' evaluation on the overall tool usage (question 1) is 1.15, which shows that the opinions have had minor differences within the user set. The test subjects were commonly eager to use similar tools for private as well as professional purposes: question 8 related to this criterion has the mark higher than the average and the rather low standard deviation of 0.74. Most of the involved persons appreciated the benefits of the system (in particular, saved costs, policy management by the end consumer and the reduction of the modeling mistakes).

The occurred usability issues define requirements for the next versions of the tool. In particular, the response delay mainly depends on the complexity of requests regarding vocabulary retrieval and validation, a missing caching mechanism of inference results and the reasoning strategy of the policy engine. The currently rather generic yet complex request of the PAT server towards the policy engine should be split in a small and simple set of specific requests to obtain vocabulary and validate usage. Furthermore, it should be investigated to find the optimal reasoning algorithm for different types of policy requests and scenarios. The user interface could be simplified by offering different levels of verbosity for editing policies. Distinct vocabularies or placements of statement slots should be offered for editing condition and conclusion triplets.

Apart from technical and usability issues, the following more socially-oriented questions should be investigated in community-driven policy modeling studies:

- How users share personal data, multiple identities, etc. Initial observations can be drawn from social networking websites (e.g., LinkedIn, Xing, etc.) where users can select whether they share a specific type of information with other users;
- Specifying, accumulating and storing arbitrary policies could result in a "policy Wikipedia" provisioning commonsense knowledge rules of what users find right and appropriate, e.g., "do not drink and drive". Such community effort would also have an anthropological effect in enabling observation of which kind of policies are shared between large communities and which policies are less popular.
- Certain policies vary by countries, cultures and time (e.g., eating any kind of food using hands could have been acceptable in certain countries in the past, but not in the present). This adds to additional technical challenges in policy versioning, matching and comparison.

Finally, we are convinced that policy acquisition from the end users is a highly important functionality for services offered in user-centered open environments, such as in the (Semantic) Web or mobile settings. Also we foresee that implementations of such ontology-based policy acquisition will become essential for any end user-oriented environment involving policies, such as business and private data management tools, ubiquitous environments, eGovernment applications.

Acknowledgements. The Telecommunications Research Center Vienna (ftw.) is supported by the Austrian government and the City of Vienna within the competence center program COMET. This work is partially funded by the EU IST projects Magnet Beyond (http://www.ist-magnet.org) and m:Ciudad (http://www.mciudad-fp7.org).

Appendix: Case Study Descriptions

Case Study 1: Eshop
You are a manager of an internet shop (Eshop). You need to model for an Eshop customer the following privacy policies in a policy editor. The customer's name is Maria.

1. Information You Give Us: we store any information you enter on our website or give us in any other way.
2. You can choose not to provide certain information but then you might not be able to take advantage of many of our features.
3. Automatic Information: we store certain types of information whenever you interact with us.
4. E-mail Communications: we often receive a confirmation when you open e-mail from Eshop if your computer supports such capabilities.
5. We also compare our customer list to lists received from other companies in an effort to avoid sending unnecessary messages to our customers.

Case Study 2: Etiquette
You need to model the following policies related to basic human behavior or etiquette in a policy editor. The policies modeled by you would be incorporated in software to help children, foreigners, robots or automated personal assistants to make choices about their behavior.

1. Restaurant. What about Doggy bags? There's nothing wrong with taking your leftovers home in a doggy bag, especially since portions are usually more than any human should eat in a single sitting.
2. Tea Party. Since it is a tea party, it's okay to eat with fingers.
3. Phone Conversation. While answering a call, do not scream or use a harsh voice.
4. Phone Conversation. In case of a poor connection or when you are abruptly disconnected, the individual who originated the call is responsible for calling back the other party.
5. Phone Conversation. If you want to leave a voice mail message on the phone, repeat your name and telephone number twice, clearly.

References

1. Berners-Lee, T., Connolly, D., Kagal, L., Scharf, Y., Hendler, J.: N3Logic: A Logic for the Web. Journal of Theory and Practice of Logic Programming (TPLP) (2007) (Special Issue on Logic Programming and the Web)

2. Bonatti, P.A., Duma, C., Fuchs, N., Nejdl, W., Olmedilla, D., Peer, J., Shahmehri, N.: Semantic web policies - a discussion of requirements and research issues. In: Sure, Y., Domingue, J. (eds.) ESWC 2006. LNCS, vol. 4011, pp. 712–724. Springer, Heidelberg (2006)
3. Collins-Sussman, B., Fitzpatrick, B.W., Pilato, C.M.: Version Control with Subversion. O'Reilly, Sebastopol (2004)
4. Crockford, D.: The application/json Media Type for JavaScript Object Notation (JSON). RFC 2647 (July 2006)
5. Davies, J., Fensel, D., van Harmelen, F.: Towards the Semantic Web: Ontology-Driven Knowledge Management. John Wiley & Sons, Chichester (2002)
6. De Roo, J.: Euler proof mechanism (2007), `http://www.agfa.com/w3c/euler/`
7. Golbeck, J., Parsia, B., Hendler, J.: Trust Networks on the Semantic Web. In: Proceedings of Cooperative Intelligent Agents 2003, Helsinki, Finland (2003)
8. Karat, C.-M., Karat, J., Brodie, C., Feng, J.: Evaluating Interfaces for Privacy Policy Rule Authoring. In: Proceedings of the Conference on Human Factors in Computing Systems (CHI 2006), pp. 83–92 (2006)
9. Kaviani, N., Gasevic, D., Hatala, M., Wagner, G.: Web Rule Languages to Carry Policies. In: Proceedings of Eighth IEEE International Workshop on Policies for Distributed Systems and Networks (POLICY 2007), pp. 188–192 (2007)
10. Kuhn, T.: AceRules: Executing Rules in Controlled Natural Language. In: Marchiori, M., Pan, J.Z., Marie, C.d.S. (eds.) RR 2007. LNCS, vol. 4524, pp. 299–308. Springer, Heidelberg (2007)
11. Riehle, D. (ed.): Proceedings of the 2005 International Symposium on Wikis (WikiSym 2005), San Diego, California, USA, October 16-18 (2005)
12. Sriharee, N., Senivongse, T., Verma, K., Sheth, A.P.: On Using WS-Policy, Ontology, and Rule Reasoning to Discover Web Services. In: Aagesen, F.A., Anutariya, C., Wuwongse, V. (eds.) INTELLCOMM 2004. LNCS, vol. 3283, pp. 246–255. Springer, Heidelberg (2004)
13. Verma, K., Akkiraju, R., Goodwin, R.: Semantic Matching of Web Service Policies. In: Proceedings of the Second International Workshop on Semantic and Dynamic Web Processes (SDWP 2005) (2005)
14. Völkel, M., Krötzsch, M., Vrandecic, D., Haller, H., Studer, R.: Semantic Wikipedia. In: Proceedings of the 15th International conference on World Wide Web, WWW 2006, Edinburgh, Scotland, May 23-26 (2006)
15. Zhdanova, A.V., Shvaiko, P.: Community-Driven Ontology Matching. In: Sure, Y., Domingue, J. (eds.) ESWC 2006. LNCS, vol. 4011, pp. 34–49. Springer, Heidelberg (2006)
16. Zhdanova, A.V.: Community-driven Ontology Construction in Social Networking Portals. International Journal on Web Intelligence and Agent Systems 6(1), 93–121 (2008)

User-Driven Semantic Wiki-Based Business Service Description

Heiko Paoli[1], Andreas Schmidt[2], and Peter C. Lockemann[3]

Abstract. A key factor for success of companies operating in a globalized market environment is a modern SOA-based infrastructure. An essential component of a SOA infrastructure is the central service registry. Current standards for organizing service registries and their implementations are driven by the technical aspects of the infrastructure. When using such technically organized service registries, business users often fail to find the needed information. With the concepts of Web 2.0 in mind, we present a new approach to the organization and implementation of the business registries that are driven by the needs of business users. The paper discusses the problems of the current technically driven approaches, presents an architecture for a business user-driven service registry and introduces an implementation of the architecture using UDDI and Semantic MediaWiki.

1 Introduction

A key factor for the success of companies operating in a global market environment is a flexible communication and information infrastructure that can be quickly and easily adapted to changing needs. Lately, service orientation has evolved as one of the more promising concepts for providing this flexibility [4]. Information infrastructures that follow the paradigm of Service-Oriented Architecture (SOA) allow information processes to be defined conveniently and with minimal effort as a succession of calls on available services [10, 12].

Judging from the many trade journals, service orientation does not yet live up to these expectations. We claim as our thesis that the failure is due to service descriptions that are of little help to the business users. Current descriptions have been written by service developers and just cover technical

Heiko Paoli, Andreas Schmidt, and Peter C. Lockemann
FZI - Research Center for Information Technologies, Karlsruhe, Germany
e-mail: {Heiko.Paoli,Andreas.Schmidt,Peter.Lockemann}@fzi.de

S. Schaffert et al. (Eds.): Networked Knowledge - Networked Media, SCI 221, pp. 269–283.
springerlink.com   © Springer-Verlag Berlin Heidelberg 2009

aspects such as service interface, formal parameters, or supported protocols. But this is not the world of the business users who initiate and control the business processes and react to numerous events in them. They need to know which services are available for which business purpose, how these services can be connected, which services have to be replaced when a business process has to be changed or whether new services are needed in order to adapt to new requirements.

As part of the solution we propose differentiating between different stakeholders. Designing information processes should be the responsibility of personnel that understands both, information systems and the business processes (we refer to them as *business analysts*). They need to know what the services have to offer to the business, and they should be able to communicate with the *business users* to map their needs to calls on the services. How these services have been technically implemented should be of little concern to them. The implementation of the services, and their connection to information processes, is the domain of *service developers*.

Service registries should address all stakeholders. Current service descriptions, though, concentrate on the service developers. To include the business aspects in a published service description would be the task of the business analysts. The objective of this paper is to discuss how the analysts can effectively be supported to carry out this task. Any solution should keep in mind that in an environment subject to frequent change, service description cannot be a one-time affair but rather a continuous and collaborative effort among business analysts and service developers [25].

Web 2.0 seems to be an appropriate interaction paradigm in which all stakeholders can be given an active part in service description. This paper presents a new collaborative and lightweight approach to describing services, and shows how business users can take an active part in it, so that a service registry would be able to cover their needs as well.

2   Problem Analysis

As discussed before, service discovery has technical and business ("semantic") facets. The technical part of a service description deals with the syntax of the service interface and is affected by the underlying SOA infrastructure. The semantic part should reflect the business objectives of the service. We examine some of the consequences of the two facets.

2.1   Capturing the Semantics of Business Aspects

The technical part of a service description has always been formulated in a way to make algorithmic processing possible. For the purpose of computer-assisted service discovery the same should hold for the semantic part.

Consequently, the business analyst must build a formal model of his or her conceptualization of the business domain, and relate the services to this model.

Take the following example. A business analyst has been given the task to build a new public information portal for flood emergency management. How will he or she find the already published services that might be useful? Suppose the analyst searches for a suitable service under the term of "flood level". Then he or she will in all likelihood miss a service for retrieving the current water level of rivers, even though this would be a good candidate for building the portal. If we had a relation from "flood level" to "water level" and used it in the discovery process, chances would be much higher that more of the appropriate services would be found.

We conclude that traditional information retrieval techniques based on descriptive terms are clearly insufficient and must be augmented by consideration of each term together with its network of somehow related terms.

2.2 Orthogonality of Technical and Business Aspects of the Service Description

Service implementations are technical artefacts and represent technical abstractions from real-world phenomena. Technical descriptions specify how they can, and must be used within a computational environment. Consequently, technical descriptions should only concern the service developers. Likewise, semantic descriptions should solely be of interest to the business analysts and users. Moreover, being an abstraction the same service implementation may be applicable in different business situations and, hence, may have more than one semantic description. Take again the water level service. It may be viewed, and employed, differently by a flood manager, the manager of a river shipping company and the manager of a hydropower plant. And finally, a service may very well have technically been implemented in different ways so that it needs different technical descriptions while the semantic description remains the same.

Consequently, both for technical and application reasons the technical and business aspects of the service description should be kept separate, something that has been known in software engineering as *separation of concerns*[1].

2.3 Support of the Dynamic and Collaborative Process of Service Description

Modern business is not a static affair. Consequently, new services may come and go, while other services must continuously be adapted or applied to new

---

[1] Progr. for Separation of Concerns,
http://www.dmi.unict.it/~tramonta/PSC07/

business cases. Continuous change to the business descriptions in the registry is, therefore, a constant challenge for which classical, waterfall-model like approaches that start with business process analysis and end with formal approval, with numerous coordination meetings in between, are ill-suited. In today's interlinked world the flexibility of SOA should be complemented by a more flexible approach where the organization of the business registry should be turned into a collaborative and continuous task along the lines of, say, the Web 2.0 concept.

2.4 Conclusion and Requirements

As we have seen in the problem analysis, a business-oriented service registry should meet three main requirements:

R1 Capture the semantics of business aspects to make services more accessible to business users

R2 Keep technical and business aspects of the service description separate for optimal support of the different user groups

R3 Support the collaborative and dynamic evolution of the service description to accommodate changing needs

3 UDDI as a Foundation

UDDI is practically the only standard for advertising services by service registries. The ambitious goal of UDDI was to establish a world-wide service registry to create a world-wide market of services and enable small and unknown companies anywhere in the world to offer their innovative services to customers on the other side of the globe. Therefore we should try to stay with UDDI as the basis of our registry unless UDDI completely fails to accommodate the requirements R1 through R3.

Figure 1 gives a condensed overview of UDDI. Central to UDDI is the UDDI registry. The registry points to the service description (WSDL) and the service itself. The description of a published service provided by WSDL should enable the service consumer to use the service via the underlying technical infrastructure. This description is therefore related to the technical interface of the service, describing syntactically its operations, formal parameters, message types, and supported protocols.

UDDI indeed provides a mechanism for augmenting the service description by metadata, although the mechanism seems fairly cumbersome for business analysts and users. The metadata take the form of name/value pairs that are stored in the technical Model (tModel) of UDDI. The name part of the

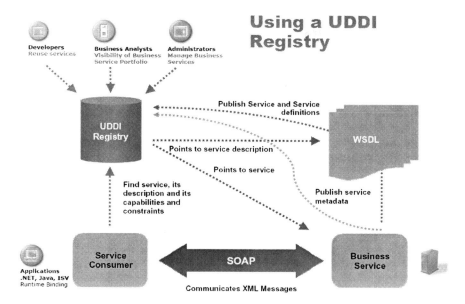

Fig. 1 Implementation and usage of a SOA with UDDI (Source [23])

tModel represents the namespace of any data structure which is to be used to characterize the service, whereas the value part is a unique pointer to the referenced data structure. The idea behind this approach is to categorize the registered services by standardized and uniformly known global category systems, such as the North American Industry Classification System (NAICS). The category system has been criticized as insufficient and the UDDI data model as very limited [20]. Indeed it seems far from satisfying requirement R1 because the category system is much too coarse to describe the services, and also too large for a user to become familiar with and to select the terms appropriate for a given situation. In addition the effort in manpower and time is inordinate to continuously develop huge and global category systems [28]. To conclude, the current UDDI concept seems little supportive of business semantics.

The concept of tModel seems to go some way towards requirement R2, though. And indeed, UDDI supports several user roles. The top left-hand corner of Figure 1 shows three user groups of the service registry: developers, business analysts and administrators. Administrators deal mainly with the technical management of the registry and the published services, and provide technical support for the other user groups but do not create or employ new services themselves. However, the two other groups, developers and business analysts, match two of our own stakeholders. But as far as service description is concerned, all user groups are treated alike: There is just one common description method.

Requirement R3 is not addressed by the concept of UDDI at all. UDDI does not care whether the global category systems remain the same or not. Therefore, the process of standardization of category systems is outside the scope of UDDI. Similarly, if a service or its application is changed, it is left to the publisher whether and how to adjust the service description. UDDI does not foresee any explicit support.

To summarize, UDDI as a concept seems well organized to support requirements R1 and R2. It offers no direct support for R3, but nor does it place obstacles in the way. On closer examination, though, even the support of R1 and R2 with the tModel as the only mechanism seems rather poor.

4 Related Work

4.1 Business Semantics (R1)

In [24] WSDL is the industry standard for describing Web services while UDDI is the industry standard for advertising them. The authors pursue the general objective of automated discovery of Web services taking semantics into account. They propose that a DAML+OIL ontology be used to annotate WSDL message parts in order to add the necessary semantics to the Web service description. WSDL itself is extended by new markup tags which allow to attach the semantic description in the form of the preconditions and effects of a Web service. By the time a service is published to UDDI, the extended WSDL description is mapped to the tModel where it becomes accessible to the discovery process. Sivashanmugam et al. also develop a three-phase algorithm for the discovery process. At the start of a process a template is generated into which the service requirements are entered. In the first phase services are matched by functionality (service operations) and then the result is ranked in the following phases on the basis of semantic similarity of input/output parameters and preconditions/effects.

The work seems to go a long way towards R1. We note, though, that requirement R2 is poorly met: The semantics are entirely embedded in WSDL and thus cannot be separated from the technical description. Further, the semantics are expressed in notations unnatural to the business user.

A bit earlier, Paolucci et al. took a similar approach [21]. They present in greater detail an algorithm for matching service requests to advertised services based on semantic descriptions in the form of ontologies. From the point of view of R2 their approach seems somewhat more advanced, since the DAML-S semantics are kept on a semantic layer. By the time a service is advertised, a DAML-S/UDDI translator constructs a standard UDDI description and stores it in the UDDI data model while the semantics for the matching algorithm is sent to a DAML-S matching engine and stored there with a reference to the constructed UDDI description. In the discovery process the stored semantics in the DAML-S matching engine is used while the DAML-S/UDDI translator provides the dependent UDDI descriptions. Unfortunately, no application experience is discussed, but it seems doubtful that business users would feel comfortable with the semantic description or would consider the approach transparent enough to evaluate the outcome of their search.

Even earlier, McIlraith et al. already employed the DAML family for semantic markup of Web Services [18]. Their objective was different, though: They wished to automate the discovery, execution, composition, and interoperation of Web services for the use in multiple agent systems. With the help of ConGolog - a programming language for robot systems - it should be possible to write generic procedures, e.g., a generic procedure to plan a business travel, without knowing which services are currently available and how they should be invoked to execute the procedures. If an agent wants to use a generic procedure, appropriate services are discovered, composed and executed automatically. Again, the discovery process is supported by ontologies. Since everything is automated the use of a common method for technical and semantic aspects is a requisite rather than an obstacle.

4.2   Separation of Aspects (R2)

The discussion in the previous section shows that separation of concerns is on everyone's mind but seems poorly executed from a business application point of view. Separation of concerns is a widely held philosophy in software engineering, but there the technical aspects predominate, and the experts involved are technical people. Still one may learn from the general model by Bergmans et al. for composing systems from multiple concerns [1]. The authors introduce a number of requirements for design-level composability, and define a category of composability problems that are inherent for given composition models. One result are criteria when separation of concerns should be applied to reduce the complexity of software by composing independent components, and when it should be avoided because of composition anomalies. We conclude that our approach does not fall into the category of composition anomalies so that requirement R2 is indeed justified.

4.3   Collaborative Service Description (R3)

Collaborative work in general, and collaborative authoring in special, is nothing new. However, since we wish to make use of standards – such as UDDI – we need to employ standards for the collaboration as well. Such a standard is MediaWiki where categories can be assigned to articles in order to support searching and navigating through its content. Krötzsch et al. extend this concept to links between articles so that they become machine-processable [17]. Links between articles can be viewed as named relations, and articles can have named attributes. Both can be used for navigation and searching by an embedded query language. The language can also be used to create dynamic articles, e.g., to have an article in which all services related to "water level" are listed. This article is automatically updated when a new article about such a service is created or when an already published article is deleted.

[17] seems to confirm that a Semantic MediaWiki is ideal as a frontend for business analysts because it is easy to use, allows adaptation to dynamic changes in a collaborative way and, moreover, is a suitable framework for the semantic needs of R1. Besides adjusting service descriptions, one mainly collaborative task is the continued development of the ontologies. [27] reports on the engineering of lightweight ontologies by using tagging mechanisms. The idea behind this approach is that interesting information is shared within a community, which is then tagged by the latter to categorize it. Concepts of a lightweight ontology can then be derived from the used tags. The ontology is constructed and changed in a collaborative and Web 2.0-like way.

5   A Comprehensive Approach to Business Service Description

5.1   Basic Architecture and Workflow

We start with requirement R3. Similar to the suggestion in Section 4.3 we make use of Web 2.0. More specifically, we take a Semantic MediaWiki-based approach to the collaborative development of the business registry. To meet requirement R2 we decide to stay with UDDI for the technical registry and to add the Wiki solution as a front end to UDDI. Finally, to satisfy requirement R1 we follow the approach of Section 4.1 and employ ontologies to capture the network of related terms. In particular, our aim is a lightweight ontology that can be easily handled by business experts without extensive training in ontology engineering. In contrast to the approach of Section 4.1 we do not extend the UDDI data model but rather use the light-weight ontology with the Semantic MediaWiki. Contents of the UDDI Registry are dynamically rendered by an extension of the Semantic MediaWiki. We refer to our solution as an *Extended Semantic MediaWiki*.

Figure 2 shows the system architecture. It consists of four main components: a UDDI-based technical registry, a Semantic MediaWiki-based business registry, an ontology server and an ontology engineering component. The figure also indicates the basic workflow within the architecture. A software developer as a service publisher can use any UDDI-compatible client to publish a new service into the registry, which may also include a technical description like a WSDL file in the case of a Web service. In addition to the technical description, the software developer may add some keywords based on the ontology in order to roughly categorize the business use of the Web service. The content of the UDDI Registry is dynamically embedded into the content of the Semantic MediaWiki, which forms the business-oriented

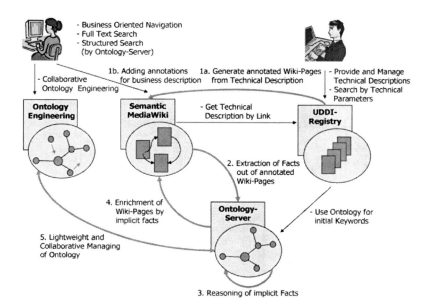

Fig. 2 System Architecture: Combining UDDI with a Semantic MediaWiki

registry. The keywords chosen by the software developer are used as an initial categorization for the service. From now on business users can search or navigate along the contents of the Semantic MediaWiki, add additional information to the dynamically generated pages, or create new pages. A Semantic MediaWiki is chosen to make the contents of the business registry machine understandable and to add implicit facts with the help of an ontology server. The ontology engineering component allows the business users to adapt the used business ontology to their needs in a lightweight and collaborative way.

We will discuss in more detail the steps that have been numbered in Figure 2.

5.2 Ontology

We observe from Figure 2 the central role of the ontology. Hence, we give a very brief outline. The left-hand side of Figure 3 gives an example of the organization of our ontology. The top level part provides the domain-independent concepts such as the terms *Concept*, *Business Object* and *Service*. These are refined to a network of concepts of the business domain of which Figure 3 just shows three examples, the terms *water*, *water level* and *water gage information*.

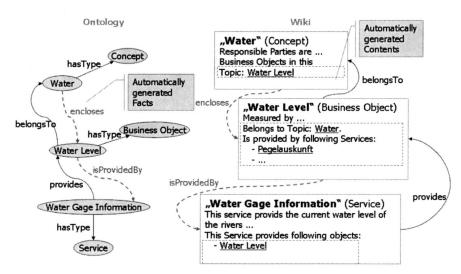

Fig. 3 Organization and presentation of the business registry together with dynamically embedded UDDI entries and implicit facts

In a collaborative environment the presentation of ontologies for effective and efficient use by the business analysts is particularly important. Presentation of the business registry is in the form of Wiki pages, with relations between concepts mapped to semantic links supported by the Semantic MediaWiki (right-hand side of Figure 3). In the example, *water* is a top level (business) concept while *water gage* is a business object concept, and *water gage information* is of type *service* and stands for a published and reusable service in the SOA infrastructure which will return a *water gage*.

We use OWL-Lite as the ontology description language. Currently we use KAON2 as a reasoner, but any other compatible reasoner should also be possible [19]. To make the ontology both persistent and generally available it is stored in a relational database from where it can be retrieved by the Semantic MediaWiki, the ontology server, the ontology engineering component, and the UDDI registry.

5.3 Annotation of Wiki Pages

A service such as *water gage information* is initially entered into the system by its developer. He or she publishes it to the UDDI registry together with a technical WSDL description, and is encouraged to augment it by intuitive keywords found in the ontology. Together with the publication a Wiki page is generated for the service, and automatically annotated with the aforementioned keywords as well as semantic links that are obtained from the relations of the general UDDI data model. The "Business Entity" element of the UDDI data model denotes the business analyst who is responsible for the business description so that the analyst may now be notified of the new service.

Business analysts can create new Wiki pages or modify existing ones (including generated Wiki pages) for the purpose of adding further annotations. The annotation of Wiki pages can be carried out by means of such Semantic MediaWiki features as semantic links, semantic attributes, and inline queries (to embed dynamic content). Many annotations can be obtained from the ontology by navigating through it and extracting further facts, or by using the reasoner to derive implicit facts or some of the semantic links. For example, on the left-hand side of Figure 3 the solid arrows represent relations that are explicitly available from the ontology (*hasType*, *belongsTo*, *provides*), while the dashed arrows represent relations that are implicitly available because of reasoning through the ontology server.

Not only does our approach satisfy requirements R2 and R3, but it clearly does so with great benefit to the two stakeholders of business analyst and service developer. A business analyst can concentrate on the business description and freely organize and annotate the Wiki pages. For example he or she may express the business context of a service, e.g., business use cases, business value etc. The business description is limited neither by the (technical) data model of UDDI nor the facilities of WSDL (that would allow us to describe a service only along its technical interface, e.g., operations, input, output parameters). On the other hand the UDDI registry remains compatible to current SOA implementations and allows developers to use their favorite UDDI tool to publish the technical description of newly implemented services.

5.4 Service Discovery

Other than in Section 4.1 we do not foresee automatic service discovery. Rather both the business analyst and the business user discover appropriate services by navigating through the ontology. This explains the emphasis we give to the presentation via Wiki pages. Take again the right-hand side of Figure 3. Note that much of the page contents for all terms is automatically generated. In particular, business object pages list all relevant services. For example, for an overview page on *water level* all *water gage* services are listed. If a new service is published which also returns a *water gage*, it will be automatically listed on the *water level* page without any additional manual intervention.

Consequently, our approach satisfies requirement R1 as well. The proposed organization of the business registry and the use of an ontology which is well known to the business analysts provides a familiar and easy-to-use environment for them. The business registry supports navigation along business objects for discovering needed services. The use of an ontology server together with the domain ontology enables a business-oriented search, e.g., a search for all services which provide a *water level*. The use of dynamic Wiki-Pages makes it possible for business analysts to build well adapted Single Points of Information for business users. In this context, a Single Point of Information means, that one dynamic Wiki-Page can contain many semantic queries related to a specific business process. Therefore all of the information necessary for fulfilling this business process can be collected automatically and presented via the same Wiki-Page.

5.5 Lightweight Ontology Engineering

In the dynamic business environment that we postulated in Section 2.3 the ontology itself is bound to frequently change as well. Rather than entrusting a central authority with modifying the ontology we rely on the combined and distributed competency of all business analysts, and perhaps even users. Accordingly, we let the ontology evolve in collaboration of the business experts whenever one sees the need. Since we cannot expect the analysts to be experts in building ontologies, the engineering of the ontology should be made as simple as possible.

We ease the task in two ways. For one the ontology is visualized as a graph, and all modifications can be easily done by dragging and dropping the nodes of the visual presentation rather than in some formal language. Second, the range of possible modifications is restricted (hence the name "lightweight engineering"). It is possible to create alternative labels for a concept and choose a preferred label for it. Concepts can only be connected via *broader-narrower* and *related* relations. By using Wiki pages all modifications to the ontology are immediately seen by all other business analysts.

6 Implementation

Figure 4 shows our implementation of the service registry. It consists of a central relational database, which holds the UDDI entries, the Semantic MediaWiki pages and the ontology. On top of the relational database we have a J2EE application server and an HTTP server with PHP support. The J2EE application server represents the technical UDDI-compatible registry, which is realized through three components: a UDDI framework to support the UDDI API (which enables technical descriptions), a SOAP Engine to support the UDDI protocol, and a UDDI browser to view the contents of the UDDI registry and publish new services. Our implementation is fully compatible to standard UDDI that explicitly allows publishers to use their own UDDI browser if they wish to (shown as the UDDI browser component at the bottom of Figure 4). The HTTP Server with PHP support represents the business-oriented registry realized through an extended Semantic MediaWiki component - the extension is necessary to support the automatic generation of content from the UDDI registry. For ontology engineering we use the

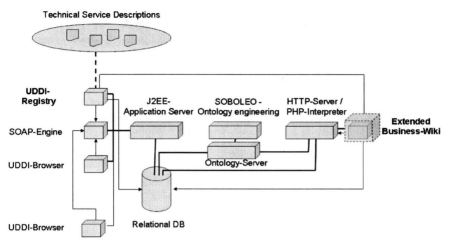

Fig. 4 Implementation of the service registry

existing tool SOBOLEO, a Web-based implementation of a Simple Knowledge Organisation System [27].

7 Experiences and Conclusions

The work presented in this paper has its origin in a project that was financed by the Ministry of Environment of Baden-Wuerttemberg. The environmental administration of Baden-Wuerttemberg has a long experience with environmental information systems in service oriented architectures. At the moment a redesign to a modern SOA-based infrastructure is planned by the State Institute for Environment, Measurements and Nature Conservation on behalf of the Ministry of Environment. The main objective is to provide all relevant parts of the system as services by a registry, and it should be possible to add a wide though unknown range of the services in the future. The system should be capable of handling hundreds of business users and service developers. To avoid duplicate work and to make all published services transparent to all business users a business oriented service registry seemed essential. The initial ontology we have used is based on an already existing and widely used taxonomy developed for the environmental information system of Baden-Wuerttemberg. The technical infrastructure as described above was developed in close communication with more than 10 representatives of business analysts and 5 representatives of developers, and was rolled out for a first testing period in April of 2007. First feedback by users sounds encouraging.

The thesis underlying our work is that service orientation will become widespread only if services can be discovered and employed with ease not just by service developers but also by business analysts. We have translated

H. Paoli, A. Schmidt, and P.C. Lockemann

the needs to three requirements, the separation of technical and semantic descriptions, natural use of the semantic descriptions by business people, and a collaborative approach to dealing with the business dynamics. First experiences seem to support our thesis for the narrow scope of environmental information systems. What is definitely needed are more systematic and wider ranging empirical studies before we can be sure that our approach is an important step in overcoming the still existing doubts on the effectiveness of service-oriented architectures.

References

1. Bergmans, L., Tekinerdogan, B., Glandrup, M., Aksit, M.: Composing Software from Multiple Concerns: Composability and Composition Anomalies. In: Proceedings of International Conference on Software Engineering (ICSE-2001), Toronto (2001)
2. Bloomberg, J.: The Four Pillars of Service-Oriented Development. zapthink (2005), http://www.zapthink.com/report.html?id=ZAPFLASH-2005418 (cited September 20, 2006)
3. Brickley, D., Miles, A.: SKOS Core Vocabulary Specification. Techreport. W3C Web Site (2005),
   http://www.w3.org/TR/2005/WD-swbp-skos-core-spec-20051102
   (cited November 30, 2005)
4. Cearley, D., Fenn, J., Plummer, D.: Gartner's Positions on the Five Hottest IT Topics and Trends in 2005. Gartner Web Site (2005), http://www.gartner.com/DisplayDocument?doc_cd=125868 (cited September 20, 2006)
5. Czarnecki, K.: Separation of Concerns - objektorientierte Frameworks und das generative Paradigma. In: OBJEKTspektrum Nr. 6. SIGS Conferences GmbH, München (1996)
6. Decker, S., Brickley, D., Saarela, J., Angele, J.: A Query Service for RDF. In: Proceedings of QL (1998)
7. Fayyad, U., Grinstein, G., Wierse, A.: Information Visualization in Data Mining and Knowledge Discovery. Morgan Kaufmann Publishers Inc., San Francisco (2001)
8. Fensel, D., Decker, S., Erdmann, M., Studer, R.: Ontobroker: The very high idea. In: Proceedings of the 11th International Flairs Conference (FLAIRS-1998), Sanibal Island (1998)
9. Garlan, D.: Software Architecture: a Roadmap. In: Proceedings of International Conference on Software Engineering (ICSE-2001) - Future of SE Track, Toronto (2000)
10. He, H.: What Is Service-Oriented Architecture. XML.Com. O'Reilly, Sebastopol (2003)
11. Hepp, M.: Possible Ontologies: How Reality Constraints Building Relevant Ontologies. IEEE Internet Computing 11(1), 90–96 (2007)
12. Huhns, M., Singh, M.: Service-Oriented Computing: Key Concepts and Principles. IEEE Internet Computing 9(1), 75–81 (2005)
13. Jones, S., Morris, M.: A Methodology For Service Architectures. OASIS Web Site (2005), http://www.oasis-open.org/committees/download.php/15071/A%20methodology%20for%20Service%20Architectures%201%202%204%20-%20OASIS%20Contribution.pdf (cited September 20, 2006)

14. Kifer, M., Bruijn, J., Boley, H., Fensel, D.: A Realistic Architecture for the Semantic Web. In: Adi, A., Stoutenburg, S., Tabet, S. (eds.) RuleML 2005. LNCS, vol. 3791, pp. 17–29. Springer, Heidelberg (2005)
15. Kiryakov, A., Popov, B., Ognyanoff, D., Manov, D., Kirilov, A., Goranov, M.: Semantic Annotation, Indexing, and Retrieval. In: Fensel, D., Sycara, K., Mylopoulos, J. (eds.) ISWC 2003. LNCS, vol. 2870, pp. 484–499. Springer, Heidelberg (2003)
16. Krafzig, D., Banke, K., Slama, D.: Enterprise SOA. Service-Oriented Architecture Best Practices. Prentice Hall, New Jersey (2004)
17. Krötzsch, M., Vrandecic, D., Völkel, M.: Semantic MediaWiki. In: Cruz, I., Decker, S., Allemang, D., Preist, C., Schwabe, D., Mika, P., Uschold, M., Aroyo, L.M. (eds.) ISWC 2006. LNCS, vol. 4273, pp. 935–942. Springer, Heidelberg (2006)
18. McIlraith, S., Son, T., Zeng, H.: Semantic Web services. IEEE Intelligent Systems - Special Issue on the Semantic Web (2001)
19. Motik, B., Sattler, U.: A Comparison of Reasoning Techniques for Querying Large Description Logic ABoxes. In: Proceedings of the 13th International Conference on Logic for Programming Artificial Intelligence and Reasoning (LPAR-2006) (2006)
20. Nickull, D., McCabe, F., MacKenzi, M.: SOA Reference Model TC. OASIS Web Site (2006), http://www.oasis-open.org/committees/tc_home.php?wg_abbrev=soa-rm (cited September 20, 2006)
21. Paolucci, M., Kawmura, T., Payne, T., Sycara, K.: Semantic Matching of Web Services Capabilities. In: Horrocks, I., Hendler, J. (eds.) ISWC 2002. LNCS, vol. 2342, p. 333. Springer, Heidelberg (2002)
22. Schaffert, S.: IkeWiki: A Semantic Wiki for Collaborative Knowledge Management. In: Proceedings of WETICE, pp. 388–396. IEEE Computer Society, Los Alamitos (2006)
23. Shen, Z.: UDDI v3.0 (Universal Description, Discovery and Integration). OASIS Web Site (2004), http://www.oasis-open.org/committees/uddi-spec/doc/spec/v3/uddi-v3.0.2-20041019.pdf (Cited September 5, 2007)
24. Sivashanmugam, K., Verma, K., Sheth, A., Miller, J.: Adding semantics to web services standards. In: Proceedings of the 1st International Conference on Web Services (ICWS-2003), Las Vegas (2003)
25. Stojanovic, Z., Dahanayake, A.: Service-Oriented Software System Engineering, Challenges and Practices. Idea Group Publishing (2005)
26. Sure, Y., Angele, J., Staab, S.: OntoEdit Multifaceted inferencing for ontology engineering. In: Spaccapietra, S., March, S., Aberer, K. (eds.) Journal on Data Semantics I. LNCS, vol. 2800, pp. 128–152. Springer, Heidelberg (2003)
27. Zacharias, V., Braun, S.: SOBOLEO - Social Bookmarking and Lighweight Engineering of Ontologies. In: Proceedings of the 16th International World Wide Web Conference (WWW 2007) - Workshop on Social and Collaborative Construction of Structured Knowledge (CKC), Banff (2007)
28. Zimmermann, O., Krogdahl, P., Gee, C.: Elements of Service-Oriented Analysis and Design: An interdisciplinary approach for SOA project. IBM DeveloperWorks (2004), http://www-128.ibm.com/developerworks/library/ws-soad1/ (cited September 5, 2007)

Facilitating Knowledge Management in Pervasive Health Care Systems

Bo Hu, Srinandan Dasmahapatra, Paul Lewis, David Dupplaw, and Nigel Shadbolt

Abstract. Realising the vision of pervasive health care will generate new challenges for knowledge management and data integration. Such challenges are fundamentally different from issues and problems that we face in centralised approaches as well as non-clinical scenarios. In this paper, we reflect upon our experiences in the MIAKT project wherein a prototype system was developed to support data integration and decision making in the breast cancer domain. While the decision making needs to rely on different clinical expertise, the MIAKT system leveraged a system ontology to glue together distributed services. Situating the MIAKT system in a highly pervasive environment reveals the inefficiency of global vocabularies via domain ontologies and the inappropriateness of "static" system ontologies with assigned system configuration instances. We examine the capability of a process calculus based language, Lightweight Coordination Calculus (LCC), in meeting knowledge management challenges in pervasive health care. The key difference in approach lies in making the representational abstraction reflect the relative autonomy of the various clinical specialisms (*eg.*, mammography or histopathology) involved in contributing to patient management. The bringing together of diverse forms of information necessary for the collective medical assessment is managed by tracking the message passing protocols undertaken by medical personnel. The scope within LCC of accommodating boolean-valued constraints allows for flexible integration of heterogeneous sources in multiple formats, which are characteristic features of a pervasive healthcare environment.

1 Introduction

"The most profound technologies are those that disappear. They weave themselves into the fabric of everyday life until they are indistinguishable from it". This is Mark

Bo Hu, Srinandan Dasmahapatra, Paul Lewis, David Dupplaw, and Nigel Shadbolt
School of Electronics and Computer Science
University of Southampton
Southampton SO17 1BJ, United Kingdom
e-mail: {bh,sd,phl,dpd,nrs}@ecs.soton.ac.uk

S. Schaffert et al. (Eds.): Networked Knowledge - Networked Media, SCI 221, pp. 285–304.
springerlink.com   © Springer-Verlag Berlin Heidelberg 2009

Weiser's vision [19] of how technologies might eventually blend in with our surroundings. Projecting this vision on to health care gives a picture wherein "smart" software agents would act on behalf of human specialists in collecting/monitoring critical life support data, extracting information from the data, jigsawing information/data together, and eventually enabling decisions and actions to be taken on the outcome of such processes. One of the most far-reaching consequences of such a vision is the emergence of a different paradigm of patient care. Currently, a person experiencing a perceptible ailment invokes the "patient-seeing-doctor" pattern, where a doctor is often an array of specialists. Instead, the new health care paradigm emphasises a degree of continuous medical surveillance, with key decisions for medical follow-ups requiring automated processing, and in a decentralised manner.

One of the fundamental questions concerning pervasive health care is how to trigger, choreograph and respond to distributed data/information resources [20]. Moreover, with the dispersal of sites of information gathering and exchange, the traceability of decisions and outcomes needs to be maintained for reasons of review, updates of protocols in the interest of improving health care provision, and for other reasons such as insurance claims. Knowledge management schemes in health care settings concentrate on the tasks of creating, discovering, preserving, delivering and exploiting the knowledge assets [12, 17]. In pervasive settings, maintaining traces of evolution of knowledge makes the passage of data, particularly those that impinge on decision making, be of vital importance. For instance, decisions about (say) increases in blood sugar levels may be drawn from information about time of day (in the context of habitual mealtimes, for instance) and forwarded to relevant medical centres where an information triage may be conducted. This simple example suggests that the "anywhere and anytime" nature of the requirement of appropriate information availability poses serious challenges for knowledge engineering.

What we report in this paper is built upon the experience and lessons of the MIAKT project that successfully developed a knowledge management framework for breast cancer screening program [3]. Using MIAKT as a stepping-stone, we investigate solutions to knowledge management issues, either general to all applications targeting the pervasive environment or specific to the health care domain. The modus operandi of the proposed solution is to view knowledge management in pervasive health care through the apparatus of interactions/conversations and examine the problem from both a behavioural and an epistemological perspective. We devise a mechanism for integration and sharing what dynamically emerges from interactions among different parties involved in providing health care services. This follows naturally from the domain requirements of recording and processing key knowledge as well as the procedures that are invoked to provide the relevant knowledge, all under the constraints of clinical guidelines and ethical concerns.

This paper is structured as follows: in Section 2 we present health care as the motivating application wherein two approaches, a system ontology-based one and an interaction-based one, are examined. Specific requirements set by the health care domain are also addressed. Section 3 reviews the MIAKT system architecture and discusses why it is not suitable for supporting pervasive health care. Section 4 presents the interaction driven knowledge management and explains, by the means of examples, the vantage of this interaction/conversation-powered framework. Section 5 concludes the paper and lays down possible future research directions.

2 Why Pervasive Health Care Is Different?

Thus far, knowledge in health care, to some extent, remains a "cottage industry" with largely tacit knowledge only explicit to isolated specialists, organisations and professional guilds. Although the necessity of collaboration has been recognised, there is little systematic knowledge sharing of clinical intervention outcomes. This is partially due to the lack of proper technologies and partially because of the division of clinical labour. Health care data is diverse in format, massive in size, and inconsistent in quality. While exposing data can be easily accommodated with the current capability (e.g. the Information Retrieval technologies [21], the database technologies [7], the standardised Resource Description Framework (RDF) [9], etc.), knowledge learnt therefrom is less "transplantable"—the process of acquiring knowledge is difficult to be standardised and prescribed. With the flux of data from heterogeneous sources, many assumptions enjoyed by conventional knowledge management becomes less applicable. Such assumptions include a centralised data repository and a globally accepted knowledge model which, although not perfect, provides a placeholder to rendezvous point upon which heterogeneous information/data can be projected. Instead, the diversity in clinical domain knowledge makes isolated knowledge islands dominant. Such an archipelagian landscape inevitably increases the cost of health care and decreases the quality of health care services. The situation is exacerbated in the pervasive health care scenario when full access to the entire domain knowledge is replaced with fragmented views that are limited by different privileges granted to the users, different usage of the data, and different hardware capacity.

An important consideration with respect to the new type of knowledge management is the distributed nature of not only the data but also the users accessing the data. With the advance of modern transportation, communication, and tele-medicine, patients are not longer restricted by physical and geographical constraints. In the situation of comorbidity (e.g. heart disease, AIDS, cancer, diabetes, or mental health), it is not a surprise to find that a patient is examined in one hospital; his/her case is reviewed by clinicians from another hospital; and he/she is treated in a third hospital by yet another group of clinicians due to speciality and availability. Data about a particular patient might be held by different departments within one hospital, from different hospitals and/or even from hospitals located in different countries. Data requests might come from members of a dedicated team logging in from their office or home, auditing committee, interns requiring information for educational purposes, and patients themselves all with different access privilege and access capabilities. Differences in work idioms in different situations evidently has the potential to significantly impinge on the quality of services that one is offered. Apart from the wide spread in geographic regions and a diverse landscape of users, the heterogeneity of clinical data is also demonstrated in the different levels of granularity of domain knowledge, different nomenclatures used in sub clinical domains, different protocols followed, different levels of details passed on in the form of medical records, and different standards reinforced by industrial manufacturers. In such an environment, knowledge which is a prime capital can only be based upon distributed and heterogeneous data/information sources and needs to be processed automatically in streaming mode. Users, therefore, need to locate the correct data providers, retrieve the most appropriate parts of the exposed data and glue together all the bits and pieces of information to make sensible conclusions. In the meantime, we need to observe the data integrity and obey regulations on data privacy and ethics. These constraints suggest that exploitation of the data should not be directly tied to the data itself but rather through dedicated "knowledgeable" services.

We emphasise the "knowledgeable" aspect of these services due to the fact that they provide added values besides the mere exposure of the encapsulated data, avoiding inter- and intra- individual variation when interpreting data, and thus knock down the entry threshold of potential data providers. We would argue that there is a clear distinction between data/information and knowledge drawn therefrom. In many applications, knowledge management is oversimplified or misinterpreted as information and/or data management. Although we appreciate the importance of organising and exposing data, we would like to argue that data/information does not equal knowledge which, though based on the (explicitly observable) data/information, benefits from procedural and conceptual transformations and re-alignment of the data and the theoretical frameworks of domain experts that are called upon in the application of knowledge in appropriate settings, clinical or otherwise [2, 5, 18]. We contend that the management of data/information concerns mainly on observation, storage and retrieval while the management of knowledge focuses more on interpretation, reuse, sharing, and revision. This distinction can help to ensure that the word "knowledge" is not abused and the exchange of data/information is not misinterpreted as the exchange of knowledge where the latter task is mainly conducted via a series of negotiation/conversation. That is to say that unlike facts and information procured through observation, knowledge is ought to stem from interaction with the physical world and other individuals and the exchange of knowledge can only be done by exchanging messages conveying what we observe with respect to our surroundings. A concrete incarnation of knowledge is, therefore, established and validated from such messages via revision, updates and transformation of the contents of a prior knowledge base. Following the same path, we believe that knowledge management is better situated when being viewed through the lens of interaction.

With the distinction between knowledge and data in mind, we reveal yet another prerequisite of knowledge management targeted at pervasive health care. Traditionally, knowledge management concentrates more on dealing with "what is", i.e. the "static" snapshot of "what" the things are, and pays less attention to the transitions among different descriptive stages in the network of the domain of discourse. Medical ontologies/vocabularies are the direct outcomes of such a static view [13]. Giving a generic definition/specification of what is in the domain of discourse is of course important, but the significance of capturing the links that intermesh isolated information islands into operable and reproducible knowledge at a large scale often gets overlooked in the actual solutions provided by tools. The relational aspects become more evident from the context-dependent nature of knowledge. In many cases, information extracted from the same data might lead to different knowledge when examined in different conversational contexts. For instance, body temperatures taken in the morning or within a certain period of the administration of febrifuge yield a totally different result from those taken in the evening or outside the time window that the medicine takes effect. The requirement of discerning contexts wherein knowledge is obtained and interpreted initiates a progress from the question "What is X?" to "How did I come to know X?" to "How can I share my experience on acquiring X?" That is to say that when managing knowledge in clinical domain, one shares also the reasoning procedures wherewith the knowledge is procured. In pervasive setting where (for example) knowledge islands might locate at different geographic regions, each participating individual needs to handle a continuous stream of monitoring data based on which its local interpretations are made. Hence, knowing the context wherein the data is to be parsed and interpreted becomes even more of a pressing issue.

3   The MIAKT Project and MIAKT System

The phenomena of calling upon a non-conventional knowledge management in pervasive health care is evident in the MIAKT project which will be used as the example application in this paper. The MIAKT project aims to support multidisciplinary meetings (MDM) for the diagnosis and management of breast cancers. MDMs include specialists from different clinical backgrounds who come together to make a diagnosis of the patient's disease based on data that have been gathered and analysed. This data may include images from a variety of imaging modalities, such as X-ray mammograms, Magnetic Resonance Imaging scans, ultrasound scans and histopathological slides cut from biopsies taken from the suspect areas. The data will also include information about the patient obtained from different sources, such as previous examinations and outcomes, medication allergy records, and family history of specific diseases.

Given the diverse nature of the expertise and the working practices within each sub-discipline expert involved, the complexity of the task of patient management in the face of potentially ambiguous data makes for multiple possible paths through the space of actions that representative agents can take. If a formulation can allow for the creation of multiple paths triggered by the requirements that are suggested by the information at hand, which is translated into messages passed to update the relevant information available to agents, then this could serve as a prototypical scenario for the pervasive world, solutions to which thus can be reused and adapted in similar applications.

Reviewing MIAKT in this light brings opportunities and challenges. The design philosophy of MIAKT system is grounded in the MIAKT architecture (as shown in Figure 1) developed primarily to allow the integration of various knowledge-based tools, that are published as services, into a knowledge management system [4].

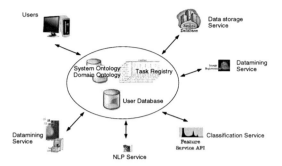

Fig. 1 MIAKT services

Exposing services instead of data has the advantage of allowing partners within the MIAKT project to retain control over their data while having their services and fragmented knowledge merged into a single knowledge management system.

Tacit knowledge underpinning MIAKT services is captured in a domain ontology and a system ontology. Different knowledge services are projected upon a common conceptualisation, the MIAKT domain ontology (BCIO) [8]. BCIO acts as a reference point. It provides handles for the relevant information concerning a particular case on different aspects and at different granularity levels. This allows a specialist to concentrate on the fragments that he/she is really familiar with and makes available his/her evidence, conclusions and the basis of the judgement to the users of other modules [8]. However, knowledge possessed by one service can only achieve its modularly accessible vision if backed up by a communication mechanism with which integration with other fragments is facilitated.

As an effort towards defining this mechanism, we developed the MIAKT System Ontology (MISO) that organises the different services and regulates data transfers within the system. Users can invoke the available distributed services annotated based on their exposed functionality. MIAKT utilises MISO to facilitate the exchange of information among different services. Analogous to other communication protocols, MISO specifies the format of messages passing from one service to another, the message initiators and recipients, the mechanism to parse and understand the contents of the messages, and the structure of the replies. Defining how the various services should work together is, however, beyond the capacity of MISO, a system ontology which is inherently static. In MIAKT, such dynamic information is scripted in a predefined, deterministic manner with descriptors drawn from MISO. Although fully compliant with the working procedure of a UK MDM, it fails to provide a mechanism to ensure this working procedure is correctly followed—such a task is unloaded to human users with "quality assurance" being laid in the hands of good will.

We contend that the service oriented architecture illustrated in Figure 1 is likely to be a typical one that is currently used to facilitate the needs of distributed health care applications. Such applications are characterised by a centralised, universal domain ontology (e.g. BCIO) and system ontology (e.g. MISO), a centralised task registry, an *ad hoc*, fixed task invocation script with centralised control mechanism, and a predefined list of available services. We, by no means, deny the usability and applicability of such an architecture; we would like to argue that the above architecture, although served well in the MIAKT scenario, cannot service the needs of the pervasive health care scenario envisaged and described in the introduction. It does not formalise *how* the data should be transferred and *how* a mutual understanding is established that underpin pair-wise communication. Such weaknesses make the MISO less favourite in domains wherein a major concern is not only what can be exchanged but also on how things should be passed and how the messages are understood by the recipients. Such behavioural and epistemological aspects are of particular importance in channeling the flow of messages. We, therefore, need a means to capture both the static aspect and the dynamic aspect of the domain knowledge. In the next section, we will retell the MIAKT story in light of process calculi.

4   From MISO to Interaction Model

The MIAKT prototype provides us an ideal platform to investigate the impacts and implications of applying semantic-rich technologies to knowledge management with respect to a truly distributed/pervasive health care system. By truly distributed/pervasive, we mean a framework avoiding centralised repositories and thus a centralised view, concrete or virtual is not endorsed. In practice, a centralised repository is credit for its effectiveness, security, and manageability. This is true as long as patients do not go beyond the catchment area of a hospital. Centralised solution becomes less attractive when one is injured while visiting another region or another country; when one needs daily care while on holiday in a retreat cabin; and when specialists are summoned up from different areas in a teleconference to discuss a rare case. Such examples all share the same characteristic of *decentralisation*. As illustrated in Figure 2, in a fully pervasive environment, we observe the relative independence of each participating agent or intelligent device.

We shall lay out the various domains of expertise that are called upon to process evidence and build up a clinically appropriate representation of a patient based on very different views. We accommodate the diversity and heterogeneity while systematically enable the choreography of individual information resources so as to combine their knowledge of a particular patient or a particular disease. While the conceptual tokens within each specialism needs to be indexed for completeness against specific sub-domain ontologies, we leave out the detailed capture of local updates of informational states that reference the conceptual tokens. Instead we take a behavioural outlook where the supporting software infrastructure gains meaning not (intensionally) via local state changes, but instead by keeping track (extensionally) of the messages and resources that were exchanged and consumed in order for the system to function as required.

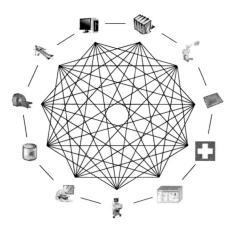

Fig. 2 Pervasive health care architecture

This shift of emphasis immediately suggests to us to take a process oriented view for system design and analysis. The formalism we use in the application is based on the Lightweight Coordination Calculus, LCC [14] which is a logic programming language based on the low-level specification of message passing and local state changes, something akin to CCS or π-calculus[11, 16]. As discussed in Section 2 a system supporting health care provided by multi-disciplinary teams requires dealing with groups of people who have seldom had their systems engineered to perform tasks together. This too, fits the paradigm of concurrency, where there is no single locus of control for task execution. Instead of the other resources existing merely to serve the control unit, these entities lead an autonomous existence and only undergo message induced transitions upon opening up access to each other—centralised control gives way to concurrent processes wherein each party accomplishes the tasks allocated to it and expose the results to accommodate the requests from the others. Moreover, this interaction based sharing of information enables a dynamic way of knowledge composition: by sharing knowledge through interactions we indirectly share data. We will demonstrate that leveraging interaction models as opposed to trying to combine knowledge in the traditional manner benefits health care knowledge management and complements the existing work in the MIAKT project.

4.1 Lightweight Coordination Calculus

LCC is a process calculus for specifying coordination among multiple participants [15]. It does so by clearly stating what role an individual plays in a messaging process. An LCC model is built upon the principle that role-playing agents should obey the laws and/or protocols that are explicitly specified against the roles that such agents are expected to take. LCC ensures the fulfillment of roles by individuals through regulating the message-flows among them. These include: the messages that should be sent and are expected to be received and what constraints should be satisfied before a message can be handled. The full picture of LCC syntax is specified in Extended Backus-Naur Form (EBNF) as in Figure 3:

$$
\begin{array}{rl}
\langle\text{Framework}\rangle := & \{\langle\text{Clause}\rangle,\}^{1+} \\
\langle\text{Clause}\rangle := & \langle\text{Agent}\rangle :: \langle\text{Definition}\rangle \\
\langle\text{Agent}\rangle := & \mathbf{a}(\langle\text{Type}\rangle, \langle\text{ID}\rangle) \\
\langle\text{Definition}\rangle := & \langle\text{Agent}\rangle \mid \langle\text{Message Clause}\rangle \mid \langle\text{Definition}\rangle\ \mathbf{then}\ \langle\text{Definition}\rangle \mid \\
& \langle\text{Definition}\rangle\ \mathbf{or}\ \langle\text{Definition}\rangle \mid \langle\text{Definition}\rangle\ \mathbf{par}\ \langle\text{Definition}\rangle \mid \\
& null \leftarrow \langle\text{Constraint}\rangle \\
\langle\text{Message Clause}\rangle := & \langle\text{Message}\rangle \Rightarrow \langle\text{Agent}\rangle \mid \langle\text{Message}\rangle \Rightarrow \langle\text{Agent}\rangle \leftarrow \langle\text{Constraint}\rangle \mid \\
& \langle\text{Message}\rangle \Leftarrow \langle\text{Agent}\rangle \mid \langle\text{Constraint}\rangle \leftarrow \langle\text{Message}\rangle \Leftarrow \langle\text{Agent}\rangle \\
\langle\text{Constraint}\rangle := & Term \mid \langle\text{Constraint}\rangle \wedge \langle\text{Constraint}\rangle \mid \langle\text{Constraint}\rangle \vee \langle\text{Constraint}\rangle \\
\langle\text{Type}\rangle := & Term \\
\langle\text{ID}\rangle := & Constant \\
\langle\text{Message}\rangle := & Term
\end{array}
$$

Fig. 3 Grammar of LCC

In an LCC interaction model, we use predicate $\mathbf{a}()$ to specify the role that an individual is playing, $\Rightarrow$ and $\Leftarrow$ to specify the direction of message flow, and $\leftarrow$ for constraints. Their use will be documented in the next subsection. *Term* and *Constant* are implementation-specific. In the current version, *Term* is a well-formed formula in Prolog logic programming language and *Constant* is a Prolog constant starting with a lowercase letter. LCC also provides constructs for parallel (**par**), sequential (**then**) and switch branching (**or**) controls.

Interpreting LCC is tantamount to unpacking LCC clauses, finding the next tasks that a set of agents are permitted to perform by definition and updating the status of an interaction accordingly. A set of clause rewriting rules are introduced to ensure LCC constructs are interpreted in a consistent manner [15]. Let C_i be an LCC clause from a model M; I_i be a set of received messages currently queueing for an individual participating in an M-based interaction; C_{i+1} be the unfolded new LCC clause; $I_{i+1} \subset I_i$ be the set of remaining unprocessed messages; and O_i be the outgoing messages generated when processing C_i. An LCC model is interpreted by exhaustively unfolding clauses as detailed in [15] to produce the following sequence:

$$
C_1 \xrightarrow{I_1, I_2, M, O_1} C_2, \ldots, C_i \xrightarrow{I_i, I_{i+1}, M, O_i} C_{i+1}, \ldots, C_{n-1} \xrightarrow{I_{n-1}, I_n, M, O_{n-1}} C_n,
$$

The interpretation of LCC constraints depends on a particular implementation. In this paper, we assume Prolog as the underlying programming language and thus interpret the constraints in terms of a Prolog logic program. Nevertheless, this by no means denies the possibility of implementing LCC constraints with other programming languages, such as JAVA.

Pooling together the rewriting rules for LCC-specific constructs and the interpretation of a Prolog program, we obtain the semantics of LCC models. For instance, in the LCC interaction model presented in Figure 4, the sequence construct **then** is unfolded by examining the first part of the sequence or, if it is closed (i.e. executed), unfolding the next part. After unfolding, the system tries to instantiate all the

$a(\text{on\_call\_doctor}, N) ::$
　$\text{routine\_check}(P) \Leftarrow a(\_A)$ **then**
　$\left(\begin{array}{l} \text{take\_temperature}(P) \Rightarrow a(\text{nurse}, S) \textbf{ then} \\ \text{take\_blood\_sample}(P) \Rightarrow a(\text{nurse}, T) \leftarrow \neg\text{blood\_test}(P) \end{array} \right)$

Fig. 4 An example of LCC

variables (e.g. P and A) to examine the satisfiability of LCC clauses. A narrative interpretation of the LCC model in Figure 4, therefore, reads "when an on call doctor receives a routine check request on a patient (P), he/she first asks an arbitrary nurse (S) to take P's body temperature. When the body temperature is done, he/she asks an arbitrary nurse (T) to take P's blood sample if P has not been given blood test before." Note that whether nurse S and T are the same person is unknown from the context.

4.2 Collaborating as LCC Role Players

In the "eyes" of LCC, knowledge management task is tantamount to negotiation. We use a few examples from the MIAKT scenario to explain how LCC interaction models are utilised. We would like to emphasise that knowledge management is built upon an awareness of the flow of information within the system, reflecting protocols and guidelines that are driven by legal and ethical concerns given the sensitive nature of clinical information. Tacit knowledge is, therefore, observed through the patterns of messages. Such a transparent knowledge acquisition procedure implies that the way that one learns can be literally copied with the same conclusion as long as the same contexts are reconstructed. For instance, in Figure 5, we define how a domain specialist could join a particular MDM event and how she could retrieve patient records from those holding the data and merge these "foreign" patient records with her local copies.

Domain specialist E's participation in an MDM starts with an invitation initiated from the meeting coordinator which is denoted as MDMC and represented using a role introducing predicate, $a(\text{mdmc}, C)$. This invitation specifies that domain specialists in an MDM should satisfy a list of restrictions given as X. In an interaction model, this is expressed as a message from the MDM coordinator (represented as an outbound double arrow leading from the coordinator to the specialist). An individual is given the full responsibility to decide whether she is capable enough to take the role of a domain specialist in a particular MDM instance. An acceptance will be sent off if she is confident of meeting all the requirements X raised by the coordinator. The source of confidence might come from her education and working experience, her knowledge about this particular patient, and/or her availability during the time this MDM event is to be held. Exactly how the constraints are satisfied and how E's confidence is interpreted are left to E herself or a software agent acting on behalf of E.

$a(\text{specialist}, E) ::$
  $invitation(E, X) \Leftarrow a(\text{mdmc}, C)$ **then**
  $accept(E) \Rightarrow a(\text{mdmc}, C) \leftarrow satisfies(X)$ **then**
  $\dots$
  $request(\text{Patient}, Y, M) \Rightarrow a(\text{mdmc}, C) \leftarrow certificate(Y) \wedge trans\_method(M)$ **then**
  $receive(\text{Patient}) \Leftarrow a(\text{mdmc}, C)$ **then**

$$\left(\begin{array}{l} get\_patient\_id(\text{Patient}, ID) \text{ \textbf{then}} \\ retrieve\_local\_record(ID, \text{Patient}_{\text{local}}) \text{ \textbf{then}} \\ \left(\begin{array}{l} align(A_l, \text{Patient}, \text{Patient}_{\text{local}}) \leftarrow find\_local\_aligner(A_l) \\ \textbf{or} \\ align(\text{Patient}, \text{Patient}_{\text{local}}) \Rightarrow a(\text{aligner}, A_r) \leftarrow \neg find\_local\_aligner(A_l) \wedge \\ \qquad\qquad\qquad\qquad\qquad\qquad\qquad find\_remote\_aligner(A_r) \end{array}\right) \end{array}\right)$$

$\dots$

$$\left(\begin{array}{l} classify(\text{Patient}) \leftarrow \neg missing\_info(\text{Patient}) \\ \textbf{or} \\ \left(\begin{array}{l} patient\_record(\text{Patient}, M) \Rightarrow a(\text{datahandler}, H) \leftarrow missing\_info(\text{Patient}) \wedge \\ \qquad\qquad\qquad\qquad\qquad\qquad\qquad found\_new\_handler(H) \end{array}\right) \text{ \textbf{then}} \\ \dots \end{array}\right)$$

$\dots$

Fig. 5 Domain specialists

For instance, a crawling tool such as semantic squirrel[1] might set off to gather all the information from E's electronic diary, her personal webpage, emails, publications, and her resume so as to compile a profile of E. Or alternatively, when safety and privacy is the concern, relevant information can be collected from more controllable sources, e.g. databases of skill sets maintain by national authorities. This can be done with or without the supervision of a human and the results could be a standalone measure or one criteria as a part of a comprehensive measure covering some aspects of E.

Upon joining an MDM, E's concern can be boiled down to several separated but closely related tasks: data acquisition, knowledge creation, knowledge sharing. Specialist E first sends a request to download the patient's record from local and remote data repositories. Together with the request, she also submits certificates Y for receiving the data and her preferred methods M for data transfer.

Each specialist only has access to a small fragment of the patient data. How an MDM team would glue the information together and build up diagnostic decision therefrom then relies on to what extent they overlay their knowledge, together with their general expertise of the field and their experiences, onto the body of a particular patient. Knowledge sharing within an MDM team should not be assumed to be on an equal basis. It might be necessary to present the conclusion together with the evidence to specialists from different background. For instance, in Figure 6 a specialist is asked to evaluate the condition of a patient and whether he/she should be recommended for an ultrasound scan. A special message is used to

---

[1] http://semantic-squirrel.org

signify the end of the communication. As long as the ending message is not received, the individual taking "$a(\text{recommendation\_request}(P,D,S),M)$" role await messages from "$a(\text{ultrasound\_expert}, S_u)$" and repetitively update its local records based on the fragments of knowledge that are available to her.

Figure 6 and Figure 7 present two alternative ways to elicit the tacit knowledge behind patient's data. In the recommendation request model, an individual takes the responsibility of liaising with the *radiographers* and *ultrasound specialists* to decide whether a patient should be recommended an ultrasound test. The flexibility of LCC is evident from the two models implementing ultrasound specialists. Figure 7(a) leaves plenty room to the users to decide how to evaluate patient's record (D) while Figure 7(b) details the situations that should be considered. In both models, tacit knowledge is extracted from patient record via communications with ultrasound specialists.

4.3 Reusing Existing Knowledge

Comparing and contrasting localised patient data against that provided by MDMC could be the first step towards establishing a common ground for exchanging knowledge. Most likely, however, patient records gathered by MDMC are not in a ready-to-use format for E. The received data must, therefore, be aligned with that kept locally by E. If an alignment has already been established and can be reused in the current task, E invokes the local *aligner* to integrate remote patient data with the local records. If, on the other hand, information in the received patient records is beyond the coverage of existing alignments, E needs to locate a dedicated aligning service and submit both the remote and local patient records for aligning. Fragments

$a(\text{recommendation\_request}(P,D,S),M)$ ::

   ...

   /* forward patient record to a field expert */
   $patient\_record(D) \Rightarrow a(\text{ultrasound\_expert}, S_u)$ **then**

   /* accumulate a final score for recommendation */
   $score(T) \Leftarrow a(\text{ultrasound\_expert}, S_u)$ **then**

$$\begin{pmatrix} null \leftarrow \neg equals(T,0) \textbf{ then} \\ null \leftarrow update(S,T) \textbf{ then} \\ a(\text{ultrasound\_recomm}(P,D,S),M) \end{pmatrix}$$

   or

$$\begin{pmatrix} null \leftarrow equals(T,0) \textbf{ then} \\ \begin{pmatrix} /* \text{ make final recommendation } */ \\ recommend\,(P,\text{``ultrasound''}) \Rightarrow a(\text{radiographer}, Rad) \leftarrow (S \geq 1) \\ \textbf{or} \\ recommend\,(P,\text{``no ultrasound''}) \Rightarrow a(\text{radiographer}, Rad) \leftarrow (S < 1) \end{pmatrix} \end{pmatrix}$$

Fig. 6 Cyclic model for ultrasound recommendation

$a(\text{ultrasound\_expert1}, X) ::$

$\quad patient\_record(D) \Leftarrow a(\text{recommendation\_request}(P,D,S),M)$ **then**

```
    /* return an overall score */
```

$\quad score(X) \Rightarrow a(\text{recommendation\_request}(P,D,S),M) \leftarrow evaluate(D,X)$ **then** $update(D)$

$\quad$ **or**

$\quad score(0) \Rightarrow a(\text{recommendation\_request}(P,D,S),M) \leftarrow already\_updated(D)$

(a) Evaluation model 1

$a(\text{ultrasound\_expert2}, X) ::$

$\quad patient\_record(D) \Leftarrow a(\text{recommendation\_request}(P,D,S),M)$ **then**

$\left(\begin{array}{l}
\text{/* assign scores to different situations */} \\[4pt]
score(1) \Rightarrow a(\text{recommendation_request}(P,D,S),M) \\
\qquad\qquad \leftarrow axillary_lymph_lump(D) \text{ \textbf{then} } update(D) \\[2pt]
\textbf{or} \\
score(1) \Rightarrow a(\text{recommendation_request}(P,D,S),M) \\
\qquad\qquad \leftarrow breast_implants(D) \text{ \textbf{then} } update(D) \\[2pt]
\textbf{or} \\
score(1) \Rightarrow a(\text{recommendation_request}(P,D,S),M) \\
\qquad\qquad \leftarrow localised_breast_nodularity(D) \text{ \textbf{then} } update(D) \\[2pt]
\textbf{or} \\
score(1) \Rightarrow a(\text{recommendation_request}(P,D,S),M) \\
\qquad\qquad \leftarrow (abnorm(D) > P3 \wedge age(D) < 35) \text{ \textbf{then} } update(D) \\[2pt]
\textbf{or} \\
score(1) \Rightarrow a(\text{recommendation_request}(P,D,S),M) \\
\qquad\qquad \leftarrow palpable_breast_lump(D) \text{ \textbf{then} } update(D) \\[2pt]
\textbf{or} \\
score(-99) \Rightarrow a(\text{recommendation_request}(P,D,S),M) \\
\qquad\qquad \leftarrow ((last_us(D) - date_of_invest(D)) \leq t) \text{ \textbf{then} } update(D)
\end{array}\right)$

$\quad$ **or**

$\quad score(0) \Rightarrow a(\text{recommendation\_request}(P,D,S),M)$

(b) Evaluation model 2

Fig. 7 Ultrasound result evaluation

of the *aligner* interaction model is shown in Figure 8. It is evident that we do not assume a conceptualisation which is globally accepted by all the participants. The existence of a domain ontology as a common reference point is not mandatory but an advantage to incorporate domain knowledge. For instance, one can align against and translate a patient record into existing standards in clinical domains, such as HL7[2], DICOM[3], etc or a purpose-built application ontology, e.g. the BCIO developed for MIAKT.

---

[2] http://www.hl7.org/

[3] http://medical.nema.org/

$a(\text{aligner}, RA) ::$
$align(\text{Patient}_1, \text{Patient}_2) \Leftarrow a(specialist, \_) \textbf{ then}$
$$\left(\begin{array}{l} \left(\begin{array}{l} global\_align(\text{Patient}_1, O) \textbf{ then} \\ global\_align(\text{Patient}_2, O) \end{array} \right) \leftarrow exist\_domain\_ontology(O) \\ \textbf{or} \\ local\_align(\text{Patient}_1, \text{Patient}_2) \leftarrow \neg exist\_domain\_ontology(O) \end{array} \right)$$
$\cdots$

Fig. 8 Data *Aligner*

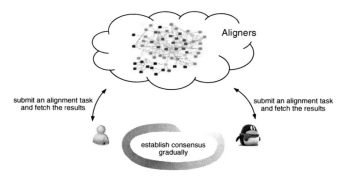

Fig. 9 Accomplishing mutual understanding

Establishing consensus (and other knowledge management tasks) is, therefore, no longer a task concerning only two parties. Instead, a group of interested peers might be involved to offer a much larger and more diverse pool of knowledge. Figure 9 illustrates a scenario whereby a consensus emerges upon consultation between those who are willing to play the *Aligner* role.

In the same vein, creating knowledge might concern more than one party who will pool their knowledge together in order to understand the problem at hand. In Figure 10, "$a(\text{ultrasound\_expert}(I, C), S_u)$" acts as a local coordinator for a group of classifiers who draw upon their knowledge based on the results of a patient's ultrasound scan using data mining or pattern recognition technique. Such knowledge is collected and composed to elicit more "knowledgeable" inspection that cannot be achieved with single knowledge source. In the same way as in the scoring example, how to compose multiple knowledge sources is left to the individual taking the role of "$a(\text{ultrasound\_expert}(I, C), S_u)$".

4.4 Being Proactive

In line with the World Health Organisation's view on "increasing the effectiveness of adherence interventions" [22], the pervasive health care paradigm offers patients more convenient and personalised health services than ever before and assists in

$a(\text{ultrasound\_expert}(I,C),S_u) ::$
$\quad \ldots$

$\quad$ /* forward ultrasound result to a group of image analyser */
$$\left(\begin{array}{l} patient\_record(D) \Rightarrow a(\text{classifier},I_h) \leftarrow I = [I_h|I_t] \textbf{ then} \\ classification(I_h,D) \Leftarrow a(\text{classifier},I_h) \textbf{ then} \\ null \leftarrow update(C,D) \textbf{ then} \\ null \leftarrow a(\text{ultrasound\_expert}(I_t),S_u) \end{array}\right)$$
$\quad \textbf{or}$
$\quad$ /* inform MDMC with the final classification */
$\quad null \leftarrow compose(C_{final},C)$
$\quad ultrasound\_result(C_{final}) \Rightarrow a(\text{mdmc},C)$
$\quad \ldots$

Fig. 10 Iterative model for Ultrasound recommendation

$a(\text{patient\_assistant},P,Status) ::$
$\quad null \leftarrow get\_sys\_time(CurrentTime) \textbf{ then}$
$$\left(\begin{array}{l} get\_body\_temperature() \Rightarrow a(sensor,S_T) \leftarrow check(Status) \textbf{ then} \\ temperature(T) \Leftarrow a(sensor,S_T) \textbf{ then} \\ null \leftarrow update(Status,T) \\ \textbf{par} \\ get\_heart\_rate() \Rightarrow a(sensor,S_{HR}) \leftarrow check(Status) \textbf{ then} \\ heart\_rate(H) \Leftarrow a(sensor,S_{HR}) \textbf{ then} \\ null \leftarrow update(Status,H) \\ \textbf{par} \\ get\_eeg() \Rightarrow a(sensor,S_{EEG}) \leftarrow check(Status) \textbf{ then} \\ eeg(E) \Leftarrow a(sensor,S_{EEG}) \textbf{ then} \\ null \leftarrow update(Status,E) \\ \ldots\ldots \end{array}\right) \textbf{then}$$
$$\left(\begin{array}{l} null \leftarrow is\_critical(Status) \textbf{ then} \\ warning(P) \Rightarrow a(family\_doctor,D) \textbf{ par} \\ alert(P,M) \Rightarrow a(visualiser,V) \leftarrow medication(Status,M,H) \wedge get\_history(H,P) \end{array}\right)$$
$\quad \textbf{or}$
$$\left(\begin{array}{l} null \leftarrow \neg is\_critical(Status) \textbf{ then} \\ null \leftarrow a(patient\_assistant,P,Status) \end{array}\right)$$

Fig. 11 Personal Assistant

patient's adherence to treatment regimens; at the same time it relieves clinicians of many tedious routine jobs and significantly reduces administrative costs. This vision can be facilitated through an interaction model dedicated to a patient. For instance, as illustrated in Figure 11, a simple personal assistant interaction model can run on mobile and/or embedded devices that continuously monitors the status of a patient. In case that a change of the patient's body temperature, heart rate, EEG signal, etc. triggers an alarm, the personal patient assistant notifies the patient's family doctor with a compiled and refined version of the patient's data while in the meantime analyses the patient's status and his/her history in order to recommend possible medication, e.g. orally taking Aspirin.

The interaction model based knowledge and data management ensure a consistent communication between patients and their doctors and an in-time feedback from patients, both positive and negative.

4.5 Handling Multimedia Data

In clinical domain, multimedia data abounds, e.g. X-ray images, MRI scans, ultrasound scans, etc. Transferring unprocessed multimedia data obviously increases network traffic, especially in pervasive health care settings where portable devices only have limited processing power, memory, and file system capabilities. Leveraging high level annotations becomes a feasible alternative to raw data and "knowledgeable" content-based multimedia data management [10] is therefore a facilitator for such data transfer. In Figure 12, we illustrate an example of how X-ray images are annotated, retrieved and visualised (by dedicated visualisers for images and texts).

$a(\text{xray\_analyser}, A) ::$

   $get\_descriptor(X) \Leftarrow a(\text{xray\_specialist}, X_S)$ **then**

   $\left(\begin{array}{l} get\_descriptor(X) \Rightarrow a(analyser_{\text{image}}, [A_i]) \text{ } \textbf{then} \\ descriptor(D) \Leftarrow a(analyser_{\text{image}}, A_i) \\ \textbf{or} \\ null \leftarrow show(X) \text{ } \textbf{then} \\ descriptor(D_{\text{local}}) \Rightarrow a(xray\_specialist, X_S) \leftarrow generate\_descriptor(D_{\text{local}}) \\ \ldots\ldots \end{array} \right)$ **then**

   $descriptor(D_{\text{local}}) \Rightarrow a(xray\_specialist, X_S) \leftarrow update(D_{\text{local}}, D)$

 $visual(show(X), image(X))$

 $visual(generate\_descriptor(X), dialogbox(X))$

(a) Xray Analyser

$a(\text{xray\_specialist}, X_S) ::$

   $null \leftarrow get\_descriptor(D)$

   $descriptor(D) \Rightarrow a(\text{xray\_provider}, R)$ **then**

   $\left(\begin{array}{l} image(X) \Leftarrow a(\text{xray\_provider}, R) \leftarrow show(X) \\ \textbf{or} \\ (not\_available(K) \Leftarrow a(\text{xray\_provider}, R) \leftarrow error(K) \\ \ldots\ldots \end{array} \right)$

 $visual(get\_descriptor(D), input(X))$

 $visual(show(X), image(X))$

 $visual(error(K), text(\text{"No image found"}, K))$

(b) Xray Browser

Fig. 12 Multimedia data handler

Expertise on image analysis is not mandatory for the individual taking the role of $a(\text{xray\_analyser}, A)$. He/she can either manually annotate the image with a set of descriptors or take advantage of his/her knowledge of other human specialists (clinical domain specialists or image analysis experts) or automated software tools to request a list of image descriptors. In both cases, tacit knowledge instead of raw data is leveraged and manipulated. When manually crafting the image descriptors, a graphic user interface could be rendered by a dedicated specialist visualiser: $visual(Data, Type)$. Similarly, in order to retrieve an X-ray image for diagnosis purposes, the Xray specialist was prompted a dialog box to input image descriptors (D). He/she then submits D to the Xray image provider. Such a request might be tackled in the local hospital or relayed to a repository at a remote site. In Figure 12(b), three different types of visualisers are invoked to provide a seamless change of visual effects in the user interface.

4.6 Enhancing Security with LCC

Among others, security is a major concern in health care applications. π-calculus and it extensions (enriched with dedicated security constructs/functions) lay down a nice framework wherein authentication, authorisation, data integrity and data encryption issues can be formally modelled and analysed [1, 6]. Although LCC is not devised as a security protocol calculus, one can implement security primitives with constraints and message passing sequences among different parties and leverage restriction scoping to ensure that secured information is only exposed to the authorised participants. For instance, upon receiving a request of patient's data, one might check whether the data requester is what he/she claim to be by asking for a authentication message, e.g. a `signature` and whether the data requester has the privilege to view the entire patient record or part of it by looking up the access policy associated with his/her ID, e.g. a compulsory match $\exists l \in L.[ID\ \text{is}\ l]$. This is under both security and clarity considerations. Certain patient information is sensible and should not be disclosed to those who are not responsible for interpreting the data. Figure 13 illustrates fragments of LCC interaction model that retrieves data based on the request submitted by an arbitrary domain specialist. It is evident that whether or not a particular specialist is qualified to receive the requested data is subject to data-specific justification using $is\_authorised(E, ID)$ where ID can be

$a(\text{datahandler}, H) ::$
  $patient\_record(\text{Patient}, M) \Leftarrow a(\text{specialist}, E)$ **then**
  $is\_authorised(E, ID) \leftarrow get\_patient\_id(\text{Patient}, ID)$ **then**
  $inform(\text{Patient}) \Rightarrow a(\text{specialist}, E)$ **then**
  $get(P, M) \Rightarrow a(\text{DataMart}, D) \leftarrow registered(D) \wedge contains(D, P)$
              $\wedge matches(P, \text{Patient}) \wedge trans\_method(E, M)$
  $\ldots$

Fig. 13 Data Handler

computed from the information in passed together with Patient. Meanwhile, this interaction model also emphasises on the customisation of data transfer methods. We use $trans\_method(E,M)$ to state that the data transfer task is specific to a particular specialist.

LCC based security also allows each individual institution to implement its local version of the security enforcement functionalities. In the *datahandler* example, each hospital can implement a local version of the *is_qualified()* constraint using password, biometrics, or public key based methods. We do not delve into this issue further since it is beyond the scope of this article.

The local implementation will be informed to all the parties that are concerned when communication is to be carried out. Expressed in a π-calculus like formalisation, we have

$$(\nu c) \left(\underbrace{\bar{c}\langle(\text{Msg}, \text{hash})\rangle}_{A(Msg)} \mid \underbrace{c(x).\text{let}\,(m,f) = x\,\text{in}\,f(m)}_{B} \right)$$

Informally, this process states that both the message and the hashing function are passed through a dedicated channel between $A(Msg)$ and B running parallel indicated with composition operator $\mid$.

5 Conclusions

In this paper we reviewed our experience in designing and developing a service-oriented system that provides knowledge management for application domains with distributed knowledge sources. Our approach is tuned in particular to handle heterogeneous knowledge in clinical domains wherein data integrity and knowledge delivery are given a strong emphasis due to the privacy and ethic concerns. Such requirements are met by concealing the data with knowledge services and composing the services based on individual applications.

It is our contention that similar data-oriented health care systems could be significantly enhanced with emerging semantics-rich technologies. In order to explore such potentials, we experimented with a system ontology to regulate what can be passed onto others by a service encapsulating the date. We took one step further to enhance this static conceptualisation with a process calculus, LCC. LCC prescribes communications with interaction models that represent the interaction/coordination procedure while leaving plenty of room for implementation specificity. It can faithfully reflect organisational and national protocols and guidelines by way of specifying the exact workflow that an event or a task should follow. The merit of an LCC empowered knowledge management system is in the fact that knowledge is not treated as a static snapshot of the domain of discourse but a dynamic and ever-changing conceptualisations of the domain. We track how knowledge is extracted and situate data in interaction/conversation of all concerned parties. This is done by leveraging the vantage of LCC as a process calculus as well as its capability of specifying constraint and its flexibility in satisfying constraints.

There are certainly many aspects needed to be further addressed with respect to an LCC-driven knowledge management. Firstly, when formalising epistemological knowledge, finding the right trade-off between speciality and generality is critical and difficult. One needs to justify to what extent the fine-grained "know-how" knowledge should be codified and delivered. Too gross might make it hard to repeat the same knowledge acquisition procedure while too fine might jeopardise the transferability and interoperability of an interaction model. Secondly, LCC can be enhanced with features such as typed variables, built-in facilities handling temporal constraints, and assertions with probability as so to better capture clinical knowledge. Although it is possible to emulate many of these with the current capacity of LCC (through calls to other programming languages implementing Constraints), explicitly introducing them can certainly increase LCC's readability and usability. Thirdly, user study is necessary to demonstrate the learning curve of LCC. Such users are preferably domain specialists who work interactively with other clinicians and intelligent software agents rather than knowledge engineers. Finally, the usability of LCC models in the clinical domain would be better demonstrated with real-life clinical protocols and guidelines. A successful health care knowledge management application might, therefore, root in a faithful representation of such protocols and guidelines, modelling of which requires trained eyes and mind.

Acknowledgements

This work is supported under the FP6 OpenKnowledge STREP project under Grant numbers IST-FP6-027253. Part of the presented work was also supported by the MIAKT project funded by UK EPSRC under Grant number GR/R85150/01.

References

1. Abadi, M., Gordon, A.D.: A calculus for cryptographic protocols: The spi calculus. Information and Computation 148(1), 1–70 (1999)
2. Ackoff, R.L.: From data to wisdom. Journal of Applied Systems Analysis 16, 3–9 (1989)
3. Dasmahapatra, S., Dupplaw, D., Hu, B., Lewis, H., Lewis, P., Shadbolt, N.: Facilitating multi-disciplinary knowledge-based support for breast cancer screening. Special issue of the International Journal of Healthcare Technology and Management (5), 403–420 (2006)
4. Dupplaw, D., Dasmahapatra, S., Hu, B., Lewis, P., Shadbolt, N.: Multimedia distributed knowledge management in miakt. In: Handshuh, S., Declerck, T. (eds.) Proceedings of ISWC Workshop Knowledge Markup and Semantic Annotation, pp. 81–90 (2004)
5. Firestone, J.M.: Key issues in knowledge management. Knowledge And Innovation: Journal of The KMCI 1(3), 8–38 (2001)

6. Fournet, C., Abadi, M.: Hiding names: Private authentication in the applied pi calculus. In: Okada, M., Pierce, B.C., Scedrov, A., Tokuda, H., Yonezawa, A. (eds.) ISSS 2002. LNCS, vol. 2609, pp. 317–338. Springer, Heidelberg (2003)

7. Gray, J., Liu, D., Nieto-Santisteban, M., Szalay, A., DeWitt, D.J., Heber, G.: Scientific data management in the coming decade. SIGMOD Record 34(4), 34–41 (2005)

8. Hu, B., Dasmahapatra, S., Dupplaw, D., Lewis, P., Shadbolt, N.: Reflections on a medical ontology. International Jour. of Human-Computer Studies 65(7), 569–582 (2007)

9. Lassila, O., Swick, R.R.: Resource Description Framework (RDF) Model and Syntax Specification. W3C February 22 (1999)

10. Marques, O., Furht, B.: Content-Based Image and Video Retrieval. Kluwer Academic Publishers, Norwell (2002)

11. Milner, R., Parrow, J., Walker, D.: A calculus of mobile processes, i. Inf. Comput. 100(1), 1–40 (1992)

12. O'Leary, D.: Guest editor's introduction: Knowledge-management systems-converting and connecting. IEEE Intelligent Systems 13(3), 30–33 (1998)

13. Rector, A.L.: Clinical Terminology: Why Is it so Hard. Methods of Information in Medicine 38, 239–252 (1999)

14. Robertson, D.: A lightweight coordination calculus for agent systems. In: Leite, J., Omicini, A., Torroni, P., Yolum, p. (eds.) DALT 2004. LNCS, vol. 3476, pp. 183–197. Springer, Heidelberg (2005)

15. Robertson, D.: Multi-agent coordination as distributed logic programming. In: Demoen, B., Lifschitz, V. (eds.) ICLP 2004. LNCS, vol. 3132, pp. 416–430. Springer, Heidelberg (2004)

16. Sangiorgi, D., Walker, D.: The Pi-Calculus: A Theory of Mobile Processes. Cambridge University Press, Cambridge (2001)

17. Shadbolt, N.: Eliciting Expertise. In: Evaluation of Human Work, ch. 8, 3rd edn., pp. 185–218. CRC Press, Boca Raton (2005)

18. Sharma, N.: The origin of the Data Information Knowledge Wisdom hierarchy (2005), http://www-personal.si.umich.edu/~nsharma/dikw_origin.htm (accessed December 2007)

19. Weiser, M.: The computer of the 21st century. Scientific American 265(3), 66–75 (1991)

20. Weiser, M., Brown, J.: The coming age of calm technology. In: Denning, P.J., Metcalfe, R.M. (eds.) Beyond calculation: The next fifty years of computing, New York, Copernicus (1998)

21. Witten, I.H., Moffat, A., Bell, T.: Managing Gigabytes: Compressing and Indexing Documents and Images. Morgan Kaufmann Publishers, San Francisco (1999)

22. World Health Organisation. Adherence to long-term therapies. Evidence for action (2003)

Integrating Semantic Technologies with Interactive Digital TV

Antonis Papadimitriou, Christos Anagnostopoulos, Vassileios Tsetsos,
Sarantis Paskalis, and Stathes Hadjiefthymiades

Abstract. Interactive digital TV is becoming a reality throughout the globe. The
most essential aspect of TV broadcasting is enhancing the interaction experience
for the viewer. To this end, we explore the potential of introducing semantics in the
distribution, processing and usage of the media content. We propose a smart iTV
receiver framework capable of collecting, extending and processing semantic metadata related to the broadcast multimedia content. A system architecture is presented
along with examples of services to illustrate the combination of semantic metadata
content, user preferences and external data sources.

1 Introduction

There is no doubt that Television (TV) is one of the most prevalent media technologies. Analog TV broadcasts have been on air for more than five decades, and their
impact has qualified TV as a massively used technology throughout the globe. Yet,
recent technological developments have created new opportunities for the evolve-
ment of the TV Broadcasting market. Technological advances include the appearance of Digital TV (DTV) and the manufacturing of Set-Top-Boxes (STBs) capable
of executing interactive applications in the TV receiver. To this end, over the past
years, there has been a coordinated effort from multiple organizations towards the es-
tablishment of standard digital transmission technologies and application execution
environments for interactive TV (iTV). This has led to the development of several
standards such as Digital Video Broadcasting (DVB) transmission specifications [6]
(for satellite, cable and terrestrial TV) and the Multimedia Home Platform (MHP)
[3] as middleware for interoperable interactive applications.

Antonis Papadimitriou, Christos Anagnostopoulos, Vassileios Tsetsos, Sarantis Paskalis and
Stathes Hadjiefthymiades
Pervasive Computing Research Group, Department of Informatics and Telecommunications,
University of Athens, Panepistimioupolis 15784, Athens, Greece
e-mail: {anthony,bleu,b.tsetsos,paskalis,shadj}@di.uoa.gr

S. Schaffert et al. (Eds.): Networked Knowledge - Networked Media, SCI 221, pp. 305–320.
springerlink.com   © Springer-Verlag Berlin Heidelberg 2009

Most interestingly, there is already an extensive network of digital TV (DTV) infrastructure based on these specifications, numbering a multitude of homes subscribed to iTV services from different broadcasters. Hence, an open market is evolving with great prospects of benefit both in terms of provider profit and user satisfaction. The next step to the future of digital TV is the enhancement of interactive applications provided to the iTV subscribers. This is critical for the consolidation of iTV and its further penetration into the home entertainment market. Towards the goal of providing rich and personalized interactive services we propose a semantics-aware iTV receiver platform and the respective business model for its exploitation. The proposed platform is named POLYSEMA and features a prototype iTV receiver system capable of collecting, extending and processing semantic metadata related to the broadcast multimedia content.

Metadata information is used at the receiver in order to offer pioneer services to users. One such service is the retrieval of web information related to the current TV program. For instance, users might have included in their profile that they are specifically interested in some movie star. Then, every time this particular actor appears on a TV program (movie, show etc.) the system can retrieve from the web information, trivia and news about that actor. Another example is allowing for a finer granularity of parental control over the media content shown on TV. Specifically, parents may designate some rules about the content their children are forbidden to watch and each time the metadata describing a video scene triggers the respective rules, the display of audiovisual content on the screen will be suppressed. Other interactive services may include smart personal video recording, semantic channel recommendation and many more.

The technical specification of the POLYSEMA platform is based on several standards widely adopted by research and industry communities. This makes the proposed system a compelling solution for STB manufacturers and TV Broadcasters worldwide. Particularly, the standards fostered by POLYSEMA include the DVB-T specifications [5], the MHP standard as an application execution environment [3], the OSGi platform for a service-oriented integration platform [1], the MPEG-7 standard as a metadata description standard [17] and Semantic Web technologies [10] as the base for semantics representation and logic-based inference.

The contribution of our work is manifold. To our knowledge, previous work on semantic applications for interactive TV focused on particular services such as channel recommendation [12]. On the contrary, we consider the development of a broad range of semantically enriched applications, all built upon an extensible set of basic services offered by the core POLYSEMA platform. Moreover, we introduce the concept of Scene-by-Scene interactive TV, which stands for the development of applications which adapt their behavior according to the content shown during each separate video scene. Another novelty of our system is that users can modify a wide set of preferences and define rules about when applications should be triggered, thus leading to an increased level of personalization. An additional critical feature of the design of the POLYSEMA receiver is that it supports the development of applications which can retrieve and display program-relevant information from the World Wide Web. This is an attempt towards the convergence of iTV with the WWW, a combination which can leverage the quality and broaden the scope of services offered in the context of iTV.

Lastly, we substantiate that our prototype system can be easily migrated to the real world of broadcasting by describing a flexible business model which allows for a multitude of value-added services from multiple providers. In this business model, media providers have to supply multimedia content, whereas service and metadata providers provide interactive services and metadata describing the content in some certain format (MPEG-7 [17]). Broadcasters multiplex interactive applications into their transport streams so that TV viewers receive both the multimedia content and the application logic required to take advantage of metadata descriptions. Receivers download metadata from the transport stream and/or the web sites of service aggregators, and process the information to match displayed content with user-defined rules and preferences.

The rest of the paper is organized as follows: Section 2 presents the work related to the POLYSEMA approach. In Section 3, we elaborate on the architecture of the proposed system, whereas Section 4 illustrates examples of interactive services and how they can be supported by the proposed iTV Platform. Section 5 is used to describe the business model which allows multiple stakeholders participate in the media content distribution value chain and thus hold a share in the revenue of the iTV market. Finally, section 6 concludes the chapter.

2   Related Work

Since the appearance of digital TV both academic and industrial stakeholders have focused on the convergence of the Web with the world of television broadcasting. The DVB consortium [6] produced the specifications of the Multimedia Home Platform [3], in order to establish the features that a state-of-the-art interactive television platform should support. The MHP specification includes a profile which supports web browsing and an email client functionality. Web browsing is realized via the introduction of the DVB-HTML [14] markup language. Unfortunately, DVB-HTML has not been adopted yet by STB manufacturers because it is a quite complex specification, thus leading to higher device costs and the rise of interoperability considerations. An alternative to DVB-HTML was recently presented in [13]. That work developed a method for transcoding web pages to MHP-compliant visual components. This meant that even STBs with lower capabilities can browse the web.

Nevertheless, convergence of WWW with DTV does not only refer to browsing the web from the television screen and remote control. It also signifies that a broad scope of diverse applications can be migrated from the web to the TV broadcasting world. Especially when it comes to personalized and smart interactive services, semantic web can be the key technology to enable the development of pioneer multimedia platforms. Two major research efforts towards this direction are the AVATAR [12] and the MediaNet [18] projects. AVATAR is a project which utilizes most of the techniques and technologies that the POLYSEMA project uses, such as video annotation, ontology-based modeling, multimedia metadata, and user profiling/personalization through semantic reasoning. The project's main objective though is to create a personalized digital TV program recommender. The system accomplishes the task by taking into account the user's profile and TV-Anytime [8] meta-information of each TV program. The POLYSEMA project, on the other hand, provides a platform which supports a variety of personalized interactive services. The MediaNET project is divided in five sub-projects, each of which covering a significant area of the multimedia content: creation, service providers, network operators, etc. As far as interactive services are concerned, it delivers the AmigoTV service and a PVR. Moreover, MediaNET offers iTV/Web convergent services such as e-voting and e-shopping via the TV device. However, the project does not investigate the benefits from using metadata during the various phases of the multimedia content lifecycle.

Another aspect in designing iTV platforms based on MHP is how applications running on the MHP runtime environment can communicate with the rest of the system. The work introduced in [21], consists of a low-level implementation of a platform, which is both OSGi and MHP compliant. It uses the notion of XbundLets, i.e. classes that implement both the interface of an OSGi bundle and the MHP application lifecycle. XBundLets feature several advantages, such as straightforward bi-directional communication between OSGi Bundles and MHP Xlets, as well as improved performance. On the other hand, to support such a dual execution environment, the reference implementation of the MHP platform needs to be modified. This is a major drawback in adopting this architectural approach, as protocol compliance is a crucial issue in Interactive Television.

3   POLYSEMA Architecture

The goal of providing semantic-awareness in iTV services demands the cooperation of several diverse information and telecommunication systems. Specifically, digital broadcasting systems have to be combined with knowledge-based systems such as reasoners and rule engines. In a task like this, the architecture of POLYSEMA platform had to be carefully designed in order to be modular enough, so that the complexity of the implementation could be kept manageable.

Another important consideration in the development of the platform was that it had to be compatible with current broadcasting technologies and standards adopted world-wide. This feature facilitates future attempts for seamless integration of semantic-aware services into the real world of Interactive TV. Nevertheless, attaining provision of innovative services without compromising compatibility with existing products is not a trivial matter.

This section is devoted to describing the modular architecture of POLYSEMA platform by presenting the novel design of the receiver, as well as the enhancements at the broadcasting server. Moreover, we explain how POLYSEMA software components cooperate with existing broadcasting and receiver solutions.

3.1 Extending the Broadcasting Server

MPEG-7 files can describe multimedia content on a scene-by-scene basis, by providing distinct descriptions for different sections of the video separated by clearly defined time markers. This model allows for a wide range of interactive applications, which adjust their behavior as the content presentation advances. A remarkable problem inhibiting such applications is that, in broadcast environments, the receiver cannot keep track of the absolute media time of individual TV programs in the incomming Transport Stream. That happens because iTV subscribers may tune to a specific program at any time during the broadcast. In such cases, applications are unaware of the absolute media time of a program event. The only way to synchronize video with an application is to transmit stream events within the stream, as described in the DSM-CC specification [2].

To overcome this problem, POLYSEMA utilizes a software module to convert MPEG-7 timestamps to stream events sent by the broadcasting server. The module parses the MPEG-7 document, which describes the content to be broadcasted, and performs a mapping of MPEG-7 time elements to stream events, at the granularity of video segments. A time event signals the transition to a new scene of the video, which is associated to an MPEG-7 segment. A stream event consists of an identifier, name and data. Data refers to any string, so it is possible that the data field may contain the MPEG-7 segment identifier. Any time the receiver tunes to a channel, the application can determine which part of the video is being displayed by listening for stream events and consulting the MPEG-7 document.

Summing up, the Transport Stream (TS) produced at the server side and sent to the receiver contains:

- AV content, MHP applications, and the broadcast file system mounted on the transport stream,
- stream event objects that convey synchronization information,
- files in the DSM-CC object carousel, which may include MPEG-7 files describing the multimedia content or links to those MPEG-7 files, so that the receiver can locate and obtain the corresponding metadata files over the Internet.

As one can see in the above list, the MPEG-7 document (or a URL link to it) is sent over the Broadcast File System. Recent work on carrying metadata in MPEG-2 Streams (see [11] for a comprehensive presentation of the respective ISO amendment) defines different methods for this purpose. Metadata can be sent either by using private sections of MPEG-2, Packetized Elementary Stream (PES) packets or the Broadcast File System. The latter approach was preferred in POLYSEMA, as it allows for prefetching of the complete metadata, so that it can be timely preprocessed by the semantic component, before the respective video scenes arrive. Moreover, the POLYSEMA receiver can retrieve the metadata from the Internet, by reading just a URL link from the broadcast filesystem. We believe that this model of metadata transmission is more appropriate, because it saves bandwidth for AV information in the TS, while the metadata can be fetched concurrently from an Internet connection available at the receiver.

3.2   The POLYSEMA Smart Receiver Architecture

Each interactive service consists of both a presentation part and a logic part which run at the Multimedia and the Semantic Component of POLYSEMA respectively (Fig. 1). Presentation refers to actions that the media player (i.e. DVB-MHP receiver) should perform when it comes to displaying the results of the logic part. Such actions are implemented in a platform-independent way by using the Java-based MHP application environment. The Semantic Component determines the actions which comprise each service, i.e. the logic behind each interactive service. The Content Retrieval and Composition Component (CRC) is in charge of retrieving and integrating any external resources to be used by invoked services. The Service Management Component is responsible for allowing TV viewers to declare their preferences and define customized services, so that the Semantic Component will know when and what actions to trigger. All components are built in an OSGi-based service-oriented fashion, in order to provide maximum flexibility in composing interactive TV services. The rest of the section describes the details of each component.

3.2.1   The Multimedia Component

The Multimedia Component of the POLYSEMA platform refers to the set of applications running in the MHP Execution Environment of the DVB receiver. There is no need for specialized hardware, as any DVB-MHP receiver can download and run the MHP applications which comprise the Multimedia Component. The only requirement is that the MHP profile running at the STB provides for Return Channel (RC) communication. This is necessary for the interaction of the Multimedia Component with the Semantic Component.

The operation of the Multimedia Component is based on a basic MHP application (Xlet) running at the receiver. Xlets are classes that run under the JavaTV application

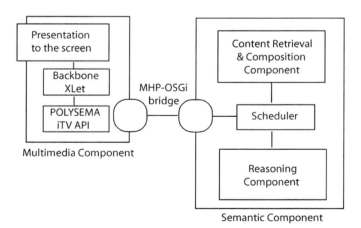

Fig. 1 Overall system architecture of the POLYSEMA platform

model, which enforces a certain application lifecycle for every Xlet and an application manager that asserts the correct execution of many Xlets in one STB. The first responsibility of the basic Xlet of POLYSEMA is reading the broadcast file system and retrieving the MPEG-7 file describing the multimedia content (or the URL link pointing at locations where the metadata file is publicly available). The link or file is then fed to the Semantic Component which is accountable for storing and processing the semantic description. Moreover, the basic Xlet subscribes and listens to incoming stream events. Upon the reception of a stream event, it inquires the Semantic Component for actions that should be carried out for the specific video scene. Each action is implemented as a separate Java class, whose instances are created by the basic Xlet.

The complete set of available actions comprises the POLYSEMA iTV Application Programming Interface (API). This set includes methods/actions such as displaying messages and information, changing channel or sound volume, recording the video or hiding the displayed content. The API is described by the Service Description XML file which is downloaded by the Semantic Component from the Service Aggregator, as will be explained in a subsequent section. This file is used by the Service Management Component to let users combine basic actions and define customized services. The POLYSEMA iTV API is also used by the Semantic Component, in order to instruct the Multimedia Component about which actions to perform. Receivers that do not have semantic processing capabilities will still be able to display the multimedia content, but will not enjoy semantics-aware interactive services. The Multimedia Component also interacts with a local HTTP Server, running as an OSGi bundle, which is used as a proxy to store prefetched information retrieved from the WWW.

3.2.2 The Semantic Component

The Semantic Component includes all processes relevant to service metadata (i.e., inference processes that drive the personalized service provisioning) and the coordination of value added services offered by the platform. A more detailed view of its internal architecture is presented in Fig.2. A basic assumption for our system is that every AV content item is described by an MPEG-7 document. In order to reason over this document we transform it to a corresponding ontology, which was based on that proposed in [19]. In fact, we have developed a stripped down version of the ontology in order to eliminate any elements not used by our system. Moreover, the user defines, through the Service Management Component, their service preferences by combining templates of possible actions and declaring rules about when such services should launch and how they should be presented. The user input is based on the TV User Ontology, which describes the user profile and preferences.

The Reasoning Component of the system uses the MPEG-7 document of the TV program and the user ontology (along with other domain ontologies such as the TV-Anytime classification schemes of MPEG-7) to infer which services should be activated during the broadcast. The Reasoning Component wraps the functionality of both a reasoning and a rule engine. Its reasoning capability is required mainly

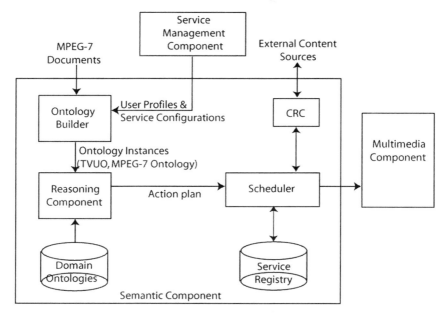

Fig. 2 Overall system architecture of the POLYSEMA platform

for classifying the multimedia content and the user preferences to predefined categories, while the rule support is used for deciding which services should be executed given the TV program metadata and the user profile. Bossam [15] is used for the implementation of the reasoning component. Once the appropriate services have been selected for execution by the Reasoning component (i.e., an action plan is formed), it is the responsibility of the Scheduler component to coordinate the execution of the respective application logic. Such logic is registered in the Service Registry module through procedures specified by OSGi.

The Scheduler instructs the Content Retrieval & Composition (CRC) and the Multimedia components what actions they should perform. The Content Retrieval & Composition component is a framework for registering and managing interfaces with external information sources. For each new source that is registered (e.g., Web site, multimedia database, RSS feed), the available content is described along with its type (e.g., text, video) and the invocation details (e.g., source URL, parameters).

3.2.3 Content Retrieval & Composition Component

The purpose of the Content Retrieval & Composition Component (CRC) is to support the interactive services that display relevant information from the WWW. Once the Semantic Component has preprocessed the semantic description and has decided that certain web information should be fetched and displayed to the viewer, it instructs the CRC component to start prefetching the relevant data.

The CRC component consists of several bundles that implement the automated access to external web sites. For instance, in the prototype system we developed bundles which access content from several well-known sites such as IMDB [7] and Wikipedia [9]. The automated access classes follow a common abstract design, which involves management of created class instances, caching retrieved data till the end of the TV program and creating output lists containing all the requested web information.

The automated access bundles of various sites differ in the actual information available from these sites. This is reflected to the methods supported by different bundles. Each installed bundle updates a registry, called CRC Registry, which is used by the Scheduler of the Semantic Component to map user requested information to bundle method calls. Every call issued to the CRC component by the scheduler results in a piece of information retrieved and being cached in the internal structures of CRC bundles. In order to make this information available for display at the Multimedia Component, the CRC component returns the information to the Scheduler which then stores it in text or image files at the local HTTP Server running within the Semantic Component. The Multimedia Component then accesses the HTTP Server to retrieve the information files and display the content to the user.

The data fetched from the WWW can be either of textual or visual format. Text data may include movies information, encyclopedic knowledge, sport statistics etc. Visual information refers to images and photos related to a subject requested by the user.

3.2.4   Service Management Component

This component implements a management console for the POLYSEMA services. Specifically, it provides the users with a graphical user interface (GUI) through which they can configure all the installed services according to their preferences. Since each residential gateway may have different services installed, this GUI cannot be predefined and delivered as a pre-built component. Hence, a dynamic GUI generation process is involved that automatically generates the GUI based on the service descriptions that accompany the service code. Moreover, the users can fill in their profiles through this component. The synthesis and rendering of the component's GUI is handled by Java Server Pages. The service configurations correspond to rules that should be applied to the displayed content and that define actions that should be taken by the other components (e.g., CRC) or by the MHP applications. Once the service preferences and user profiles are defined by the users, they are automatically translated into ontological instances and/or rules. These knowledge elements are used as input to the Reasoning Component and drive the service personalization process.

3.2.5   System Integration with OSGi

The software components of POLYSEMA and particularly of the Semantic Component are deployed on the OSGi platform in order to take advantage of the flexibility of OSGi bundles. Each of the aforementioned components is developed as an OSGi bundle, which can access other bundles' services through their registered interface in the OSGi registry.

The Multimedia Component could not be developed as an OSGi bundle, because it consists of applications running in the MHP receiver. The OSGi and MHP environments have different properties because they follow different design principles (see [21] for a relevant discussion). In order to retain full compatibility with industry standards such as MHP and DVB, and still harvest the service management flexibility of OSGi, we decided to build a delegate bundle in the OSGi platform that conceals the nature of MHP applications from the rest of the system. The MHP-based Multimedia Component communicates with the delegate bundle via the IP return channel of the receiver. The remaining OSGi-based components of POLYSEMA interact with the Multimedia Component by accessing the respective MHP-delegate OSGi service.

4   Provision of Interactive Services

This section lists the basic services we implemented for the POLSYSEMA prototype and gives a sample interactive application to highlight the way interactive services are executed in POLYSEMA. The services which are currently supported are:

- *WWW information retrieval service*: This service displays information gathered from the Internet relevant to specific elements of the media (e.g. actors, directors, scene locations etc.). The Multimedia Component receives a list of information objects retrieved from the WWW by the CRC Component. The list may contain pieces of text or images, displayed through a scrolling window interface. The

Fig. 3 Displaying information from the WWW regarding the actor

user can navigate through the available information by their remote control or disable the execution of the service (Fig. 3).

- *Parental control service*: Parental control implements a control mechanism for blocking media access to inappropriate content. Options regarding the censoring of the inappropriate content can be one of a) Video off, b) Sound off, c) Video and sound off and d) Channel change. All options cause the suppression of inappropriate content for a time duration specified by the Reasoning Component of the system. The last option tunes the receiver to another channel in the Transport Stream defined by the user. After a time period indicated by the Reasoning Component, the receiver returns to its former state.
- *PVR service*: The PVR service allows the recording and playback of broadcast content. The service is activated by the receipt of a message by the Semantic Component, which instructs the automatic recording of the multimedia content for a specified time period. The user can browse a list of recorded multimedia, from which the individual recordings can be deleted, played-back, etc.
- *Alerting service*: This service enables the asynchronous display of messages in the form of an information alert (emergency news, sports event, etc.). These alerts can originate from different sources. An interesting case is displaying the contents of an RSS feed (Fig. 4). In this case, the CRC component is instructed to listen to specific feeds and store the respective information. Upon a change in the information included in the feed, the updated information is sent to the TV screen in form of an alert.
- *Channel recommendation service*: Channel recommendation aims to provide the viewer with suggestions for channels of interest to them. To accomplish this task, an external information flow is used to determine the program of the available channels. The program is described in an XML format according to the xmlTV

Fig. 4 Displaying an RSS Feed from BBC

reference specification and can be provided by the Broadcaster of the TV program. The xmlTV files contain enough information to form the basis of semantic processing at the POLYSEMA's Semantic Component. During the semantic processing, the provided information is matched against the user preferences and channel suggestions are created. The implementation of this service uses the alerting service, since the channel suggestions are displayed as special alert notices.

To outline how services are implemented in the POLYSEMA platform, we give an example of a service and its execution. Consider a user who includes in their profile their interest in vehicles in general. It could be requested that, in case of appearance of a vehicle in a TV program the system should collect information about it from the Web (e.g. Wikipedia). Fig. 3 depicts a screenshot of such an application.

The MPEG-7 document is assumed to contain detailed metadata that annotate the scene in which an actor drives a small motorbike (Fig. 3). A URL link to the MPEG-7 document, located at the Service Aggregator server, is transmitted through the transport stream. At the beginning of the film, the MHP application running at the receiver requests that the Semantic component downloads the MPEG-7 over the Internet. Afterwards, the Semantic component creates an MPEG-7 ontology from the document, processes the descriptions of the film's scenes to match them with the user rules and produces the respective actions. As the broadcast advances, each film scene annotated by the MPEG-7 metadata, is eventually displayed on screen. The corresponding stream event is triggered at the receiver and the MHP application requests from the Scheduler component to return the desired actions, i.e. the outcome of the preprocessing of the Reasoning component for that particular scene. Subsequently, the Scheduler supplies the desired content to the Multimedia Component through certain layout templates gathered and formatted by the Content Retrieval & Composition Component. Then, the MHP code corresponding to the specific service is invoked in order to present the relevant information to the user.

5   Business Modeling Issues

This section aims at presenting a business model which would allow the involved stakeholders to gain revenue from the introduction of semantics-aware Interactive Services by exploiting the POLYSEMA system. The existence of a sustainable business model is essential because it can leverage the practical application of our proposed system in the real world of broadcasting. To this end, we illustrate an appropriate business model in Fig. 5. In this figure, one can see the revenue (R_i, i=1..3) and information (I_i, i=1..5) flows between the core players and roles of this model.

The central entity in this figure is the Broadcaster. The Broadcaster is responsible for assembling any video content, application logic and metadata resources into a single digital television Transport Stream to be broadcasted to all subscribers. MPEG video files are supplied by the Media Content Provider. Application logic refers to the MHP applications required to run at the receiver in order to make semantic interactive services feasible. These applications could be provided by

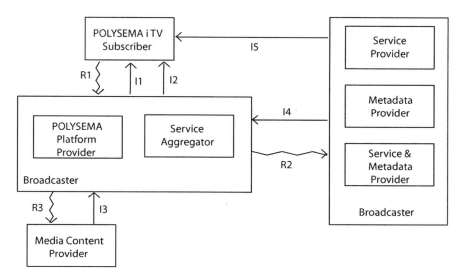

Fig. 5 Business Modeling for POLYSEMA

another market player, namely Platform Provider, but we simplify the diagram by having the Broadcaster assume the responsibility of this role. Additional responsibility of the Platform Provider is to directly provide initial installation assistance and further technical support to the subscribers.

The business model includes a class of market players called Service & Metadata Providers. There can be a wide variety of Service & Metadata Providers, which would allow for a broad scope of diverse applications to run on top of the basic interactive platform offered by the Platform Provider. It is not necessary that all Service Providers will provide Metadata and vice versa. Separate Players can produce only services or metadata. Nevertheless, having a single player assume both responsibilities will facilitate the creation of more meaningful metadata descriptions, as the descriptions will be written in correspondence to the context of specific interactive services.

The possibility of existence of several Service & Metadata Providers indicates that a Service Aggregator player should manage the input from them. In the figure, this role is also assumed by the Broadcaster. This is because it is a good practice to let subscribers deal with a single entity, which is then responsible to distribute the revenue to the rest of the players. One of the activities of the Service Aggregator is to provide access control management. Each subscriber will have to contact the public server of the Service Aggregator to download the catalogue of services available by the platform. Without this catalogue, the receiver will not be able to provide any advanced interactive services. This catalogue can be dynamically created by the Service Aggregator, based on accounting information specific to each subscriber, thus offering a flexible, internet-based access control management scheme.

Table 1 Description of revenue and information flows

Flow ID	Description
I1	This flow represents the Transport Stream that also contains the DVB Carousel. It consists of a) the video content, b) typical broadcast information, c) the MHP-related service logic and d) a link to the Service Aggregator public server.
I2	This flow represents the information sent to the subscriber upon login to the platform. It consists a) of catalog of services that corresponds to its subscription, and b) the MPEG-7 metadata file that may also depend on its subscription (in case some services come with their own MPEG-7 annotations).
I3	The video content to be delivered by the broadcaster.
I4	a) service data and b) MPEG-7 metadata for the broadcast program.
I5	a) service OSGi part (e.g., CRC bundles), b) service-specific domain ontologies.

A detailed description of the information flows between the various players of the business model is summarized in Table 1.

Finally, regarding the revenue flows we should note that various payment models and subscription types can be supported. For example, a user may subscribe for a fixed set of services or for unlimited use of services (flow R1). Alternatively, users can pay for some basic set of services and be additionally charged depending on their service usage. The flow R2 can then be implemented in a pay-per-install manner, which gives income to the Service Providers depending on the installations (for a limited period) of their services by the users. Another, Internet-oriented, revenue sharing method among the Broadcaster and Service Providers is pay-per-impression. According to this model, each provider receives commission every time a user uses its service a fixed number of times (e.g. a thousand times).

6   Conclusion

The research work carried out in the context of the POLYSEMA project is driven by the great importance of metadata in providing future iTV services and the need to manage them efficiently. Moreover, the POLYSEMA platform supports applications which adapt their behavior as the content presentation advances, allowing thus for innovative iTV services. Additionally, we believe that more effective personalization can only occur if the preferences of each user are known. This can only be achieved if semantic reasoning process takes place in the end-user premises. Future research may include an even more generic framework for designing services and integrating a variety of external web resources into the TV watching experience.

In this paper we have described the design and implementation of the POLY-SEMA platform as well as some sample services and some relevant business modeling issues. However, the POLYSEMA project deals with more issues peripheral to the topic of semantics-aware multimedia. Firstly, some algorithms for semantics extraction from subtitles have been designed. Such extraction involves automatic video classification from subtitles [16], ontology learning from subtitles and video summarization from subtitles. Moreover, in order to be able to test the platform with real MPEG-7 descriptions, we had to develop a new tool for MPEG-7-compliant video annotation [20]. This tool exploits ontologies for the annotation process and exports both MPEG-7 documents as well as MPEG-7 ontological instances (i.e., populates our MPEG-7 ontology). More updated information on the project results can be found in [4].

Acknowledgements. This work was partially funded by the Greek General Secretariat for Research and Technology (GSRT) under the Operational Program "Information Society", co funded by the EU.

References

1. About the OSGi Service Platform - Technical Whitepaper Revision 4.0. Technical report, Open Services Gateway Initiative
2. DVB - Implementation guidelines for Data Broadcasting, v1.2.1. ETSI TR 101 202
3. DVB - Multimedia Home Platform (MHP) Specification 1.1.1. ETSI TS 102 812 V1.2.1
4. POLYSEMA Project Web site (2008), http://polysema.di.uoa.gr
5. Implementation guidelines for DVB terrestrial services; Transmission aspects. ETSI TR 101 190 (2004)
6. The Digital Video Broadcasting Project (DVB) (2007), http://www.dvb.org/
7. The Internet Movie Database (2007), http://www.imdb.com/
8. The TV-Anytime Forum (2007), http://www.tv-anytime.org/
9. Wikipedia, the free encyclopedia (2007), http://en.wikipedia.org/wiki/
10. Antoniou, G., van Harmelen, F.: A Semantic Web Primer. MIT Press, Massachusetts (2004)
11. Lopez, A., et al: Synchronized MPEG-7 Metadata Broadcasting over DVB networks in an MHP Application Framework. In: Proceedings of Inrernational Broadcasting Convention (IBC 2003), Amsterdam, Netherlands (2003)
12. Fernandez, Y.B., Arias, J.J.P., Nores, M.L., Solla, A.G., Cabrer, M.R.: Avatar: an improved solution for personalized tv based on semantic inference. IEEE Transactions on Consumer Electronics 52(1), 223–231 (2006)
13. Ferretti, S., Roccetti, M.: MHP Meets the Web: Bringing Web Contents to Digital TV for Interactive Entertainment. In: Proccedings of the eighth IEEE International Symposium on Multimedia (ISM 2006) (2006)
14. Gil, A., Pazos, J., Lopez, C., Lopez, J., Rubio, R., Ramos, M., Diaz, R.: Surfing the Web on TV: the MHP approach. In: Proceedings of IEEE International Conference on Multimedia and Expo (ICME 2002) (2002)
15. Jang, M., Sohn, J.: Bossam: an extended rule engine for the web. In: Antoniou, G., Boley, H. (eds.) RuleML 2004. LNCS, vol. 3323, pp. 128–138. Springer, Heidelberg (2004)
16. Katsiouli, P., Tsetsos, V., Hadjiefthymiades, S.: Semantic video classification based on subtitles and domain terminologies. In: SAMT Workshop on Knowledge Acquisition from Multimedia Content (KAMC), Genova, Italy (December 2007)

17. Martinez, J.: Standards: Overview of MPEG-7 description tools. IEEE Multimedia 9(3), 83–93 (2002)
18. Travert, S., Lemonier, M.: The Medianet Project. In: Proceedings of Image Analysis for Multimedia Interactive Services, Portugal (April 2004)
19. C. Tsinaraki, P. Polydoros, and S. Christodoulakis. Interoperability support for Ontology-based Video Retrieval Applications. *Springer Lecture Notes in Computer Science, Content-Based Image and Video Retrieval*, pages 582–591, 2004.
20. Valkanas, G., Tsetsos, V., Hadjiefthymiades, S.: The POLYSEMA MPEG-7 Video Annotator
21. Vilas, A.F., Diaz Redondo, R.P., Cabrer, M.R., Pazos Arias, J.J., Solla, A.G., Duque, J.G., Nores, M.L., Fernández, Y.B.: MHP-OSGi convergence: a new model for open residential gateways. Wiley, Software: Practice and Experience 36(13), 1421–1442 (2006)

Marrying Game Development with Knowledge Management: Challenges and Potentials

Jörg Niesenhaus and Steffen Lohmann

Abstract. The game industry has long been neglected as a market and research area for knowledge management and semantic technologies. However, as the budgets for game projects are growing and game development is subject to an increasing professionalization and specialization coupled with outsourcing and offshoring, new needs and potentials for continuous knowledge management and the use of semantic technologies emerge. This chapter starts with a description of the current situation and examines typical game development activities and involved parties that could benefit from a continuous knowledge management support. Subsequently, it provides a general framework architecture and implementation examples that show how knowledge management and semantic technologies can be employed to support game development.

1 Game Development as an Application Area for Knowledge Management and Semantic Technologies

The game industry has gone through an overwhelming economic growth within the past years and analysts foresee a strong growth in the nearby future as well. The branch of game development and publishing is already a major industry with its strongest markets in North America, Japan, and Europe. In 2007 the nine European core markets[1] sold video and computer games worth € 7.3 billion (excluding hardware sales) [15]. Games software sales in the U.S. recorded € 6.9 billion (9.5 billion

Jörg Niesenhaus and Steffen Lohmann
University of Duisburg-Essen
Interactive Systems and Interaction Design
Lotharstrasse 65, 47057 Duisburg, Germany
e-mail: {joerg.niesenhaus, steffen.lohmann}@uni-due.de

[1] The European core markets are Great Britain, Germany, France, Italy, Spain, the Netherlands, Switzerland, Sweden, and Finland.

S. Schaffert et al. (Eds.): Networked Knowledge - Networked Media, SCI 221, pp. 321–336.

dollar) in 2007 and the Asia-Pacific market earned € 7.4 billion in 2006[2]. The industry's annual growth ranges from 17 percent (U.S. sales between 2003 and 2006) [7] up to 21 percent (German sales between 2006 and 2007) [4]. Pricewaterhouse-Coopers predicts a doubling of the German market size of 2006 for computer and video games reaching € 2.6 billion in 2010 [4]. One reason for this steady growth is the diversification of gaming products converting more and more casual gamers into active consumers[3]. Furthermore, people that were socialized with games often tend to keep on gaming resulting in an age above 30 years for the average American or British gamer[4] [15].

1.1 Application Area for Knowledge Management

The ongoing economic growth is accompanied by an increasing professionalization in the development of digital games. In addition, higher budgets and larger development teams cause a growing specialization. Large scale projects involve up to 1,000 participants nowadays and accumulate development costs up to tens of millions of euros[5]. Due to the outsourcing of development parts, the whole process also tends to be more geographically dispersed. Specific components (e.g., game engines, graphic assets, level artifacts, or development environments) are frequently bought from third-party developers [20].

Along with this professionalization, specialization, and globalizaton, the documentation and maintenance of knowledge in game projects gets a higher priority. Complex dependencies have to be handled and not only data but also knowledge has to be transferred between project partners. Furthermore, a huge amount of knowledge on game-specific aspects is generated, such as the design of the game world, the storytelling, the basic game mechanics, and technical specifications (cp. [19]). In simple terms:

Nowadays, an essential part of game development is knowledge management.

However, any attempt that aims to support knowledge management in the development of digital games is faced with specific challenges as the development activities are characterized by highly agile and creative processes. Thus, appropriate

---

[2] Official numbers of 2007 were not available at time of writing.

[3] Popular examples for this diversification are music games such as "Singstar" (a popular competitive karaoke game by Sony) and "Guitar Hero" (a video game that is delivered with plastic guitar controllers enabling gamers to playback famous songs) or "brain training" games (compilations of several math games and puzzles which are meant to train the human brain) such as "Brain Age" or "Big Brain Academy" as well as intuitive control schemes offered by modern consoles.

[4] According to the Entertainment Software Association (ESA).

[5] An example for a large scale project is *Grand Theft Auto IV* by Rockstar Games [22].

knowledge management solutions must satisfy a number of criteria (which we abbreviate with the term "ALIGN" according to their initials):

- Adaptable: being easily adaptable to changing project demands
- Lightweight: following the principles of simplicity and ease-of-use
- Immediate: adding immediate benefit to the project and all its participants
- Generic: providing a general solution for various projects
- Nonrestrictive: not dictating strict procedures but fostering creativity

1.2 Application Area for Semantic Technologies

The complex dependencies that connect the artifacts involved in game development and the knowledge structure of the whole project can benefit from a semantic grounding. In general, semantic technologies attracted significant attention in software engineering recently. Initiatives such as the W3C's Ontology Driven Architecture (ODA) [24] or the OMG's Ontology Working Group [25] testify the growing interest for ontology-based approaches in software engineering and related disciplines. Furthermore, some lightweight approaches connecting the so-called Social and Semantic Web such as Semantic Wikis [21] or Social Semantic Annotation [9] are promising candidates for application in game development, since they are normally easy to use, require only little effort and can simultaneously provide a semantic structure that is of great help in large software development projects [18].

However, not only the development process can benefit from the application of semantic technologies. As game worlds are often large in size and complex in their dependencies, they are normally represented in huge class hierarchies or ontology-like strutures. Authentic storytelling or the behavior of non-player characters (NPCs) are typical cases where the game logic uses rules and inferences upon a formal representation of the game world. Research in the area of interactive digital storytelling has shown many opportunities to organize huge amounts of data in order to create believable story worlds. For instance, Crawford suggests an "inverse parser", a dramatic sublanguage for storytelling applications where users select predefined words and terms to communicate with the story world [6]. As a basis for the sublanguage Crawford proposes the use of systems like WordNet[6]. Another approach in this area is the system "HEFTI"[7] by Ong and Legett that uses a knowledge base as an combination of "story components" that represent certain time segments in the story and are composed by a "story builder" instance [16]. The knowledge base also includes "contextual sets" to categorize different plot entities, events, scripts, actions, and characters within the story components.

Other approaches analyze games on a more general level. For instance, the aim of the "Game Ontology Project" [26] is the development of an ontology that

---

[6] *WordNet* is a lexical database which groups substantives, verbs, adjectives and adverbs into sets of cognitive synonyms, each expressing a distinctive concept; http://wordnet.princeton.edu/

[7] *HEFTI* stands for Hybrid Evolutionary-Fuzzy Time-based Interactive Storytelling System.

Fig. 1 Challenges, requirements, and promising solution concepts in the integration of game development, knowledge management, and semantic technologies

describes central structural elements of digital games and their interrelations. All these approaches reflect the growing demands for semantic technologies to formally describe games and game worlds and generate intelligent character behavior and authentic storytelling.

Figure 1 summarizes the challenges, requirements, and potential solution concepts regarding the integration of game development, knowledge management, and semantic technologies. In the following, we focus on the support of knowledge-intensive activities in the game development process as a promising application area of semantic technologies. Based on a description of the state of the art in Section 2, potentials for integrated knowledge management are identified in section 3. In Section 4 we propose a framework aiming to support knowledge management in game development on three distinct levels. We close with a conclusion and outlook on future work in Section 5.

2 State of the Art

So far, no established or even standardized continuous support for qualitative knowledge management exists in the area of digital games. Due to many differences between game projects and work-related software projects, established methods and tools from software engineering cannot be transferred one-to-one. Tools currently used by game development teams are mostly designed for single use cases and are combined by the teams in order to support more complex tasks.

2.1 Data and Document Management

Normally, development studios use version control systems (such as *Perforce*[8], *Alienbrain*[9]) in order to administrate data and documents. Knowledge about technical and game mechanical issues is primarily stored inside the single files of these version control systems that are updated in variable time intervals. Outdated knowledge partially exists for any length of time; contrary and inconsistent conclusions across several documents are not rare.

Usually, it is differentiated between runtime files and design documents. The game is generated out of the runtime files (e.g., engine, scripts, graphics) in its different design stages. The design documents describe guidelines and design decisions concerning the use of technology and game mechanics and define responsibilities. Though it is common practice to set references between the documents, this is primarily done manually so that references often become obsolete or fragmentary as time passes by. In many cases, most files are insufficiently or not at all commented due to a lack of time. Thus, the function of a file can often only be derived from its location in the file system, the version control structure, and its name. The absence of a global knowledge management is clearly noticeable in this structure.

In addition to the version control systems, web-based project management platforms (such as *Mantis*[10], *Jira*[11]) are used to support, for instance, bugtracking, task coordination, and change requests. Due to the lack of a comprehensive platform solution for game development, some studios and publishers develop their tools in order to customize the workflows according to the needs of their project.

It is essential for the project to establish well-defined workflows and knowledge exchange policies that are easy to handle and not too time-consuming. Another vital point is the establishing of a project vocabulary or use case examples in order to give team members an idea on how to contribute to the system and to ensure a consistent form of knowledge storage.

2.2 Outsourcing Content and Code Production

Outsourcing content and code production in game projects places special demands on knowledge management. External teams and partners have to match the existing

---

[8] *Perforce* is a software configuration management system;
http://www.perforce.com
[9] *Alienbrain* is an asset management system supporting graphic intensive projects;
http://www.softimage.com/products/alienbrain
[10] *Mantis* is a web-based bugtracking system; http://www.mantisbt.org/
[11] *Jira* is a browser-based bugtracking system with customizable workflows;
http://www.atlassian.com/software/jira/

production standard and integrate their code pieces, graphic assets, and further multimedia objects into the given game structure.

In order to share data with external partners, development studios often use web-based document sharing platforms (such as *Microsoft Sharepoint*[12]). In addition, specific communication channels protected by security protocols (e.g., Virtual Private Network (VPN)) provide connections for critical data transfers between partners. Further information is often communicated via e-mail or the above-mentioned web-based project management platforms.

Given this infrastructure, the transfer of data is well supported but when it comes to communicate knowledge between the commissioning studio and external partners problems start to emerge. Loose design documents and knowledge scattered across several e-mails or postings on web-based platforms result in inaccuracies and misconceptions in the development process which are time-consuming for both sides. The commissioning studio outsources development tasks with the goal of saving manpower and money but the additional communication and coordination overhead is often underestimated and not considered in the calculations.

2.3   Media Exchange and Consistent Game Design

The realization and communication of a consistent game design is not only challenging for the cooperation between commissioning studios and external partners but also for the studio's internal communication. Much of a game's knowledge is implicitly provided by artworks, videos, mockups, and other media formats. To ensure a consistent design, it is therefore important to set up relations between these different media artifacts and to provide sufficient metadata. Since the development of the game design is a creative and complex process that normally involves several departments of a development studio, the communication of technical and artistic aspects is nearly equally important. Metadata extensions for graphic files (such as XMP [1]) are a first step in the direction of coupling data and knowledge exchange. However, metadata is yet not widely used within game development. To sum up: Developing a consistent game design and communicating it to the project participants, particularly to outsourcing partners, is a crucial task for the success of a game that is not sufficiently supported at the moment.

3   Potentials for Integrated Knowledge Management

By analyzing typical activities and involved parties in the development of digital games, the potentials for continuous knowledge management support become apparent.

---

[12] *Microsoft Sharepoint* is a browser-based collaboration and document management platform; http://office.microsoft.com/en-us/sharepointserver/default. aspx

3.1   Knowledge Management with Respect to the Development Process

At the beginning of the game development process in the phases of *pitching* and *pre-production* (see Figure 2) central activities are the generation of ideas and the creation of new game concepts: Numerous ideas are developed and discarded leading to a permanent change of the game's shape. Converting these agile processes into permanent knowledge is of great value for a development team as it allows to reconstruct at a later time why ideas were discarded, which challenges occurred, and how a problem was finally solved. During development of a game the team occasionally returns to an earlier point of discussion and reconsiders decisions on the basis of a new understanding of the game context. In many cases, previous experiences are included in the considerations so that it is furthermore useful to activate knowledge of projects that have been successfully accomplished in the past. Knowledge should always be connected with project structures, files, and program code to enhance the chance of reusing already existing components [19].

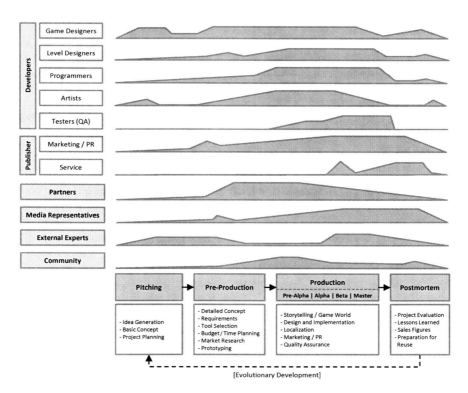

Fig. 2 Typical game development process. The curves describe the distribution of knowledge-intensive activities.

With the beginning of the *production* phase, the demand for constant documentation increases with every generated version of the game. In this phase, priority should be given to the interconnection of file structures, data, and knowledge to make every development step traceable and – if required – revocable at a later time. The production phase is usually initiated by the core team of the development studio that sets up the technical infrastructure of the project, such as the engine technology and development environment, and defines the fundamental game design. Then, the team is gradually expanded by members of the artistic and programming departments. In later production phases specialists such as story writers, sound and level designers, voice actors, or testers complement the team. Being not able to share all knowledge right from the start of the project, it is apparent that people who join the team at a later development stage need to quickly catch up all relevant knowledge about the game.

After the completion of a game project, the review and final feedback discussions start. This phase has come to be called *postmortem* in developer jargon. At this stage, the processes, problems, and experiences of the completed project are discussed in order to draw conclusions for future developments. In addition to the feedback given by the developers, the experiences of the service units and publishers as well as feedback of external experts are brought into discussion. Furthermore, reviews of magazines and community feedback are discussed in order to identify critical issues that need to be patched immediately or in a future version of the game. It is differentiated between features that were very well received by the community and features that earned a lot of criticism. Afterwards, the criticized aspects of the game are analyzed in detail and the team decides whether to invest time and resources in the improvement of these aspects or to skip them. The outcomes of this review and discussion phase define the basis for subsequent developments and should therefore be saved in structured form allowing easy access in the future.

3.2   Knowledge Management with Respect to the Involved Parties

The exact composition of the involved parties is, of course, subject to variation and depends largely on a project's size and goals. Typically, the following groups take part in the development of digital games (see Figure 2):

The group of *developers* includes all participants actively involved in product development, such as game designers, programmers, graphic artists, level designers, etc. All developers contribute with their personal experiences and expertise to the project. Due to an above-average fluctuation of participants between development teams or departments it is eminently important to externalize project-relevant knowledge of the individual team members in order to prevent a loss of knowledge when a member leaves the team.

The *publisher* is responsible for the finance, placement, marketing, and distribution of the game. Besides the continuous dialogue with the developers, all knowledge-intensive processes converge at the publisher making it possible to launch a product successfully. The coordination of marketing and public relations, the localization of a game for different markets, and the organization of the distribution are only a few examples for these processes. Publishers often distribute several products at a time; therefore, it is vital for them to know when a product is ready to ship in order to calculate the costs, start the marketing campaigns, sign contracts with the retailers and present the products at trade fairs. If problems occur and a project fails to match the gold master date[13] the publisher has to update the planning and contracts. Based on the given information, publishers decide on additional human or financial resources for the project to finish it as fast as possible or they cancel it if the success of the project is doubtful or if it can be foreseen that an project extension will consume far too many resources.

The cooperation with outsourcing and offshoring *partners* requires a particularly intensive form of knowledge exchange since it is essential to create a shared understanding of the project and to ensure that all externally developed components fit seamlessly into the game (see Section 2.2). Typical outsourcing partners are studios or freelancers that specialized on graphical assets, sound design, and voice acting or specific game contents such as level design or artificial intelligence programming. Some of the tools that have been developed by the studio are also used by partners. Therefore, it is important to share not only updates or new versions of a tool but additionally all relevant knowledge about it (e.g., bug reports, new features, how-tos, etc.).

The group of *media representatives* consists of journalists, editors, and producers who work for media formats dealing with digital games. Often, members of this group get the chance to test an early version of the game in order to prepare previews. The feedback of these previews is of great help for the developers since media representatives are often among the first external persons that review the game. Due to their broad experiences with digital games they often give valuable advice regarding bugs or weaknesses of the tested version. Moreover, journalists have to match their own deadlines before magazines go to press or television reports are produced; therefore, it is important for publishers and development studios to know these deadlines when releasing news, demo versions, or media of an upcoming game that shall have a specific media coverage (e.g., a cover page on a magazine).

Furthermore, *external experts* are often involved in game projects to assist the developers, for instance, in technical, usability, or child-welfare issues. It is of great importance to the developers and publishers to receive early feedback on possible obstacles in order to still have an influence on changes. A high age rating by a public rating agency[14] (e.g., 'Mature' or 'Adult-Only') can have a negative impact on the sales numbers of a game and is an important factor for the calculation of a publisher. Frequently, developers are supported by domain experts when designing products for special target groups or application areas. For instance, pedagogues might support the developers in the areas of serious games [14] and game-based learning [17]. Sometimes, external usability engineers are hired in order to evaluate the game interfaces, identify usability issues or conducting large-scale playability tests [11].

---

[13] The final version of a game which goes into manufacturing is called *gold master*.

[14] In many countries *public rating agencies* define generally binding age restrictions for media products.

The *community* that builds around published or announced game products is often characterized by a high activity and engagement when it comes to the critical review and discussion of a game or the generation of ideas for improvements and extensions. There are numerous well-established community portals, boards, and weblogs that focus on digital games (e.g., Gamespot[15], IGN[16]). Thus, we consider it valuable to integrate the community as far as possible in the development process in order to gain new ideas, helpful suggestions, and feedback from outside. Some projects go as far as to give away basic decisions of the game design to the community and let the game fans actively participate in the development process (e.g., *Top Secret*[17]). More common, some game developers offer polls or questionnaires to the community which might have an impact on certain features or give the developers a better idea of the most wanted features for a game. Another method to involve the community in the development process is to create a "closed beta" (for preselected community members) or a "public beta" (for everyone who is interested) phase where gamers are allowed to play parts of the nearly finished game for free. Closed or public betas are normally combined with ingame or online questionnaires, board discussions, and feedback forms. In addition, the developers evaluate game logs in order to balance features or find bugs.

4   Continuous Integration of Knowledge Management

In order to serve the demands for continuous knowledge management support in game development we propose a framework architecture consisting of a collaboration environment, embedded feedback channels, and knowledge extraction mechanisms. All these components are connected by a central repository that uses semantic technologies for knowledge representation (see Figure 3). In the following, we describe the framework's architecture in more detail and illustrate possible types of support by implementation examples.

4.1   Knowledge Repository

A knowledge repository forms the central access point for all knowledge management activities in the development process. It stores the project's knowledge in structured form and incorporates the following features:

- *Best Practices knowledge:* Initially, a basic set of ontologies provides conceptualizations that proved to be successful in previous projects. These fundamental structures describe the project on a rather general level by pointing to

---

[15] One of the biggest game portals for the American and European market;
http://www.gamespot.com

[16] Big American and British portal for interactive entertainment and new media;
http://www.ign.com

[17] *Top Secret* is a massive multiplayer online racing game which is developed under the lead of David Perry together with about 60.000 community members.

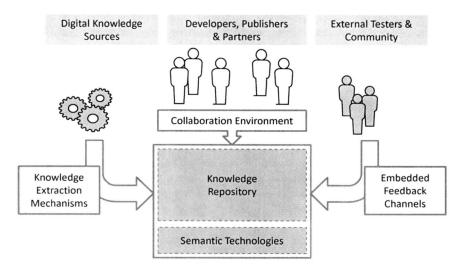

Fig. 3 Framework for supporting knowledge management in game development

important concepts that should be filled with project-specific knowledge during development. The Best Practices knowledge must, however, not only be based on a team's own background. The knowledge repository also enables the reuse of existing ontologies that have been developed in third-party projects or were published specifically for application in software engineering [23, 8]. However, the Best Practices knowledge should only be considered as a helpful starting point instead of being misunderstood as a structure the project's knowledge management must adhere to; it should not hamper creativity and innovation.

- *Shared understanding:* The initially provided knowledge base is collaboratively adapted and extended by all project participants during development. It acts as a shared conceptualization that consolidates the different perspectives of the involved parties so that it adequately represents the project's consensual knowledge at any time.
- *Evolutionary conceptualizations:* The knowledge repository is continuously updated in accordance to the project's evolution. History and version control mechanisms allow to track, review, and selectively rollback changes.
- *Context-sensitive integration:* A large part of the knowledge in the repository is semi-automatically derived from the project's context. For instance, if user feedback refers to specific components of the game (see Section 4.3) a reference to these components is stored along with the feedback in the knowledge repository allowing for future retrieval and reconstruction of the contextual setting. Vice versa, the knowledge management support is adapted to the development context ideally providing "the right features at the right time in the right way".

- *Hybrid formality:* The knowledge repository supports different degrees of expressiveness: Some parts of the project's knowledge might already be in a highly structured form while others are less formal and structured. Correspondingly, sophisticated techniques such as automated reasoning can only be applied to parts of a knowledge base that offer sufficient formality.

The knowledge repository is based on semantic technologies, in our case on the XML-based knowledge representation formats RDF, RDF Schema, and OWL [3]. By using these Semantic Web standards, ontologies that are available on the web can easily be added to a knowledge base of the repository. The application of Semantic Web standards is additionally motivated by the fact that the implementation of the framework's components is also mainly based on web technologies in our approach. This enables access for distributed development teams simply by using a web browser, without the need of installing specific knowledge management software on local devices.

In addition, the knowledge repository provides interfaces for syndication and further processing of parts of the knowledge base (e.g., via web services or news feeds). That way, a developer weblog (see Figure 4a) or mailing list can easily be connected to the repository. The other way around, external knowledge (e.g., provided by hardware producers) can also be integrated via appropriate interfaces according to the access rights.

4.2 Collaboration Environment

A collaboration environment provides comprehensive access to the knowledge repository. It is designed according to the principles of simplicity [13] and quick collaboration [5]. Besides the developers, the publishers and partners have separate access rights and are enabled to adapt and update parts of the project's knowledge base. Typical community features such as commenting and rating are combined with semantic technologies allowing for an enhanced knowledge retrieval.

Figure 4b shows an implementation example of the collaboration environment that is based on the *OntoWiki* system [2] and has been developed in the context of the SoftWiki project[18]. The user interface provides features for intuitive, web-based editing and updating of knowledge bases and allows easy interlinking of knowledge pieces or referencing on the underlying topic structure. In addition, participants can 'tag' parts of a knowledge base with freely chosen keywords, resulting in an emerging 'tag space' that represents the participants' vocabulary with respect to the developed product [18]. The effort and formal overhead of expressing knowledge, modifying the knowledge base, or setting relations between knowledge instances are minimized due to the adoption of the Wiki paradigm [5].

---

[18] SoftWiki Distributed, End-user Centered Requirements Engineering for Evolutionary Software Development; http://softwiki.de

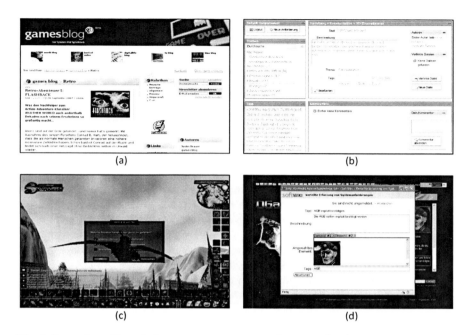

(a)   (b)

(c)   (d)

Fig. 4 Examples of knowledge management support for game development

4.3 Embedded Feedback Channels

The central collaboration environment is extended by decentralized feedback channels that can be embedded directly into the development or run-time game environments. Depending on the type of project and state of development, different groups are equipped with appropriate feedback channels (e.g., QA-team, community, external experts).

Figure 4c shows a possible realization implemented in the scripting language LUA [10]. It can be seamlessly integrated into the run-time environment of the game, in this case, of the online game *World of Warcraft* (WoW)[19]. The tool provides user-initiated input forms that are available at any time in the game and equip the WoW user with an opportunity to report encountered problems or suggestions for improvement. Moreover, it can be used to trigger predefined questionnaires at certain time events or situations.

Figure 4d shows an alternative implementation that can be easily embedded in web browsers in order to elicit feedback on browser games without a change of the environment. In addition, the tool captures contextual information by linking feedback to artifacts of the game environment, if possible (e.g., location, time, application status, etc). This contextual knowledge helps in later analysis as it allows for a more systematic exploration and facilitates the understanding of feedback.

---

[19] http://www.worldofwarcraft.com or http://www.wow-europe.com

Embedded feedback channels have a high potential when it comes to foster community engagement. As mentioned in Section 3.2, communities that build around games are characterized by an above-average activity and commitment. Providing participation opportunities and incentives that stimulate community engagement can be highly valuable to product improvement and innovation generation. With the right tools, communities might be actively involved in development, leading to games that better meet the users' desires.

4.4 Knowledge Extraction Mechanisms

Next to these forms of knowledge management support requiring active participation of the involved participants, the framework also considers project-related knowledge that is passively provided by available sources. Examples are user statements on weblogs, community portals, and discussion boards or documents and product descriptions from previous projects. Relevant information available in these sources is semi-automatically extracted and integrated into the project's knowledge base in order to get a comprehensive impression on how a game product is perceived by others. Due to an increasing use of XML- and RDF-based formats for content and knowledge representation, automated processing is increasingly feasible.

A central requirement for extraction mechanisms is that the integration process remains always in the control of the developers in order to not swamp the project's knowledge base with unstructured data. With the *Semantic Integrator* we proposed a tool that supports semi-automatic knowledge discovery from large datasets and integration in an existing ontological structure [12]. Document sources can be mined for project-related contents by composing search queries with relevant concepts from the project's knowledge base. The results are presented in structured form; project-related terms and paragraphs are highlighted. Statements that are considered as relevant for the project can be extracted and integrated into the knowledge base according to the conceptual structure (e.g., as feature requests, ideas for improvement, etc).

5 Conclusion

We analyzed and systematized typical knowledge-intensive activities and involved parties in game development and proposed a general knowledge management framework aiming to serve the demands of this application area. In particular, we tried to point out that a continuous integration of knowledge management support and semantic technologies into the development process of digital games is not only crucial for the success of large and distributed projects but also results in several benefits for the involved parties. These include easier adherence to the timetable and lower dependency on the knowledge of individuals reducing the risks and costs of development. Continuous knowledge management support and semantic representation also facilitate the development of game series and secondary or downstream exploitation, for instance, if a similar game structure or the same engine are reused in subsequent projects. Besides its function as documentation, the knowledge base can also serve as a source of inspiration for these subsequent projects.

Similar to other application areas of software engineering, ontologies can help to create a shared understanding in game projects and are able to support the classification and interoperable exchange of game artifacts and knowledge. The game industry is well-known for being a driver of new technologies. Therefore, game development might be a promising testbed of semantic technologies. Vice versa, semantic technologies and knowledge management solutions that proved to be successful in game contexts might be suitable candidates for application in non-gaming environments. Altogether, the interplay of game development, knowledge management, and semantic technologies offers a lot of potential in different directions that need to be further explored and evaluated. Our general goal was to take a first step towards a continuous knowledge management support for agile and creative development processes. It has become clear that this goal is faced with several specific challenges that require combined research and development efforts.

References

1. Adobe Systems Inc.: Adobe XMP: Adding intelligence to media (2008), `http://www.adobe.com/products/xmp/index.html` (cited December 1, 2008)
2. Auer, S., Dietzold, S., Riechert, T.: OntoWiki – A Tool for Social, Semantic Collaboration. In: Cruz, I., Decker, S., Allemang, D., Preist, C., Schwabe, D., Mika, P., Uschold, M., Aroyo, L.M. (eds.) ISWC 2006. LNCS, vol. 4273, pp. 736–749. Springer, Heidelberg (2006)
3. Allemang, D., Hendler, J.A.: Semantic Web for the Working Ontologist: Effective Modeling in RDFS and OWL. Morgan Kaufmann, San Francisco (2008)
4. German Trade Association of Interactive Entertainment Software (BUI): Facts (2008), `http://www.biu-online.de/fakten/` (cited December 1, 2008)
5. Leuf, B., Cunningham, W.: The Wiki Way: Quick Collaboration on the Web. Addison-Wesley Longman, Amsterdam (2001)
6. Crawford, C.: Chris Crawford on Interactive Storytelling. New Riders (2005)
7. Entertainment Software Association (ESA): Industry Facts (2008), `http://www.theesa.com/facts/` (cited December 1, 2008)
8. Gaevic, D., Djuric, D., Devedic, V.: Model Driven Architecture and Ontology Development. Springer, Heidelberg (2006)
9. Good, B., Kawas, E., Wilkinson, M.: Bridging the Gap between Social Tagging and Semantic Annotation: E.D. the Entity Describer. Nature Preceedings (2007), hdl: 10101/npre.2007.945.1
10. Ierusalimschy, R., de Figueiredo, L.H., Celes, W.: The Evolution of Lua. In: Proceedings of the 3rd ACM SIGPLAN Conference on History of Programming Languages, ACM, New York (2007)
11. Isbister, K., Schaffer, N.: Game Usability: Advancing the Player Experience. Morgan Kaufmann Press, San Francisco (2008)
12. Lohmann, S., Heim, P., Ziegler, J.: Semantic Integrator: Semi-Automatically Enhancing Social Semantic Web Environments. In: Auer, S., Bizer, C., Müller, C., Zhdanova, A. (eds.) The Social Semantic Web 2007 – Proceedings of the 1st Conference on Social Semantic Web (CSSW). LNI, vol. 113, pp. 167–172. Köllen (2007)

13. Maeda, J.: The Laws of Simplicity. The MIT Press, Cambridge (2006)
14. Michael, D.R., Chen, S.L.: Serious Games: Games that Educate, Train, and Inform. Thomson Course Technology (2005)
15. Nielsen Games Survey: Video Gamers in Europe 2008. Interactive Software Federation of Europe (2008),
    `http://www.isfe-eu.org/tzr/scripts/downloader2.php?`
    `filename=T003/F0013/8c/79/w7ol0v3qaghqd4ale6vlpnent&`
    `mime=application/pdf&originalname=ISFE_Consumer_Research_`
    `2008_Report_final.pdf` (cited December 1, 2008)
16. Ong, T., Leggett, J.: A Genetic Algorithm Approach to Interactive Narrative Generation. In: Proceedings of the 15th ACM Conference on Hypertext and Hypermedia. ACM, New York (2004)
17. Prensky, M.: Digital Game-Based Learning. Paragon House Publishers (2007)
18. Riechert, T., Lohmann, S.: Mapping Cognitive Models to Social Semantic Spaces - Collaborative Development of Project Ontologies. In: Auer, S., Bizer, C., Müller, C., Zhdanova, A. (eds.) The Social Semantic Web 2007 - Proceedings of the 1st Conference on Social Semantic Web (CSSW). LNI, vol. 113, pp. 91–98. Köllen (2007)
19. Rollings, A., Morris, D.: Game Architecture and Design: A New Edition. New Riders (2003)
20. Saltzman, M.: Game Creation and Careers: Insider Secrets from Industry Experts. New Riders (2003)
21. Schaffert, S., Bry, F., Baumeister, J., Kiesel, M.: Semantic Wikis. IEEE Software 25(4), 8–11 (2008)
22. Bowditch, G.: Grand Theft Auto producer is Godfather of Gaming. Times Online UK (2008), `http://www.timesonline.co.uk/tol/news/uk/scotland/article3821838.ece` (cited December 1, 2008)
23. Torres, F.R., Calero, C., Piattini, M. (eds.): Ontologies for Software Engineering and Technology. Springer, Heidelberg (2006)
24. W3C: Ontology Driven Architectures and Potential Uses of the Semantic Web in Systems and Software Engineering (2006), `http://www.w3.org/2001/sw/BestPractices/SE/ODA/060211/` (cited December 1, 2008)
25. OMG: Ontology Working Group (2008), `http://www.omg.org/ontology/` (cited December 1, 2008)
26. Zagal, J., et al.: Towards an Ontological Language for Game Analysis. In: Proceedings of DiGRA 2005 Conference on Changing Views – Worlds in Play (2005)

Author Index

Anagnostopoulos, Christos   305
Auer, Sören   1, 61

Bessler, Sandford   249
Brüggemann, Stefan   187
Bürger, Tobias   113

Dantcheva, Antitza   249
Dasmahapatra, Srinandan   285
Dieng-Kuntz, Rose   155
Dietzold, Sebastian   61
Dögl, Daniel   113
Du, Ying   25
Dupplaw, David   285

El Ghali, Adil   155
Erling, Orri   7

Gabner, Rene   249
Ghidini, Chiara   95
Giboin, Alain   155
Gindl, Stefan   217
Granitzer, Michael   95
Grebner, Olaf   25
Gruber, Andreas   113
Grüning, Fabian   187

Hadjiefthymiades, Stathes   305
Heese, Ralf   235
Heino, Norman   61
Hinkelmann, Knut   79
Hu, Bo   285

Kern, Roman   95
Kohlhammer, Jörn   205

Kouba, Zdeněk   171
Křemen, Petr   171

Landefeld, Rico   129
Lewis, Paul   285
Ley, Tobias   79
Liegl, Johannes   217
Lindstaedt, Stefanie N.   79, 95
Lockemann, Peter C.   269
Lohmann, Steffen   321
Luczak-Rösch, Markus   235

Maier, Ronald   79
Martin, Michael   61
Mikhailov, Ivan   7

Nejkovic, Valentina   141
Niesenhaus, Jörg   321

Ong, Ernie   25

Paoli, Heiko   269
Papadimitriou, Antonis   305
Paskalis, Sarantis   305
Pellegrini, Tassilo   1

Riss, Uwe V.   25, 79

Sack, Harald   129
Schaffert, Sebastian   1
Scharl, Arno   217
Scheir, Peter   95
Schmidt, Andreas   79, 269
Shadbolt, Nigel   285
Shafiq, Omair   113
Spiliopoulos, Vassilis   45

Tifous, Amira   155
Tochtermann, Klaus   1
Toma, Ioan   113
Tosic, Milorad   141
Tsetsos, Vassileios   305

Valarakos, Alexandros G.   45
von Landesberger, Tatiana   205

Voss, Viktor   205
Vouros, George A.   45

Weichselbraun, Albert   217

Zeiß, Joachim   249
Zhdanova, Anna V.   249

Subjet Index

aggregation operator 53
AJAX 138, 142
ALC 175
algorithmic description 220
Architecture 49, 65
associative networks 97
associative retrieval 97
automated detection of erroneous
  instances 200
automated extraction of opinions 217

Black-Box techniques 172
business intelligence 12
business model 316
business processes 270

career guidance 88
case study 56
centralised data repository 287
channel recommendation service 315
cognitive psychology 102
collaboration environment 332
collaborative authoring 129
collaborative service description 275
collaborative tagging 89, 142, 145
COMA++ 58
Common Information Model 199
communities of practice 155
concept-property selector 53
consistency checking algorithm 189
consistency of metadata 32
context-dependent nature of knowledge
  289
corporate IT infrastructure 237
corporate ontology 240
corporate ontology engineering 235
Corporate Ontology Lifecycle Methodology
  240
Corporate Semantic Web 236

DAML+OIL 274
data quality management 187

DBC/JDBC 7
DBMS 13
decentralization 291
decision-centered visualization 209
DICOM 297
Digital TV 305
Digital Video Broadcasting 305
DILLIGENT 236
discrete steps 47
DOLCE 124
domain of discourse 288
domain specific language 194
DVB-HTML 307
D2RQ 16

effectiveness of retrieval systems 96
electoral behavior 218
Erfurt API 65
explicit semantics 143

framework 331
functional requirements 244
functions and features 256

game-based learning 329
game development 322, 327
game industry 321
Glass-Box techniques 175
globally accepted knowledge model 287
Globus toolkit 114
GRDDL 13
grid computing 113
grid service 118
Grid Service Execution Environment 122
Grid Service Modeling Language 117

Haystack 40
HCOME 236
HL7 297

IMDB   222, 313
improving health care provision   286
inconsistency detection   188
index layouts   10
informal learning   85
information maps   63
Information retrieval & evaluation   104
information visualization   207
Intelligent Content Objects   113
interaction protocol   143
interactive TV   305
IkeWiki   133, 246

JAVA   293
JENA   12, 52
JSON   255
JspWiki text formatting rules   146
J2EE   280

KAON2   278
knowledgeable services   288
knowledge artifacts   26
knowledge-based system   210
knowledge capitalizing problem   156
knowledge extraction mechanism   334
knowledge-intensive activities   327
knowledge management criteria   323
knowledge maturing   81
knowledge representation paradigms   62

LCC syntax   293
lifecycle   241
Lightweight Coordination Calculus   292
Linking Open Data   20, 76
.Lucene   99

Maariwa   131
machine learning methods   219
MaknaWiki   132
management phases   24
mapping association tree   50
mapping relational data   16
MarQL   136
maturing services   86
Maximum Entropy Model   222, 228
metadata annotation   195
METHONTOLOGY   236
minimal unsatisfiability preserving
   subterminology   172
monotonic logic   177
MPEG-7   309, 316

multimedia data   300, 310
Multimedia Home Platform   305

Natural Language Processing   218
NEPOMUK   27
NeOn Methodology   236
North American Industry Classification
   System   273
N3   255

OGSA   115
Ontology Alignment Evaluation Initiative
   46
ontology-based web publishing   134
ontology-based information system   235
ontology concepts   159
ontology engineering methodology   236
ontology language   64
ontology mapping   46
ontology metamodel   131
OntoWiki   61, 246, 332
On-To-Knowledge   236
OSGi   308
OWL   64 ,189, 199, 253, 332
OWL API   57
OWL-DL   176, 201

pairs selector   54
parental control service   315
parser   51
Part-Of-Speech detection   222
Pellet   200
Personal Information Model Ontology   33
personal learning environment   80
personal task management   34
PlatypusWiki   132
process model   83
policy engine   255
pOWL   67
PVR service   315

quality assurance   290

RapidOWL   236
RDF   195, 287, 332
RDFa   13, 73, 151
RDF API   37
RDF Graph   22
RDF inferencing   11
RDF mapping   17
RDF Reactor framework   39
RDF/S   64,157, 255

RDF Sponge   13
relevance judgement   106
requirements   48, 63
revision control system   138
Rhizome   132
Rich Internet Application   136
Rise   132

scientific workplace   2
security protocol   326
semantic   98
semantic annotation   100, 136
Semantic Desktop   96
Semantic MediaWiki   88, 132, 276
semantic policy storage   251
semantic task management architecture   36
Semantic Web   1, 7, 46, 62, 96, 113, 130,
    142, 156 171, 250
Semantic Web Services   113
sentiment detection   217
SER model   84
SeRQL   37
service discovery   279
Service Oriented Architecture   122, 269
SHIN   172, 177
SIOC   16
SKOS   281
Similarity   46, 50
similarity matrix   51
similarity method   52
SKOS   74
SOAP   31, 86
SOBOLEO   281, 90
SoftWiki   73
Social Semantic Desktop   27, 29
social network management   34
social software   250
social task management   34
Social Web   151
SoftWiki   332
SPARQL   10, 16, 37, 106, 133
SPARUL SPARQL extension   12

SQL data manipulation   22
SweetWiki   133
SWOOP   245

tag cloud   144
Tagged Linguistic Units   220, 229
Tagging   89
task-analytic approach   206
Task Model Ontology   33
task pattern lifecycle   91
Textual   99
tf-idf measure   101
triple storage   8
TV User Ontology   311
typed relationship   143

UDDI   86, 272
UDDI Registry   276
Unified Activity Management   40
user-driven policy construction   251
user-generated content   250
user-generated policies   250
user interface   72
user study   259

Visual Analytics Framework   214

WebSphere Policy Editor   254
weighting of annotations   101
Wikipedia   130, 313
WordNet   323
work activities   26
WSDL   86, 272
WSML   118
WSMO   114
WSMX   115
WYSIWYG   133, 151

XBRL   13

Zend Framework   68

LaVergne, TN USA
17 August 2009
154756LV00003B/46/P

9 783642 021831